HERE

**COACHELLA VALLEY
WILD BIRD CENTER
46-500 VAN BUREN ST.
INDIO, CA 92201**

D0982491

The Growth and Development of Birds

The Growth and Development of Birds

Raymond J. O'Connor
Director,
British Trust for Ornithology

A Wiley–Interscience Publication

JOHN WILEY & SONS
Chichester · New York · Brisbane · Toronto · Singapore

Library of Congress Cataloging in Publication Data:

O'Connor, Raymond J.
   The growth and development of birds.
   A. Wiley-Interscience publication.
   Includes Index.
   1. Birds—Growth.   2. Birds—Development.
   I. Title.
QL698.026   1984   598.2′3   84-3724

ISBN   0 471 90345 0

British Library Cataloguing in Publication Data:

O'Connor, Raymond J.
   The growth and development of birds.
   1. Birds—Growth
   I. Title
   598.2′31   QL698

ISBN   0 471 90345 0

Typeset in Great Britain by Photo-Graphics, Honiton, Devon, England.
Printed by St. Edmundsbury Press Ltd., Bury St. Edmunds, Suffolk.

Contents

Contents

Preface

This book is concerned with how birds grow, with emphasis on the selective advantages in growing in one way rather than in another. Avian growth patterns and the factors regulating them form a fascinating subject that has insights to offer ornithologist and general ecologist alike. However, the field is one which has developed with astonishing speed over the last decade or so and no review of the major ideas and recently discovered facts is yet available outside the technical literature.

The present book has three origins: the comparative studies of breeding adaptations in birds developed by David Lack, the new perspectives in growth studies introduced by Robert Ricklefs, and the experimental studies by Patrick Bateson into the role of imprinting and sensitive periods in behavioural development. The pioneering study of Cullen (1957) on the development of the Kittiwake has also influenced my thinking. She showed that several behavioural and morphological features of young Kittiwakes deviated from the ancestral ground-nesting gull patterns in ways that seemed to accommodate the chicks in their life on a narrow cliff ledge. Here I have attempted to show that her findings are quite typical of avian development in general, with birds growing in particular ways because of the selective advantage to them in doing so, given the way of life of their particular species. In the book I have tried to bring a diversity of findings from physiological, behavioural, and ecological studies within the single framework of evolutionary significance. As all biological patterns are ultimately explicable in terms of the benefits resulting therefrom under natural selection, one expects this to be possible. The surprising thing is that it has not been attempted before.

The book is structured around the life-cycle of the developing bird. An initial chapter provides an overview of the ecological significance of growth patterns. This is followed by a chapter on nests and again places emphasis on selective advantages, perhaps more so than is usual with this topic. The next six chapters deal with the physical development of the young bird and the following three focus on those aspects of social behaviour of particular importance to the developing young. Mortality is discussed at this point because it can be measured only at population level, not at that of the individual. The three chapters following deal with various behaviours, with song and migration discussed as separate topics because our understanding of their development in young birds has undergone major change in the last few years. A final chapter looks at the role of experience in birds' lives.

I started this book whilst on the staff of the Zoology Department of the University College of North Wales and completed it whilst working on rather different duties at the British Trust for Ornithology. I am grateful to the Trust for a short period of scientific leave to complete a critical stage of the writing. Several of my colleagues there read chapters in draft and their comments in getting the text into more readable form are appreciated: Adrian Cawthorne, David Glue, Bob Hudson, Peter Lack, John Marchant, Robert Morgan, Mike Moser, and Kenny Taylor. I am grateful to Christopher Perrins and to Robert Ricklefs for crucial encouragement to persevere with the book. My wife Deirdre put up with the loss of many of our evenings and weekends and helped with copyright clearances. Finally, my thanks are due to the staff at John Wiley and Sons, who tolerated an extremely long incubation period for this book.

Glossary

Allometric
A term used to describe a relationship between two variables X and Y of the form $Y = aX^b$, where a and b are constants. Such equations are widespread in growth and metabolism studies and are also known as power functions, since Y varies with the bth power of X. The relationship has the convenient property that a plot of $\log Y$ against $\log X$ is linear.

Altricial
Describes nestlings born blind and naked and that remain for some time in the nest and dependent on their parents for food and warmth.

Cursorial
Birds that habitually run.

Endothermy
The ability to generate adequate body heat internally by metabolic processes.

Energy density
A term used to denote the amount of energy contained in each gram of body tissue.

Energy loading
The amount of energy put into the egg by the laying female, usually in the form of yolk.

Full-term
A term used to describe an embryo that has completed its period of development in the egg and so is ready to hatch.

Gular flutter
The method of panting used by birds in very hot conditions. The roof of the mouth is fluttered in a shallow manner to move air across the moist surface there, thus removing water and the heat it carries.

Homeothermy
The ability to maintain a constant body temperature over a range of environmental temperatures.

Metabolic intensity
The rate of metabolism per unit body weight.

Nidicolous
Adjective describing young birds that remain in the nest for further development (= nest-dwelling).

Nidifugous
Adjective describing young birds that leave the nest shortly after hatching.

Ontogeny	Technical term for development.
Pipping	The first externally visible stage of hatching, when the first weak cracks appear on the surface of the shell.
Poikilothermic	A term usually equated with 'cold-blooded', describing young that require an external source of heat, e.g. parental brooding, to maintain their body temperature above environmental levels.
Precocial	Young that can leave the nest shortly after hatching and live independent lives, though usually with some form of parental care (brooding, guarding, food-finding) available to them.
Ptilopaedic	Covered with down.
Superprecocial	Describing chicks such as those of megapodes that can lead fully independent lives from the time of hatching, with no further parental care available to them.
Thermogenesis	The generation of heat by internal means such as chemical metabolic processes or by shivering.
Thermolysis	The ability to dissipate heat in hot conditions.
Thermoneutral zone	The temperature range over which a bird's metabolic rate is lowest, with temperature regulation adjusted by physical means, e.g. sleeking down the plumage to promote heat loss, and conversely. Below this range the bird must increase its metabolic rate to compensate for the heat loss across its maximum plumage insulation, and above this range it must pant and lose water to avoid over-heating.
Zugenruhe	A German word widely used in migration studies to describe the 'migratory restlessness' shown by captive migrants during the time they would normally migrate.

CHAPTER 1

Patterns of Avian Development

Development patterns among birds are classified according to the relative maturity of the young bird at hatching. Some young birds, such as domestic chicks and most ducklings, hatch from the egg in a state of advanced maturity. They have their eyes open at hatching or very shortly thereafter, they are covered with down at birth, and they are capable of walking or swimming from a very short time (perhaps as little as 2 hours) afterwards. Such young, which normally lead a life to some extent independent of their parents, are said to be precocial in their development. (Megapode chicks are totally independent of their parents and are sometimes described as superprecocial.) At the other extreme is a pattern of hatching in a very immature state – blind, naked, and incapable of self-care to the slightest degree, with the young fully dependent on their parents for periods of perhaps as little as a week but usually more. This pattern is typical of most small songbirds and is called altricial development. Such young are also termed nidicolous (nest-dwelling) because they remain in the nest, in contrast to the nidifugous (nest-fleeing) behaviour of precocial species.

CLASSIFICATION OF DEVELOPMENT MODE

In practice there exist many patterns of development intermediate between these extremes. Nice (1962) has classified these patterns into four major classes, with a number of subclasses to some of them. Figure 1.1 shows these categories in relation to the main features of development used by Nice, notably the presence or absence in the hatchlings of down and of vision, the mobility of the young, and the extent to which they feed themselves or depend on their parents. Although some of these features – for example, whether the eyes are open or not at hatching – are of doubtful significance in defining the developmental status of a nestling, it is clear that there exists a broad continuum of development patterns among birds and that this gradation effectively parallels a gradient in nestling independence of parental care.

The continuum suggested by the layout of Figure 1.1 conceals the fact that certain categories of young do not occur among birds. Amongst birds, only one instance (the Finfoot *Heliornis*) is known of naked young leaving the nest, though hoatzins, Opistocomidae, are unusually mobile despite being

1

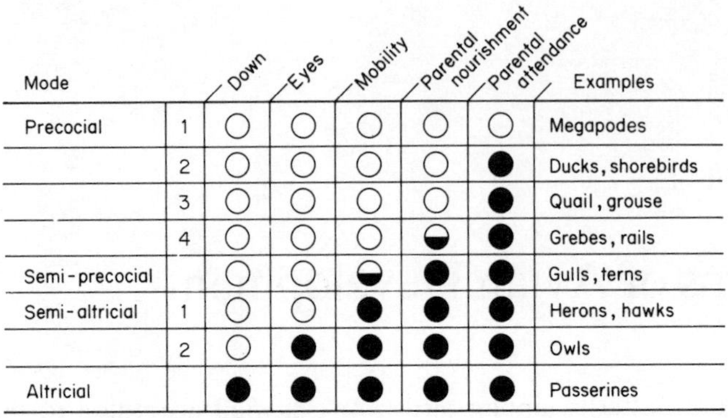

Figure 1.1 Characteristics of the developmental modes in Nice's (1962) classification of avian growth. *Key*: ○, Precocial characters; ●, altricial characters. Redrawn with permission from Ricklefs (1984).

Table 1.1 Development mode and taxonomic class in birds. From Nice (1962)

Taxonomic class	Development mode
Sphenisciformes: Penguins	Semi-altricial 2
Struthioniformes: Ostriches	Precocial 3
Rheiformes: Rheas	Precocial 3
Casuariiformes: Cassowaries, Emus	Precocial 3
Apterygiformes: Kiwis	Precocial 2
Tinamiformes: Tinamous	Precocial 2 or 3
Gaviiformes: Loons	Precocial 4
Podicipediformes: Grebes	Precocial 4
Procellariiformes: Albatrosses, Petrels	Semi-altricial 1, 2
Pelecaniformes	Semi-precocial
Phaëthontidae: Tropic-birds	Semi-altricial 2
Pelecanidae: Pelicans	Altricial
Sulidae: Boobies, Gannets	Altricial
Phalacrocoracidae: Cormorants	Altricial
Anhingidae: Anhingas	Altricial
Fregatidae: Frigate-birds	Altricial
Ciconiiformes	
Ardeidae: Herons, Bitterns	Semi-altricial 1
Balaenicipitidae: Whale-headed Storks	Semi-altricial 1
Scopidae: Hammerheads	Semi-altricial 1
Ciconidae: Storks, Jabirus	Semi-altricial 1
Threskiornithidae: Ibises, Spoonbills	Semi-altricial 1
Phoenicopteridae: Flamingos	Semi-precocial
Anseriformes: Swans, Geese, Ducks, Screamers	Precocial 2
Falconiformes: Hawks, Vultures, Secretary Birds	Semi-altricial 1

Table 1.1 (*continued*)

Taxonomic class	Development mode
Galliformes	
Megapodiidae: Megapodes	Precocial 1
Cracidae: Guans, Curassows	Precocial 4
Tetraonidae: Grouse	Precocial 2, 3
Phasianidae: Quails, Pheasants	Precocial 3, 4
Meleagrididae: Turkeys	Precocial 3
Opisthocomidae: Hoatzins	Semi-precocial?
Gruiformes	
Turnicidae: Button-quails	Precocial 4
Gruidae: Cranes	Precocial 4
Aramidae: Limpkins	Precocial 4
Rallidae: Rails, Coots, Gallinules	Precocial 4
Rhynochetidae: Kagus	Semi-precocial
Eurypygidae: Sun-bitterns	Semi-precocial
Cariamidae: Seriemas	Semi-altricial 1
Otididae: Bustards	Precocial 4
Charadriiformes	
Jacanidae: Jacanas	Precocial 2
Haematopodidae: Oystercatchers	Precocial 3, 4
Charadriidae: Plovers, Turnstones	Precocial 2
Scolopacidae: Snipe, Woodcock, Sandpipers	Precocial 2, 4
Recurvirostridae: Avocets, Stilts	Precocial 2
Phalaropodidae: Phalaropes	Precocial 2
Dromadidae: Crab-plovers	Semi-altricial 1
Burhinidae: Thick-knees	Precocial 4
Glareolidae: Pratincoles, Coursers	Precocial 4
Stercorariidae: Skuas	Semi-precocial
Laridae: Gulls, Terns	Semi-precocial
Rhynchopidae: Skimmers	Semi-precocial
Alcidae: Auks, Auklets, Murres	Precocial 2
Columbiformes	Semi-precocial
Pteroclidae: Sand Grouse	Precocial 2
Columbidae: Pigeons	Altricial
Psittaciformes: Parrots	Altricial
Cuculiformes	
Musophagidae: Touracos	Semi-altricial 1
Cuculidae: Cuckoos	Altricial
Strigiformes: Owls	Semi-altricial 2
Caprimulgiformes	
Steatornithidae: Oilbirds	Altricial
Podargidae: Frogmouths	Semi-altricial 1
Nyctibiidae: Potoos	Semi-altricial 1
Caprimulgidae: Goatsuckers	Semi-precocial
Apodiformes: Swifts, Hummingbirds	Altricial
Coliiformes: Colies	Altricial
Trogoniformes: Trogons	Altricial
Coraciiformes: Kingfishers, Rollers, Hornbills	Altricial
Piciformes: Jacamars, Toucans, Woodpeckers	Altricial
Passeriformes: Perching Birds	Altricial

born without down. There must therefore exist functional constraints on the possible combinations of neonatal characters, and this is reflected in a rather close correlation between mode of development and taxonomic class (Table 1.1). For most orders developmental class is rather constant, but the Gruiformes and the Charadriiformes are exceptionally diverse: the former contains five precocial families, two semi-precocial families, and one semi-altricial family, whilst the later has precocial development in nine families, semi-precocial development in four, and semi-altricial in one (the aberrant Crab-plovers, Dromadidae). The auks, Alcidae, are the only case in which several developmental modes are present within a single family.

EGG SIZE AND COMPOSITION

The degree of independence and precocity shown by a bird at hatching is the result of prior provisioning of the egg by the adult. Since precocial chicks achieve greater development within the egg than do altricial young, one expects to find the precocial–altricial gradient in development paralleled by an equivalent gradient in egg size or in yolk (= energy) content. Table 1.2 shows that this is indeed the case, with egg weight in precocial species averaging some ten times heavier than in altricial species and with semi-altricial and semi-precocial species at correspondingly intermediate positions. However, much of this spread in egg weight between developmental modes is due to associated differences in adult body size. Rahn *et al.* (1975) found that within most orders of birds larger species lay larger eggs; the relationship is not linear but is allometric (that is, linear when both axes are transformed logarithmically), with an exponent of 0.770. Hence as adult weight doubles, egg weight increases by $2^{0.770}$ or 70.5 per cent. Figure 1.2 illustrates the relationship for various semi-altricial raptors. Rahn *et al.* (1975) compared relative egg size in different taxonomic orders, by using fitted regressions of this type to compute within each order what weight of egg a theoretical 100 g member of that order would lay. These weights are summarized in relation to development mode in the rightmost column of Table 1.2 and show that, even after allowance for adult size, precocial species still lay relatively larger eggs

Table 1.2   Egg weights in young of various developmental modes

Mode	Mean egg weight $\pm$ S.E. (N)[a] (g)	Egg weight of 100 g bird[b] (g)
Altricial	8.62 $\pm$ 3.70 (24)	4–10
Semi-altricial	34.92 $\pm$ 5.75 (5)	11–14
Semi-precocial	48.96 $\pm$ 9.03 (11)	
Precocial	91.57 $\pm$ 30.90 (22)[c]	9–21

[a] Based on data in Carey *et al.* (1980).
[b] Computed from Rahn *et al.* (1975) by Ricklefs (1984).
[c] Data has skewed distribution so the mean is not the optimal measure of location.

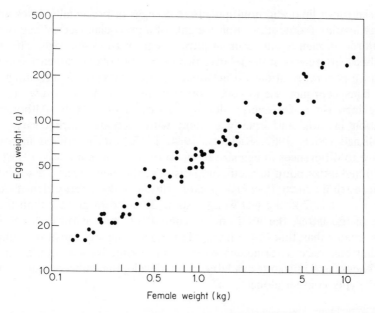

Figure 1.2 Egg weight in relation to female weight in Nearctic and Palaearctic raptors. Based on data in Newton (1979).

than do altricial species. Although the spread in egg weights between groups is much reduced after adjustment for adult size, the significance of the larger absolute egg sizes in the more precocial species should not be overlooked. The smaller an egg is, the easier it is for a female of given size to produce it. Hence the evolution of the altricial mode of development, even in association with egg size alone, would have permitted the evolutionary reduction of body size and the degree of adaptive radiation shown within the passerines and other small land-bird orders (Dawson and Evans, 1960).

The altricial–precocial gradient of development pattern and hatchling independence is paralleled by a gradient in egg composition as well as in egg size (Table 1.3). Eggs of altricial species contain more water and less solid

Table 1.3 Egg composition in species of various developmental modes. Based on Carey *et al.* (1980)

	Altricial	Semi-altricial	Semi-precocial	Precocial
Water (% content)	84.3	81.7	76.5	74.7
Lipid (% content)	5.9	6.3	9.5	10.3
Other solids (% content)	9.8	12.0	14.0	15.0
Energy density (kcal/g)	1.14	1.24	1.63	1.91
Yolk content (% content)[a]	24	26	33	41

[a] Based on samples of species different from those for other data.

material (both lipid and non-lipid) than do eggs of precocial species and the energy loading (content per unit weight) of a precocial species' egg is nearly two-thirds as high again as in an altricial egg of the same size. This is due largely to differences in the relative proportion of yolk within each egg and is part of a general relationship between energy loading and percentage yolk: each 1 per cent increase in yolk content in an egg is associated with a 0.033 kcal/g increase in its energy density (Carey *et al.*, 1980). Although this variation in yolk and energy content with increase in precocity is well established (Nice, 1962; Ricklefs, 1974, 1977a), its restricted importance relative to differences in egg size tends to receive less emphasis. Carey *et al.* (1980) make this point in their comparison of the eggs of the Cassowary and of the estrildid finch *Poephila guttata*: although the energy density of the former is, at 1.87 kcal/g wet weight, some 36 per cent greater than the 1.37 value of the latter, the total energy content of the Cassowary's egg is 851 times greater than that of the finch. Hence, ignoring any constraints imposed by adult body size, altering egg size is a more effective way of adjusting the total energy and nutrients available to the developing embryo than is altering relative yolk content alone.

ECOLOGICAL CONSTRAINTS ON DEVELOPMENTAL MODE

An obvious contrast between altricial and precocial development is the different timing of the peak demand which each makes on the parent birds. Altricial species produce eggs that are small and of low yolk content, each of which features make a rather smaller demand on the female than is subsequently made by the task of feeding helpless young. Conversely, precocial species invest heavily in their production of large energy-rich eggs but benefit subsequently from the considerable independence of the resulting chicks (Ar and Yom-Tov, 1979). Consequently, we can expect altricial and precocial species to differ in their susceptibilities to ecological factors affecting either the timing and intensity of egg production or the cost and efficiency of post-natal growth. Two ecological factors – food availability and predation intensity – are especially important here. If food is readily available, both to females for egg production and to independent young for growth, precocial development is feasible, but if food is difficult to obtain for egg formation, then the small eggs of altricial development are favoured. Similarly, predation risk must be influential, with altricial young vulnerable to brood discovery by predators over a short period of intense growth and precocial young vulnerable only to individual chick discovery. However, precocial young are vulnerable for longer, since they grow at a slower rate than do altricial chicks, and their vulnerable period is longer the more precocial they are.

Case (1978) presents a simple word model of how food availability interacts with predation to determine the optimal growth rate for a species. When food is obtained easily, the parents need be away from a nest only briefly and at short distances; the costs of quickening growth are then small and growth rate is maximized. But as food becomes scarce, so the time taken to meet the increment in feeding rate needed for a small increase in growth becomes disproportionate, thus exposing nestlings to mortality (from predation or adverse weather) more severe than that saved by faster growth. An optimum intermediate growth rate would then exist. Precocity of growth must therefore be optimized in relation both to the availability of food for adults and young and to the intensity of predation on the young.

Few experimental studies exist to demonstrate a direct link between growth rate and food supply. Harris (1978) fed young Puffins daily with 50 g of sprats as a supplement to the food brought to them by their parents; he found that their growth rates were substantially elevated, though not as much as in captive young fed in the laboratory. Högstedt (1981) supplemented the food available to breeding Magpies and found that brood weights were about 12 per cent higher in the experimental territories. In similar experiments with Crows, pairs with excess food bred earlier and more successfully but fledgling weights were unaffected (Yom-Tov, 1974). For most species any increase in food availability is likely to be manifested in extra eggs (larger clutches), with growth rates adjusted only as a fine-tuning of nest productivity to food supply (p. 78). Those species responding positively to food supplements are probably single-chick species (like the Puffin) or those with irregular or unpredictable food sources. For long-lived species, on the other hand, refraining from breeding in years of poor food may be a better strategy than attempting to rear chicks through poor growth conditions (Andersson, 1976; Drent and Daan, 1980).

Table 1.4 shows the general relationship that exists between growth rates within various taxonomic families and some relevant variables: maturity at hatching, parental mating system, parental participation in rearing, and feeding frequency (Case, 1978). Rapidly growing birds are usually fed frequently and by both parents, irrespective of precocity. For example, the semi-precocial gulls, terns, and skuas grow more rapidly than the offshore-feeding (and consequently infrequently fed) altricial pelicans, boobies, and frigatebirds. Figure 1.3 illustrates this relationship quantitatively for seven species of seabird breeding off the Northumberland coast of England. Similar considerations underlie the diversity of developmental strategies among the auks (Alcidae).

The production of energy-rich eggs by precocial species is favoured by a ready availability of the foods required for egg formation by females. It is no accident that adults of precocial species are predominantly primary consumers, as in waterfowl and in gallinaceous species (Ar and Yom-Tov, 1979). For precocial chicks, however, such foods often lack components essential to post-natal growth and at least a proportion of animal material is taken by precocial young. Thus, most cursorial land-bird chicks feed on items which are easily captured but widely dispersed and difficult to find (Nice, 1962). Self-feeding from an early age is then possible, provided that parental

Table 1.4 Development mode and parental rearing characteristics in various bird groups. Data are listed in decreasing order of growth rate, with groups of equal rate linked. From Case (1978).

	Maturity at birth[a]	Mating system[b]	Parental participation by Male	Parental participation by Female	Feeding frequency (times per day)[c]
Apodidae	A	M	+	+	5– 30
Raptors	SA	M	+	+	6– 40
Passeriformes	A	M	+	+	16–400
Columbidae	A	M	+	+	3– 22
Phalacrocoracidae	A	M	+	+	10– 16
Ardeidae	SA	M	+	+	6– 25
Stercorariidae	SP	M	+	+	No data
Laridae	SP	M	+	+	4–100
Alcidae	A	M	+	+	1– 2
Pelecanidae	A	M	+	+	3– 5
Ciconiidae	SA	M	+	+	5– 10
Sulidae	A	M	+	+	1– 3
Procellariidae	SA	M	+	+	0.5–1.0
Phaethonidae	SA	M	+	+	2– 3
Anatidae	P	M, Pr	$(+, 0)^d$	$(+, +)^d$	Self-feeding
Charadriidae	P	M	+	+	Self-feeding
Scolopacidae	P	M	+	+	Self-feeding
Tetraonidae	P	M, Pr	+	+	Self-feeding
Diomedeidae	SA	M	+	+	0.5
Fregatidae	A	M	+	+	No data
Phasianiidae	P	M, Pg, Pr	+	+	Self-feeding
Meleagrididae	P	Pg	0	+	Self-feeding
Rallidae	P	M	$(+, 0)^d$	$(+, +)^d$	Self-feeding

[a] P, precocial; A, altricial; SA, semi-altricial; SP, semi-precocial (Nice, 1962).
[b] M, monogamous; Pr, promiscuous; Pg, polygamous (Lack, 1968).
[c] Numbers present the range of feeding rates recorded among species of the family.
[d] In some species the male participates, in others he does not.

guidance to food is available. Such species may also be driven to an optimum growth rate if chick growth and chick mortality are linked. For example, if foraging chicks are more likely to be taken by predators than are those under cover whilst resting, if body size increase outstrips the learning of hunting techniques, a mortality cost to growth will be present. Here selection for parental guarding also takes place, reducing such cost. Indeed, the difficulties

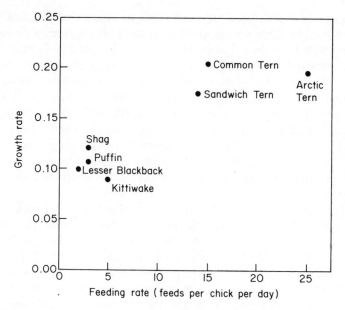

Figure 1.3 Growth rate (summarized as rate constants from fitted Gompertz equations) in relation to feeding rate in various seabirds. Based on data in Pearson (1968) and Ricklefs (1973).

of providing adequate care at this stage has been shown to limit reproductive success in some waders (Safriel, 1975), a discovery which might account for the greater growth rates of those precocial species in which both parents guard the brood (Case, 1978).

Another point of precocial–altricial distinction resides with brood energetics. Since precocial species provision their eggs with the necessary nutrients for hatchling maturity, their subsequent energy costs are low. Brood size limitation in precocial species is thus primarily through parental ability to produce eggs (though, as noted above, in some species demand on parental guarding ability may be limiting). In contrast, altricial species are limited by the food-gathering abilities of the parents (Lack, 1968) and brood energetics therefore call for efficiency in energy management.

A final correlate of the precocial–altricial dichotomy is that brood sizes in precocial species are generally larger than in altricial species (Ar and Yom-Tov, 1979). With stable populations this immediately implies that precocial broods suffer higher mortality. On the other hand, since precocial young can escape predators, few broods fail to produce some young, in contrast to altricial broods in which predation often results in complete failure (Ricklefs, 1969a). Precocial development is thus a better strategy in environments affording only a restricted breeding season, whilst altricial development is adaptive with prolonged breeding seasons that provide opportunities for laying repeat clutches following earlier failure.

These four features of the precocial mode of development – slow growth, easily obtained food for young, low requirement for energetic efficiency, and high (but rarely total) brood mortality – are well adapted to Temperate Zone characteristics and poorly adapted to the tropics. First, the seasonal food flush of higher latitudes provides abundant food for egg production (particularly by herbivores) and imposes less constraint on the inefficiencies of food (= energy) gathering by precocial young. Second, since the bulk of energy expenditure on precocial young occurs in egg formation and incubation each chick loss is expensive; such losses are fewer in the reduced predation regime of high latitudes. Finally, the cooler climate of temperate regions increases the cost of parental brooding, thus favouring the early homeothermy of precocial young.

Exactly the converse arguments should favour the altricial mode of development in tropical species. Seasonality is not well marked and allows breeding over an extended period, though probably rarely in conditions of superabundant food. Predation is intense, so that many nests fail, but this is compensated by the greater opportunity for repeat attempts; in some species such opportunity can be increased by moulting during the nesting period and prolonging the breeding season into the normal moult season (Foster, 1974a). Finally, the warmer climate necessitates lower energy expenditures.

These expectations are supported by available information. World-wide, about 80 per cent of all birds develop as altricial young, largely because the small eggs needed with altricial development permit small body size and small body size offers more ecological opportunities in, for example, reducing competition (Witschi, 1956). However, proportionately more species outside the tropics develop as precocial or semi-precocial young than is the case within the tropics (35 per cent against 15 per cent). The imbalance is still more marked for certain groups, particularly amongst those exploiting fruits and flying insects. Parrots are almost exclusively found in the tropics and subtropics, pigeons and doves are three times as abundant in the tropics as elsewhere, and various aerial insectivores are three to six times better represented in the tropics than elsewhere. Carnivorous birds are the main group of altricial or semi-altricial species more abundant in temperate than in tropical zones, probably because of the greater abundance of ground-living small mammals outside the tropics (Ar and Yom-Tov, 1979).

INTRASPECIFIC VARIATION IN GROWTH PATTERNS

Just as there can be a considerable variation of developmental mode within a single family when ecological considerations demand it, as in the auks (Alcidae), so too there can be a considerable variation in growth pattern within a single species. Only in recent years has the extent of such individual variation been studied extensively and related to ecological factors (O'Connor, 1978a,b; Dunn, 1979; Ross, 1980). These studies suggest that the degree and type of variation in any species is adapted to the overall breeding strategy of the adults, at least in those species which feed their young.

Birds should rear as many young as they can on the prevailing food levels, with the optimal level of investment in the current brood being a balance between expected success now and expected success of future breeding opportunities (Ricklefs, 1977b; Dunn, 1979). Whatever the optimal level of parental investment may be, the female's contribution to it by way of eggs laid has to be made several weeks before the brood's requirements for food (or parental guarding) become limiting. (Lack, (1954, 1966) reviews the evidence that many species are limited in productivity by their ability to feed their young.) Females must therefore determine their clutch size long before they can measure directly how much food is available for the resulting brood. How, therefore, may they optimize their clutch size?

For altricial species, the nature of the food supply is apparently critical in answering this question (O'Connor, 1978a). For some, the eventual abundance of nestling food is predictable (directly or indirectly) at laying. Thus for Great Tits the spring temperatures are correlated with the subsequent abundance of the caterpillars on which the nestlings are fed, so by responding to spring temperature levels the Great Tits can effectively predict the later food supply. In other species no such correlation may be available to the bird. Therefore, one can envisage the food supplies of different species as varying along a gradient of predictability, with each species engaging in a different breeding strategy according to its position on the gradient. If the food supply for nestlings is highly predictable at laying, then each female will lay just that number of eggs that is optimal for her (allowing for the quality of her territory, her age and breeding experience, and so on). The growth pattern of such a species will then tend to be highly stereotyped and probably close to optimal as a result of this clutch size adjustment.

A second feature of the nesting season food supply that is of importance to growth pattern is that of its stability over the nestling period. Such stability was assumed implicitly above but it can vary (to a degree – see below) independently of predictability. If food levels fluctuate from day to day during the nestling period (for example, because insect availability varies with the weather), some young are at risk of dying during the poor periods. If such periods are normally only of short duration, one would expect to find selection for adaptive modification of growth pattern in such conditions; then additional young could be reared on the surplus available on most days of the nestling period. One obvious way to modify growth patterns to meet such conditions is for the nestlings to store fat and other nutrients and to use these reserves to see them through temporary shortfalls in food supply. For this reason the strategy is termed the resource storage strategy (O'Connor, 1978a) but, in fact, the strategy encompasses a complete suite of adaptive growth characteristics. Thus, (1) growth (weight gain per unit time) must be labile, to permit compensation of poor spells during the good spells; (2) fat deposition must progress at least in step with metabolic rates, to ensure constancy of survival prospects during an interruption in food availability at any stage of development; (3) thermoregulation must be achieved early in the development of the nestling, to free the adult from having to feed young at a time when even feeding themselves may be difficult; and (4) some ability to adjust brood size to low food availability during any period of sustained food shortage must be available, for the incidence of such periods is necessarily significant with an erratic food source. These several features are indeed found in the growth patterns of aerial insectivores and other species with erratically available foods (O'Connor, 1978a).

BROOD REDUCTION STRATEGIES

Although food sources whose levels are predictable (in the context of rearing young) are necessarily stable, the converse is not true. A food supply can be extremely stable over the duration of a nestling period, yet be totally unpredictable by any measurable parameter at the time of egg laying. Relatively short-lived flushes of food typically come into this category, as with aphid infestations, chironomid emergences, and tree fruiting. A female of a species reliant on such foods cannot predict precisely how many nestlings she will in fact be able to rear. Always to anticipate poor conditions by laying a small clutch would usually be wasteful, since in most years she would be able to rear more young than she would then have. Anticipating the best conditions, on the other hand, would normally leave her with more young than she could rear at optimal rate without detriment to her own survival. Prolonging the nestling period in the poorer years cannot work because, at a constant mortality risk, the lengthened exposure to mortality (especially predation) quickly offsets the gain in clutch size (Ricklefs, 1969b; O'Connor, 1978a). The way around these various difficulties appears to be for the female to lay the large clutch but to start incubating the eggs before the clutch is complete. This ensures that the clutch hatches asynchronously to yield a brood of young differing in age and therefore in size. If the parents cannot find enough food to rear all the young hatched, a competitive feeding hierarchy based on age and size differences between siblings will ensure that some chicks are fed consistently, thus maximizing the chances of part of the brood surviving. Were food distributed evenly among the young, all might become so weakened that none survived. Asynchronous hatching may therefore be regarded as an insurance against food supplies failing in the course of the nest cycle, after the female has committed herself to a particular brood size by laying the corresponding size of clutch. Such an idea is based firmly on Lack's (1954) hypothesis that clutch size in many species is adjusted to correspond to the number of young for which the parents can later expect to find food. An extension of the idea is due to Dawson (1972), who suggests that the female may profit by laying more eggs than she can on average expect to raise. Should the food levels during the rearing period prove unusually good, the extra nestlings will survive; if not, they will die quickly since they are subordinate members of the feeding hierarchy produced by asynchronous hatching. The laying of such 'insurance' eggs is known amongst the penguins of the Sub-Antarctic (Warham, 1975). In these, two eggs of very different size are laid in the normal clutch, typical volumes in the case of the Rockhopper Penguin being 140 ml and 204 ml for first and second laid eggs respectively; the chick from the smaller egg survives only if the larger egg fails to hatch.

Dawson's idea would work best for species in which (a) egg production is energetically cheap, (b) nestling food supplies vary substantially in time and space, and (c) hatching asynchrony is reinforced by other adaptations enhancing the efficiency with which surplus nestlings are disposed of (O'Connor, 1978a). House Sparrows satisfy these criteria (Dawson, 1972; O'Connor, 1978a), but there are many other species which can be expected to do so (Ricklefs, 1965).

Brood reduction in this way can be viewed, at first sight, as a simple parental manipulation of the feeding regime of the brood, to allow the adults rear at least some offspring in conditions of scarcity. There are theoretical arguments for this being the case (Alexander, 1974), but they are a matter of some controversy and in any case are not consistent with the patterns of parental feeding in birds. Adults do not at first recognize their young individually (p. 171) they feed the young bird responding most quickly to their arrival (Löhrl, 1968); and they respond to experimental increases in nestling demands by themselves foraging more intensively (von Haartman, 1949; Drent and Daan, 1980). These points are more easily accommodated in a theory (O'Connor, 1978c) based on Hamilton's (1964) idea of inclusive fitness than on ideas of parental manipulation.

Figure 1.4 summarizes the results of this theory. Three categories of birds are involved in the brood reduction process, each of them with different genetic interests. First, there are the adults who control the extent of any pre-adaptations towards brood reduction (such as when incubation of each egg starts). Second, there is the victim of the brood reduction process. Third, there are the survivors, whose enhanced survival in the reduced brood is the outcome of the death of the victim. From the viewpoint of each adult, every young carries half the parental gene complement and the loss of a chick from a brood of size B represents a proportional loss of 1 in B young. If the proportional increase in survival of the remaining young is greater than this, the adult will benefit by a reduction in brood size. For the victim, its own death zeros its individual fitness but, because it shares genes with each and every sibling, selection can nevertheless still act on its inclusive fitness. If more of the victim's genes are likely to survive in those of its siblings remaining after brood reduction than will survive if the larger brood size persists, one gets selection for suicide! Naturally, the loss of both halves of its own gene set means that the differential required for a net benefit to accrue to a victim by dying is always above the level at which its parents benefit (Figure 1.4). Finally, seen from a survivor's viewpoint, nestlings at the top of the hierarchy lose a half-share in the gene complement of the victim (as does the adult), but both halves of a survivor's genes benefit from the improved success of the smaller brood (whilst the adult benefits through the one

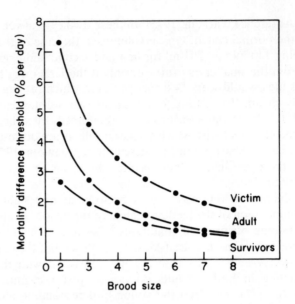

Figure 1.4 The change in mortality on reducing brood size by one needed to have a net gain in fitness on the parts of the survivors, the parent and the nestling dying, as a function of brood size. Redrawn from O'Connor (1978c).

half-share it has with each survivor). The threshold at which a nestling at the top of a competitive hierarchy benefits from brood reduction thus lies below the adult level (Figure 1.4). It should be fairly obvious that the differences between these thresholds will decrease in larger broods (O'Connor, 1978c).

Testing of this theory against observed patterns of mortality reported in the literature supported the ideas presented above (O'Connor, 1978c). Nestling mortality, for example, averaged 37.1 per cent amongst 13 species for which the theory predicted brood reduction, but averaged only 17.7 per cent in eight species for which brood reduction was not predicted.

Apart from this theoretical support for Lack's interpretation of the significance of hatching asynchrony, there exist several lines of evidence for his view (Hahn, 1981). These fall into four categories: (1) observations of aggressive interactions and fratricide between siblings, with older and larger siblings dominant; (2) evidence that growth rates and survival are lower amongst the late-hatched chicks of asynchronous broods; (3) evidence that brood asynchrony is associated with unpredictable food supplies, including evidence of intraspecific variation in asynchrony with seasonal change in breeding conditions; and (4) experimental manipulation of hatching asynchrony, leading to greater mortality in artificially synchronous broods.

Sibling aggression and fratricide

Sibling aggression has been reported in at least 25 species, most of them birds of prey (13 species, nine of them eagles) and herons and cranes (eight species); the remainder are seabirds of various types (two pelicans, two boobies, one skua) (O'Connor, 1978c; Hahn, 1981).

The factors governing aggressive behaviour towards siblings have been examined experimentally in the case of the South Polar Skua (Proctor, 1975). Hunger is responsible for the release of this aggression, the intensity of which is related to the degree of hunger. Attacks are first observed about a day after the second chick hatches, possibly because the adults attend particularly to the helpless newly-hatched chick. Aggression can be induced experimentally in second chicks but is normally inhibited by the greater size of its sibling. Normally the younger chick simply moves away from the nest. If the older chick is well fed, however, it does not attack its younger sibling, which then gets its share of parental attention.

Amongst eagles the importance of sibling aggression varies from species to species (Meyburg, 1974). In some, such as the Lesser Spotted Eagle, two chicks usually hatch but only one fledges. If the young hatch significantly asynchronously, the older becomes adept at taking meat offered by the adult and the younger is offered progressively fewer pieces of meat by the adults and then only for brief presentations. The feeding behaviour of the adult adjusts to the requirements of the older chick and the younger dies. But in other cases the young hatch more nearly synchronously and the older attacks the younger whenever they are unbrooded. The younger then retreats to the edge of the nest where it may die of cold or starvation of may fall over the edge. In these species, therefore, the second chick is attacked by the older only if it represents a significant threat to the latter's dominance, but is otherwise doomed anyway. Presumably the second egg of such species is there purely for insurance. In other eagles, such as the Spanish Imperial Eagle, aggression occurs between siblings but their survival is linked directly to the availability of prey, i.e. the idea advanced by Dawson (1972).

In some species sibling aggression becomes fratricidal, though direct evidence of this has rarely been obtained. In Pink-backed Pelican broods in Uganda the larger sibling always feeds first and frequently attacks the younger, which moves to the edge of the nest to escape; even then it may be ejected over the edge of the nest (Din and Eltringham, 1974). Ingram (1959, 1962) provides direct evidence for fratricide and subsequent cannibalism in Short-eared Owls. He describes how the adults stockpile small mammals or other prey at the edge of the nest, where the older siblings can reach them. In this way a hungry owlet can feed immediately, inhibiting its aggression towards smaller siblings. Should food prove inadequate, however, a smaller sibling may be attacked and eaten, as reported by Ingram (1962).

Examples of adults feeding dead nestlings to surviving siblings have been reported of eagles (Meyburg, 1974). The adults may themselves consume some or all of the carcase, an adaptive response if food is in short supply. Progenicide – the killing of young by their parents – has been reported less frequently but does occur, e.g. in Pink-backed Pelicans (Din and Eltringham, 1974) and in Double-crested Cormorants (Siegel-Causey, 1980). In these cases the breeding attempts were abandoned shortly afterwards and might therefore have been due to adults becoming unable to cope with their commitment.

What happens if one manipulates the degree of asynchrony of hatching experienced by birds? Hahn (1981) undertook just such experiments with chicks of the Laughing Gull; by swapping eggs between clutches, she obtained samples of synchronously hatched clutches to compare with asynchronous ones. Lack's explanation for the occurrence of asynchrony led her to expect two results. First, some of the artificial synchronous broods should fail because all three young were inadequately fed, whilst this should be rarer in asynchronous broods. In agreement with this, Hahn found that 23.1 per cent of her experimental broods failed, against only 10.4 per cent of the staggered broods. Second, partial brood success should be more frequent in asynchronous broods, since parents of the former can more frequently salvage part of the brood. This prediction was also supported, for Hahn found that asynchronous nests were 50 per cent more successful in fledging some young than were her experimental broods (and this after allowing for total failure).

Correlates of brood hierarchies

Hatching asynchrony on its own will not always lead to the establishment of competitive hierarchies within the brood. In Blue Tits, for example, large broods hatch over a couple of days and the later young benefit from the increase in parental feeding behaviour which takes place over that time (O'Connor, 1975a). Hence part of the size difference induced by asynchrony between siblings is negated by the additional feeds of late young. It is easy to envisage similar effects due to egg weight, growth rate, and so on (Howe, 1976). Birds using brood reduction to adjust to nestling food availability therefore need additional adaptations serving to maximize sibling differences and thus to enhance the efficiency of the strategy.

There appear to be six such adaptations (O'Connor, 1978a). Chick size at hatching is usually correlated with egg size (p. 32), so sibling size differences (and hence differences in competitive abilities) are maximized if there are large differences of egg weight within the clutch. Second, hatching asynchrony should be greater the more need there is for brood reduction efficiency. Third, the more mature a hatchling is, the more readily it can respond to parental feeding. Fourth, a high growth rate maximizes the value of the head-start provided by asynchronous hatching. If such growth is additionally a function of nestling body size (at least during the early days), the resulting size differences between siblings increase multiplicatively (O'Connor, 1975b). Finally, there is little point in nestlings of this type laying down fat reserves, since to divert energy by doing so could allow a competitively inferior sibling to erode the size difference. For the dominant sibling, its rank in the brood ensures that it will have priority in consuming what food is brought to the nest for as long as it needs it. When these points were compared and contrasted in three passerine species, each feature was quite pronounced in the House Sparrow, a species whose ecology is appropriate to brood reduction tactics, and contrarily modified in species expected to minimize brood reduction tactics (O'Connor, 1978a).

OVERVIEW

It should be clear from the discussion above that the growth pattern of any species is determined by life-history constraints imposed primarily by the feeding ecology of adult and young and the associated mortality risks. Movement along the continuum from superprecocial to altricial strategy involves reducing the female's investment in egg provisioning, but a corresponding reduction in the maturity and independence of the hatchling involves a greater parental responsibility for the maintenance and regulation of essential life functions. These trends have two important evolutionary consequences. First, the smaller altricial egg can be produced by females of small body size, thus allowing altricial species to exploit a great variety of niches not open to larger animals, particularly in the tropics. Second, because the largest element of offspring provisioning by altricial species occurs at the nestling stage, the male can contribute a larger share of the total reproductive effort needed. Altricial species can therefore work to a food-gathering limit more often set by the joint food-gathering abilities of the pair than by the egg-production abilities of the female. This new limit can be pressed still more closely in species unable to predict at laying how many young a pair can expect to raise, by resorting to systematic brood reduction to bring the number of young into line with what the adults can adequately feed. This new limit is, however, a further step away from the female limit at egg production, being set no longer by parental interests alone but by considerations of inclusive fitness in which nestling interests feature explicitly.

CHAPTER 2

Nests

One of the chief ways in which birds care for their young is through their construction of nests. All but a few aberrant species use at least a rudimentary nest, the major exceptions being the brood parasites, nightjars, auks, and the Fairy Tern.

NEST SIZE AND PARENTAL INVESTMENT

As with most forms of parental care, effort expended in nest building involves a cost to the parents, with a corresponding benefit in the form of better egg or chick survival. Hence the extent of nest-building effort is determined by a trade-off between parental costs and reproductive benefits, and this is reflected in the size and complexity of the nests built. Many species build quite complex nests but expend little energy in doing so. Indeed, when circumstances dictate, a nest may be completed in less than a day (Skutch, 1976). But in other species the costs of nest construction may be limiting. One clear example of this is provided by the Eastern Phoebe, which nests commonly in the eastern United States, building a large nest on rocky bluffs, in mouths of caves, or on overhangs on stream banks, and on man-made equivalents (buildings, bridges, and culverts) (Weeks, 1978). Nests are relatively large and of two types: (1) nests supported from below ('statent' nests) and weighing 200 g or less, and (2) nests without support from below ('adherent' nests) and generally exceeding 400 g in weight. Nest construction takes several days in both cases, but the larger adherent nests are additionally costly because the female is unable to perch alongside during the early stage of construction and has to hover while applying nest material. Consequently, birds building adherent nests take longer to recover and initiate egg-laying than do birds building statent nests, and they subsequently produce significantly smaller clutches. Eastern Phoebes also repair old nests (of both types) whenever these are available, renovating them by the addition of moss to the rim and of a new lining, and will at times modify old Barn Swallow nests for the purpose. Figure 2.1 shows the selective advantage of such repairs: birds availing themselves of pre-existing nests averaged a 5.6 per cent gain in clutch size over those birds building afresh. Thus, energy expended on nest construction can detract from the parents' capacity for subsequent investment in eggs and young.

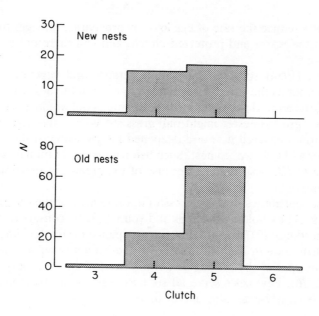

Figure 2.1 Clutch size distribution of Eastern Phoebes, *Sayornis phoebe*, building new nests (top) or adapting pre-existing nests (bottom). Data from Weeks (1978).

EVOLUTION OF NESTS

Evolutionary pathways of behaviour patterns can often be traced by comparing the same behaviour in primitive and in advanced taxa. However, amongst birds the variety of nests and functions thereof, even within a single taxon, is such that the styles of nest building have little evolutionary significance and provide no assistance in tracing the origins of their construction. Three explanations have been advanced, each of them related to the survival of eggs and young. The first is based on the descent of birds from reptiles, some of which lay their eggs in hollows excavated in the ground and subsequently warmed by the Sun. Such is the tactic of the Australian megapodes and, in times of danger to egg or young, of various ground-nesting plovers. The second explanation is that, during courtship, displaying birds are frequently in an ambivalent mood, for males between attacking the female as intruder and mating her as partner, for females between fleeing the male's aggression and approaching him as mate. Re-direction of aggression, as in the grass-pulling of Herring Gulls in territorial disputes, may result in accumulation of vegetation around the laying site. Similarly, a balance of attack and flight tendencies may lead a tern to circle the prospective mate at constant distance whilst the latter itself turns to face it constantly, thus creating a circular scrape. Any such development – of vegetation ring or shallow scrape – can subsequently reduce the rate of egg loss, by preventing the eggs from rolling away from the warm and protected clutch, and will therefore evolve as an adaptive trait.

Makatsch (1950) suggests that simple scrapes and vegetation rings developed further as the sitting birds tended to deepen the nest-cups by adding to the rim, the natural position to which the sitter's bill would add material. In vegetation a ground-nester might pull grass stems over the nest and into its walls to form a covered nest and deepened interweaving of the walls might form a door and fully woven nest. Such behaviours would then be selected for and evolve to full nest building because of the greater production of young which results.

The nest contributes to the successful development of the offspring in any of six ways: (1) by holding the eggs and young; (2) by conserving warmth of eggs and nestlings; (3) by concealing the occupants; (4) by shielding the clutch or brood from rain or hot sunshine; (5) by restraining predator attacks; and (6), in some species, by serving as dormitories for some or all of the family (Skutch, 1976). Few nests serve all six purposes and in many instances the nest form is clearly geared to one of them.

NESTS AS CONTAINERS

As containers, nests are functional only in creating a stable platform to hold eggs and young in trees or shrub, against cliff-faces, on soft ground, or even floating on water. Such a basic function to nest building occurs mostly amongst the non-passerine birds. Indeed, many non-passerines do not use a nest construct at all but rely on simple scrapes on the ground or on a ledge. Of those that do build, the majority construct simple nests of earth, stones, or vegetation. Passerine birds, on the other hand, build nests of complex form and structure, to compensate for their small size in relation to weather and predation (Welty, 1975).

Some species have lost the habit of nest building. Most of the 39 species of tern nest colonially on the ground, but a few tropical species have taken to tree-nesting, even in the absence of ground predators. The Black Noddy builds a small platform of leaves on which to lay, yet loses many eggs and chicks during high winds. The Fairy Tern is even more vulnerable, laying on a bare branch! Houston (1979) suggests that such habits may be associated with the avoidance of ectoparasites. Nesting in trees results in nests being further apart in the three-dimensional space afforded by the foliage than they would be if placed on the ground below. For Black Noddies on Cousin Island in the Seychelles, Houston found that nearest neighbour distances (in three-dimensional space) averaged 5.1 m in trees: such a ground density would have left them only 0.8 m apart on the flat. Nest material also harbours ectoparasites, so reduction in nest material can reduce the possibility of disease and of parasite transmission. Houston found that none of the 17 Fairy Tern chicks he examined bore feather lice but 18 of the 25 Black Noddy chicks examined were infected. Statistically, such a difference is highly unlikely to occur by chance. It has also been shown for Bank Swallow colonies that ectoparasite infestation increases with the density of nests in the colony (Hoogland and Sherman, 1976). If losses to ectoparasites are high enough (see Chapter 12), and if nests need not be constructed against environmental stress or against predation, it may be selectively advantageous for the Fairy Tern to accept the losses of eggs and chicks consequent on the evolutionary loss of nest construction.

NESTS AND MICROCLIMATE

Nests positioned in different places experience dissimilar environmental regimes. For the eggs to experience stable conditions, therefore, nests must either be constructed differently in each place or be positioned only in places with an acceptable microenvironment. Nests may also be positioned to ameliorate the conditions experienced by the sitting bird (and thereby the cost of providing the thermal conditions required by the eggs and young). Thus the microclimates of individual Antarctic Skua nest sites studied in Antarctica were rather milder than the general environment, indicating a degree of selection by the adults (Le Morvan et al., 1967). Such choice of nest site can show a dividend in energetic terms. For example, hole-nesting species dwelling inside the trunks of living trees experience a more stable and higher temperature regime than do birds nesting on branches, and this is reflected in earlier onset of breeding (Gofman, 1977). Indeed, even within a single hole-nesting species nest temperatures may vary from place to place. In Wytham Woods, Oxford, certain nest-boxes are laid in by Great Tits relatively early each season, and these proved to have above average insulation properties. The females in these boxes therefore conserved more energy than did conspecifics elsewhere and were able to accumulate the energy required for egg formation sooner (O'Connor, 1978d). Skowron and Kern (1980) review some of the reported explanations of this type advanced for individual nest construction practices. Thus, as a broad generalization, nest-site selection is directed to ameliorating the energetic stress on the sitting bird. The better insulated a nest is, the lower is the thermoregulatory cost to the sitting bird, but the more the better insulated nest costs to build in the first place. In practice, nest conductances vary between species but in cold climates they are broadly in line with adult plumage conductances (Whittow and Berger, 1977). That is, even in cold climates nest-building effort is limited to providing insulation comparable to that of the sitting bird's own plumage, so that heat applied to the eggs through the brood patch is lost through the nest only at the same rate as metabolic heat is already lost through the adult's plumage.

Nest orientation

The orientation of the nest may play a significant role in the shelter and warmth afforded its contents. The Cactus Wren of the Arizonan desert is one species for which nesting success has been proved to be related to nest orientation. The species builds a closed retort-shaped nest with a lateral entrance in a cholla cactus or other spine-bearing shrub. Early nests are usually placed with the entrance facing away from the prevailing winds whilst later, in the hotter part of the breeding season, nests are oriented into the wind (Ricklefs and Hainsworth, 1969). Examination of breeding success in these nests showed that nests correctly oriented in the preferred direction produced more young than did nests in other directions. Hatching success did not vary in this way, so it must be the survival of the young in nests protected from excesses of heat which is crucial (Austin, 1974).

Nest construction

Grebes have a different method of using the Sun to provide heat for their developing eggs. They build floating nests anchored to aquatic vegetation, with a large mass of submerged material breaking the surface as a shallow disc on which the eggs are laid. On sunny days the adults do not sit but leave the dark wet covering material to absorb heat from the Sun. Wet material is a good heat conductor and egg temperatures are therefore maintained (Bochenski, 1961). After sunset, though, the nest cools rapidly and must be tended by the adults. Similar solar heating is exploited by the Egyptian Plover, which nests on river sandbars and covers its eggs with a layer of sand to improve absorption.

A second mode of energy conservation available to grebes and other species using putrefying nest material is the use of the heat of decay of such material. In freshly built nests putrefaction is slow, but after the adults have incubated their clutch for a few days the elevated temperatures experienced by the vegetation in the nest-cup induces rapid decay. Nest temperatures therefore rise through the incubation period, even in the absence of the adult (Bochenski, 1961). Great Crested Grebes in fact complete the building of their nests during incubation and the nest mass available for putrefaction thus grows greatly just as putrefaction accelerates.

PROTECTION FROM PREDATORS

Nests provide protection from predators by virtue of their inaccessibility, impregnable construction, or concealment, through colonial nesting, or through association with aggressive animals capable of driving off predators.

Nests are generally made inaccessible by their location on islands, on cliff-edges, in burrows or cavities, at the tips of tree branches, and on or over water. Each tactic excludes major classes of predators, though not all, and birds may be particularly vulnerable to those animals which succeed in overcoming the defence offered by the birds' choice of site. Seabird islands provide spectacular examples of inaccessible nest sites, with many thousands of birds nesting perhaps only a bill-length apart from each other.

A few species provide their eggs and young with such protection by creating their own islands. The Horned Coot of the Bolivian and Chilean Andes constructs a conical island of stones and pebbles in the shallows of a mountain lake, a mound extending to perhaps a metre in height and comprising a tonne or more of material, and builds a nest of vegetation on this. Grebes, Moorhens, and other waterfowl similarly build islands of floating or semi-floating vegetation on which to lay, usually at the edge of emergent vegetation in lake or river.

Cavities, burrows, and cliff-edges offer such good protection from most predators that birds dependent on them may be limited by their availability. In the Dinas reserve in Wales the supply of additional sites in the form of nest-boxes for the hole-nesting Pied Flycatcher was followed by an immediate increase in breeding density there. In Michigan the success rate of Eastern Bluebirds in natural nests (55.6 per cent) was identical to that of pairs in nest-boxes (55.5 per cent) but the proportion of nests in natural holes decreased as the season progressed, due largely to competition from other cavity-nesting species such as the Starling, Tree Swallow, and House Wren (Pinkowski, 1977).

In trees, inaccessibility is best obtained by suspending nests on trees without access from vines or branches on adjacent trees, as with typical orapendula nests, or by nests suspended from the ends of twigs or vines dangling over water, a typical site of the Northern Royal Flycatcher of tropical America (Skutch, 1976) and of the Penduline Tit of Europe. Such sites are sometimes natural, at other times the result of the bird's activity. Red-headed Weavers in tropical Africa spend much time snipping leaves from the supporting branch and surrounds of its nest, as does the Strange Weaver. The effect is to prevent the entry of snakes into the down-pointing entrance of the nest (Skutch, 1976).

Thorns may provide the desired inaccessibility for some species. The effectiveness of thorny vegetation in protecting nests against predation is shown by Lack and Lack's (1958) study of the Long-tailed Tit. Breeding success in nests placed in brambles and other thorny shrubs ranged from 12 to 47 per cent, but nests in thorn-free oak or ash trees were always totally unsuccessful.

The ovenbirds *Furnarius rufus* and *Furnarius cristatus* of the South American grasslands provide a much cited example of impregnability as a defence. The nest is in the form of a globe containing 3–5 kg of mud made up of sand and cow dung mixed with grass and hair. Given adequate supplies of mud and hair, a pair can complete the nest in 4–5 days. The nest contains two chambers, one of them an entrance passageway following the curve of the outer wall. The inner chamber, protected behind the inner wall of the passageway, is lined with grass and a few feathers. Once the mud has hardened the walls are extremely strong and resistant to many forms of attack, both from animals and from weather.

Other species limit such defences to the reinforcement of natural sites. European Nuthatches, for example, add mud to the rim of any over-sized tree cavities, until the access hole is suitably small and impassable to larger predators. Again, the Red-billed Hornbill of the African savannah walls the female into the nesting cavity, leaving only a small slit readily defensible by her massive bill through which to provision the female and small young.

Concealment of eggs and young

Concealment of eggs, young, and adults is a major anti-predator function of nests in many species. For some the nest structure is completely cryptic, with the outer part composed of fine lichens or other vegetation appropriate to the surroundings. For other species the eggs are highly cryptic and are held together with a sparse rim of shells or small stones drawn from the surroundings. In extreme cases, such as ground-nesting Nightjars and Wood-cock, the sitting bird is itself so patterned as to be marvellously camouflaged and the incubator can then sit tight in the face of even a very close approach by the potential predator. But for other species the concealment is offered by the siting of the nest. Early-nesting Yellowhammers, for example, build in low (initially the densest available) hedgerow vegetation in early spring and rather higher as the season and vegetation profile develop (Peakall, 1960).

Colonial nesting and predation

Colonial nesting may reduce predation in two ways. First, the sheer violence of attack by a large number of birds may deter the predator from continuing to hunt in the colony. Kruuk (1964) found that Carrion Crows and Herring Gulls showed avoidance reactions to attacks by Black-headed Gulls. The crows especially behaved timidly when attacked on the ground and were often seen to make large detours around the colony rather than pass over it. Second, predators may be so distracted by the attacks of the birds that their hunting efficiency whilst in the colony decreases. Kruuk found that predation success decreased with frequency of attack by gull. By distracting the predator's attention with their attacks, the gulls make it harder to spot the camouflage of the egg and chick patterning.

Association with aggressive animals

Table 2.1 shows some of the reported associations between nesting birds and an aggressive species, in most cases with a presumed and in some cases a demonstrated benefit to the weaker species. Such defence may also be combined with other defensive measures. For example, Magpies suffer

Table 2.1 Some examples of protective associations between nesting birds and other species

Nesting association between species	Species	Nesting associate	Benefit
Mixed species colonies	Terns	Black-headed Gulls	Mutual augmentation of colonial defence
Association with aggressive species	Eider	Black-headed Gull colony	Anti-predator behaviour of gulls
	Snow Goose	Snowy Owl nest	Owl is main enemy of arctic fox, the chief risk to the goose
	Merlin	Fieldfare colony	Merlin raptorial capabilities deter thrush predators but colonial mobbing by the thrushes increases Merlin nest success
	House Sparrow	Imperial Eagle nest	Sparrow nests built into huge structure of the eagle nest are effectively unapproachable by predators
Association with stinging insects	Long-tailed Tits	Red ants, *Formica rufa*	Insect stings deter approach by nest predators
	Black-throated Warbler	Hornet nests	
	White-collared Kingfisher	Tree termites	
	Dull-coloured Seed-eater	Wasps, especially *Polistes canadensis*	
	Chestnut-headed Orpendula	Wasp and bee nests	Wasp/bee swarms exclude botflies, a major nesting parasite
Association with vertebrates	Water Thick-knee	Crocodile nest	Brood guarding by crocodiles excludes egg predators
	Magpie	Man (dwellings)	Carrion Crows, an egg predator of Magpies, are shy of humans
	Masked Weaver	Man (traffic streets)	Heavy traffic precludes nest approach by predators

considerable interference from Carrion Crows, in some cases being evicted from their nest or territory by the latter species, in other cases losing eggs and young. But because crows keep away from the immediate vicinity of human dwellings whilst Magpies are less shy, one defence is for the Magpies to build their nests near houses. A second is to build a dome over their nest. Each of the two measures contributes to the final reproductive success of pairs employing both. These defences were adopted (at least in part) as a result of experience: 66 per cent of the Magpies nesting near human habitations were experienced breeders whilst only 42 per cent of those nesting in remote areas were experienced (Baeyens, 1981).

COMMUNAL NESTS

Although most nests are tended by a single pair, examples exist of communal nests in which more than one female lays. Some such nests have the character of apartment houses, with multiple nest chambers within a single nest mass. The Sociable Weaver of Southern Africa is one example, the birds building massive nests several metres across and containing perhaps 80 individual nesting chambers. Other nests of this type have a single clutch into which several females lay. Examples are the anis and the babbling thrushes, in which several pairs combine to construct the nest in which eggs are incubated and nestlings reared communally. Amongst the magpie-larks, Grallinidae, of Australia similar behaviour occurs, but only the more dominant females in the group lay eggs (Welty, 1975). Ostriches also nest communally, with the several females of the male's harem laying into a single nest and incubation being performed by the male and dominant female (Bertram, 1980). Over 130 species engaging in some form of communal nesting are listed by Skutch (1961) and for Africa alone Grimes (1976) lists 52 species.

NESTS AS DORMITORIES

Niche nests, covered nests, and cavity nests all lend themselves to use as dormitories in which adults and offspring may spend the night in shelter and relative safety from predators. Open nests provide little protection of this type and are used for slumber mainly by waterbirds, notably coots and gallinules. The male Moorhen builds special nest platforms away from the egg-nest and broods damp and tired chicks there. Young Magpie-geese construct similar platforms on which to rest in their Australian swamps, doing so first when only 2–3 days old.

Skutch (1976) reviews the possible pathways by which dormitory use of nests might evolve. A first step is exemplified by the Sulphury Flatbill, an American flycatcher in which the female spends all her nights in her nest – from building through nest cycle to nest dilapidation – as long as it is usable, roosting in foliage thereafter. The male never shares the nest in this way. A second step is for both parents to sleep in the nest throughout the breeding period, the case for a wide variety of closed-nest species. Since only one adult can incubate, the other is present only for comfort or convenience. This is the case also where nest-helpers sleep in the nest, as with the Black-eared Bushtits studied by Skutch in Guatemala. The third stage of dormitory use of nests is where fledglings return to the nest to sleep, often alongside the young of later broods. This is typical of many swallows, swifts, and some other migratory species. Such sleeping habits end in these birds with the onset of autumn migration. Few examples of the behaviour in resident species have been well enough studied to know the cause of its cessation. In single-brooded species the offspring abandon the used nest once fully able to fly.

THE MOUND-BUILDERS

Quite the most extraordinary 'nests' of any birds are the mounds of fermenting organic matter used by the Megapodidae of Australasia. One genus of the family, *Macrocephalon*, lays in holes dug in the sand near the shore, incubation heat subsequently coming from solar heating, and certain species in the other five megapode genera also resort to solar or volcanic heat. Most members of the group, though, rely on the fermentation of vegetable and organic matter in nesting mounds raked together by the adults (Clark, 1964). The nesting behaviour of the Mallee Fowl has been particularly well studied by Frith (1959). Work on the mounds begins in winter, the birds spending as much as 11 months scratching leaves, twigs, and sand from a large area (many square metres) around the growing nest pile. If the weather is favourable, this material ferments and then cools towards incubation temperatures (in this species about 32–36°C, rather lower than for many bird species). The birds dig out a central laying chamber in the fermentation zone and lay eggs into it through their breeding season. The birds regularly attend the mound and regulate its temperature by adding or removing material as required or, late in the season when fermentation rates have fallen, by opening the mound to allow the Sun to heat the eggs. The eggs receive little additional attention, not even being turned, and, in keeping with the relatively low incubation temperatures, develop only slowly, over 50–100 days. The chicks themselves are quite incredibly independent, kicking (not pecking) open their shell, digging themselves free from their burial chamber, and capable of weak flight on the day of hatching! The escape of the hatching chick is aided only by the generally loose texture of the mound, consequent on its constant working over for thermoregulation by the adults (Clark, 1964).

SUMMARY

Nest construction is a reproductive activity which may detract from other forms of reproductive investment but which has evolved because of the greater production of young which results with nests. Nest construction is more strongly developed amongst passerines than amongst non-passerines and is directed principally towards the amelioration of climatic or predation burdens. The energy costs of caring for eggs and young can be reduced by orientation of the nest to take account of the daily path of the Sun and by building nests of materials whose heat and fermentation of decay will supplement incubation heat. Protection from predators may be achieved principally through concealment of the nest or by placing it in a position inaccessible to predators. These defences may be reinforced by nesting colonially or in association with aggressive animals so that attacks on potential predators are heavier than a pair could provide alone. Amongst precocial species nest use is principally restricted to the incubation period, but in altricial species nests may continue in use as dormitories long into the post-fledgling period.

CHAPTER 3

From egg ...

The earliest stages of development of young birds are spent in the egg. To a large extent, what happens to the young at this time is determined by the biology of the adults. The colour, size, and endowment of the egg are all features over which the embryo has no direct control, but these features are open to selection for their effects on the survival of the egg and of the embryo it contains. Similarly, the incubation regime to which the egg is subjected is set by parental behaviour and is not under the immediate influence of the embryo, except perhaps when nearing hatching. In the present chapter the general patterns of egg size and composition and of incubation behaviour are reviewed, with discussion of embryonic development left for Chapter 4. Processes of egg recognition are also considered, since they determine egg care by the parents, but other aspects of egg biology less directly concerned with the development of young – egg laying, size of clutch, and so on – are not discussed; they are reviewed in depth by Lack (1968).

EGG SIZE AND COMPOSITION

Shape

The majority of birds' eggs are oval and their shape has no particular significance. Many shore-birds lay three or four pyriform (top-shaped) eggs which they arrange compactly with the narrow end inwards: incubation heat is dissipated less rapidly from such a group. Similarly shaped eggs laid by auks are adapted for survival on bare cliff-tops: if disturbed, they roll about their narrow end in a circle, rather than off the nest ledge (see below).

Colour and pattern

Bright yellow is about the only colour unrepresented in the eggs of one species or another (Skutch, 1976). The egg may be uniform in colour or may be overlaid with patterns of spots, streaks, or blotches of one or more additional colours. These markings are applied to the egg by pigment glands in the wall of the uterus and their pattern is determined by the motion of the egg at application. There can therefore be substantial variation in patterning

29

amongst eggs even of the same clutch, though the main characteristics are under fairly tight genetic control.

Two of the selective forces for egg colour have been identified experimentally. Holyoak (1969) darkened the pale eggs of Jackdaws and found that 85 per cent of the darkened eggs were lost by the birds putting their feet through them, against a maximum of 9.8 per cent of control eggs which disappeared before hatching. White eggs are thus a safer bet for a hole-nesting species (Lack, 1958). The other trend Lack noted was towards cryptic colouration of more exposed eggs. Tinbergen *et al.* (1962) found that plain white and plain brown eggs near a gull colony were predated by crows more heavily than were similar numbers of eggs with the spotted natural pattern, thus showing the advantage of the gull patterning. But this applies only if the crows are prevented from spending a long time hunting for the eggs, and normally the crows are under attack by the gulls. As is often the case, the advantage is paid for by a disadvantage in another direction: patterned eggs exposed to direct sunlight reach lethal temperatures sooner than do lighter but less cryptic eggs (Montevecchi, 1976a).

A further and intriguing trend in egg colouration is that the third (and normally the final) egg to be laid in gull and tern clutches is often uniquely patterned (Chamberlin, 1977; Gochfeld, 1977). Since third eggs in gull clutches are those least likely to produce fledglings because of sibling competition for generally limited food (Parsons, 1970), it is possible that this unique patterning serves the adaptive end of concentrating predation on the least valuable egg of the clutch. Under intensive mobbing by the adults, predators are likely to take the most conspicuous egg and to carry it away to consume in peace. Montevecchi (1976b) found that small gull eggs were particularly vulnerable to American Crow predation because they could be carried off; larger eggs needed to be broken and the contents eaten on the spot by the crows whilst under attack by the gulls.

Many ground-nesters are themselves extremely cryptically coloured and sit close until discovered by the predator. In such species, e.g. Nightjars, the eggs are often less cryptic than the adult bird. Another tactic is to cover the eggs when leaving them unattended. Such behaviour is well known in Temperate Zone species of grebes and waterfowl (Anatidae) as well as in South American screamers (Anhimidae) (Kear, 1970a) and tinamous (Tinamidae) and in at least 13 species of waders (Charadrii) in the tropics (Maclean, 1975). Maclean believes the relative frequency of egg covering in tropical ground nesters is correlated with a greater diversity of predators there, but he shows that in at least some cases the covering serves to protect the eggs against overheating in the hot sunlight when left unattended. Covering also serves a thermoregulatory function – but against heat loss, not gain – in grebes (Bochenski, 1961) and in ducks (Chapter 8).

Egg size and adult weight

Statistical analysis of the egg weights of over 800 species show that within each order or family egg weight is related to body weight by the function

$$E = a \ W^{0.67} \ , \tag{3.1}$$

where E is egg weight, W is body weight, and a is a constant for each order or family (Rahn et al., 1975). Figure 1.2 illustrates the results for falcons. The two-thirds exponent in this result means that egg weights are more or less proportional to the surface area of the female, within any order or family. Small species produce proportionately larger eggs for their size than do larger species.

There are substantial differences in egg size between different orders, once body size is allowed for. A 100 g falcon, for instance, lays an egg of some 14 g whilst a 100 g pigeon lays only a 7 g egg (Rahn et al., 1975). What accounts for such differences? Table 3.1 shows that egg weights for birds of a specified body weight are high in those orders producing young covered with down at birth (i.e. ptilopaedic chicks) and low in those hatching naked chicks. It seems that development of down whilst within the egg is feasible only if the female puts more nutrients into the egg at the outset. Note that all the nidifugous groups have heavy eggs for their size, as expected. However, other groups, whose young are downy but remain at the nest, also start from unusually large eggs, so the correlation is with the development of down and not with the independence of the young. In most of these downy nidicoles (species in which the young remain at the nest) the young are left unattended for long periods by parents searching afar for food. Seabirds, owls, and raptors all come into this category. The one order with heavy eggs but naked chicks is

Table 3.1   Relative egg weight in various orders of birds in relation to the presence or absence of natal down

	Egg weight for 100 g female[a] (g)	Downy young?
Cuculiformes	4.5	No
Psittaciformes	6.6	No
Piciformes	6.6	No
Columbiformes	6.8	No
Passeriformes	7.7	No
Galliformes	9.2	Yes
Coraciiformes	10.3	No
Strigiformes	11.5	Yes
Gruiformes	12.9	Yes
Falconiformes	13.7	Yes
Anseriformes	14.2	Yes
Charadriiformes	17.4	Yes
Procellariiformes	21.1	Yes

[a] Calculated from equations in Rahn et al. (1975).

the Coraciiformes, reflecting an analysis based on the kingfisher and hornbill families. These are unusual in certain respects, such as the unique behaviour of walling of the female hornbill into the nest for the incubation period and for some or all of the nestling period (p. 24). The unusually small eggs of the Cuculiformes (Table 3.1) are the result of their parasitic habits. Small hosts have small eggs, so even large cuckoos must reduce their egg size to match.

The absolute range of egg weight between different species is huge, ranging from about 0.3 g for some hummingbirds to about 1600 g for the Ostrich. Eggs of 8 and 9 kg are reported from the extinct Aepyornithidae (Rahn et al., 1975). This range corresponds to eggs of from 25 per cent down to 1 per cent of female body weight.

Variation in the size of the egg has implications for certain aspects of development. Thus the incubation period I of an egg is related to its weight W by the equation

$$I = 12.03 \ W^{0.217} \tag{3.2}$$

in days (if W is in grams) (Rahn and Ar, 1974). Hence doubling egg size involves a 16 per cent increase in incubation time, with consequent increases in exposure to predation and the risk of bad weather. A species' egg size is thus a compromise between the greater development of the embryo possible in a relatively large egg and the disadvantages of the longer incubation required to achieve this, but this interaction is also affected by the ability of the female to mobilize the nutrients needed for eggs of various sizes.

Within a species there can be considerable variation in egg weight, both between females and between the clutches and eggs of any one female. In general, eggs are smaller whenever a female has difficulty in acquiring enough food for egg formation, whether this difficulty be through inexperience or through seasonal variation in food levels: eggs are larger in good feeding conditions.

Are there particular advantages in hatching from a large egg? In some species, and at certain times, the answer is yes. To begin with, large eggs may hatch more successfully, as happens among Woodpigeons (Murton et al., 1974). But there exists a general correlation between the size of the egg laid and the weight of the chick hatched from it (Figure 3.1). In some species a heavy chick is simply a generally larger chick which has developed to a greater extent while in the egg. In other species the larger egg produces a hatchling of constant body size but with a larger yolk sac, and such chicks can survive without food for longer periods. All sorts of intermediate conditions between these extremes occur. In the Herring Gull large egg size is correlated with larger fat reserves at hatching, reserves which the chick may depend on until the parent–offspring bond is strongly formed (Parsons, 1970; Hahn, 1981). In the Laughing Gull large eggs yield chicks with bigger fat reserves which are probably subsequently the source of the metabolic water needed to counter the salt-laden diet (Ricklefs et al., 1978). In Japanese Quail the large eggs

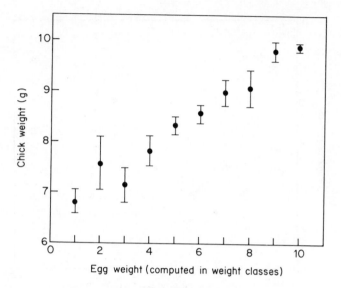

Figure 3.1 Mean weights of Japanese Quail, *Coturnix coturnix*, chicks hatched from eggs of different weight (range 8–14 g). Vertical lines show one standard deviation on each side of the mean. From unpublished data (R. J. O'Connor and B. Owen, manuscript in preparation).

yield chicks with heavier coats of down which are therefore better insulated against heat losses (R. J. O'Connor and B. Owen, manuscript in preparation). Thus different species lay larger eggs than usual in response to their specific ecological needs.

Schifferli (1973) found that Great Tits which hatched from large eggs in late nests fledged more successfully than did young from small eggs. The former grew faster and reached larger maximum weights. In early broods the nestling food supply (various defoliating caterpillars) is plentiful and the small-egg chicks catch up in size before fledging, but for late nests the caterpillar supply has dwindled as the insects pupate and the small chicks retain their handicap. In a separate study by Perrins (1965) it was found that Great Tit survival was greater for heavy nestlings than for their lighter contemporaries (Figure 3.2).

Egg composition

Figure 3.3 shows the gross morphology of the newly laid egg of the domestic fowl. The egg consists essentially of a fertilized ovum surrounded by various life support tissues and by a hard shell. The shell provides the mechanical strength of the egg and its thickness therefore varies between species. Thicker shells give greater strength but make it harder for the developed embryo to cut itself free at hatching: shell thickness in each species is thus a compromise

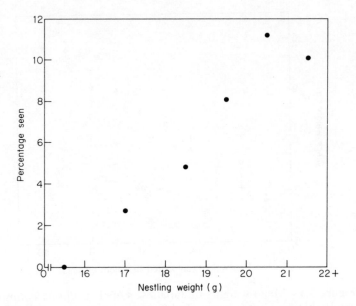

Figure 3.2 Percentage of Great Tit, *Parus major*, nestlings seen a month after fledging in relation to their weight at 15 days of age. Data from Perrins (1979).

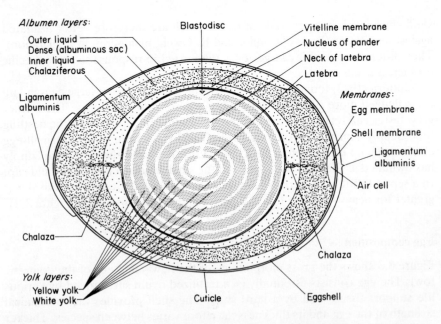

Figure 3.3 The gross morphology of the egg of the Domestic Fowl, *Gallus domesticus*. Redrawn from Romanoff and Romanoff (1949).

between opposing selective forces. In many species calcium is drawn from the shell during the skeletal growth of the embryo, leaving the shell thinner towards hatching than at laying. Abnormal egg shell-thinning in the wild often stems from the presence of pesticide residues in the diet of the laying female, and such eggs suffer heavy breakage losses during incubation (Newton, 1979).

The shell is permeable to gases, which penetrate the shell through a mosaic of pores. As the new-laid egg cools, air is drawn into its interior to form a small air space between the two shell membranes, usually at the blunt end of the egg. Movements of oxygen, carbon dioxide, and water between embryo and air space and between air space and environment provide for the respiratory needs of the developing embryo, and the size of the air space therefore enlarges during incubation. The air space probably serves an important water-conservation end. Most eggs lose about 18 per cent of their initial weight during incubation, largely as respiratory water (Rahn and Ar, 1974), and this water has to be provided from within the egg. Simkiss (1974) found that the existence of an air space within an egg provides a slightly cooled zone in which expired water is partially re-condensed, with a net reduction in losses to the environment. Water is the main solvent for biochemical processes within the egg, so its retention is important: dehydration at critical points during incubation is a significant source of embryonic mortality (Romanoff, 1967).

The yolk contains the main reserves of food substances for the developing embryo. Its size varies between species, with generally larger proportions of yolk in the more precocial species (Table 3.2). Part of this extra yolk feeds the greater developmental maturity of these embryos and part provides a food reserve at hatching. Ovovitellin is the main protein of the yolk and is particularly rich in sulphur and phosphorus. Proteins account for about one-third of the dry matter of the yolk, but lipids form a still greater proportion and explain the high energy contents of eggs of precocial species (Table 3.2). Altricial and precocial species also differ in respect of the water content of the yolk (about 10 per cent higher in the former) and in the amount of the phosphatide lecithin present. This latter acts as a growth stimulant and

Table 3.2 Yolk and albumen content of eggs of various developmental modes. Calculated from data in Ar and Yom-Tov (1979), Ricklefs (1977a) and Ricklefs and Montevecchi (1979)

Developmental mode	Yolk content (%) (mean ± s.e.)	Albumen (%) (mean ± s.e.)
Altricial	21.8 ± 0.65	70.4 ± 1.38
Intermediate	29.9 ± 1.66	62.3 ± 1.63
Precocial	36.6 ± 0.82	53.7 ± 0.96

its higher content in altricial yolk accounts for the accelerated development of such young (Rolnik, 1970). The main carbohydrate in yolk is glucose.

The water reserves of the developing embryo are largely contained in the albumen. Total water content of altricial eggs is a few per cent higher than in precocial eggs of the same weight.

EGG RECOGNITION

Most birds do not recognize their own eggs, despite being able to recognize their own nest or nest site. Selection for egg recognition is of course generally lower than for chick and nest recognition, for once deposited in the nest eggs usually remain there. Experimental studies on Adelie Penguins have shown that many birds, especially experienced pairs, will accept whatever objects appear in their nest about their date of egg laying but will reject plastic eggs introduced before then (Frederickson and Weller, 1972). Birds of such species may go on incubating an empty nest whilst their eggs lie in full view before them. Other species, especially those breeding in crowded colonies without substantial nests, do recognize their own eggs as such and will incubate them away from the original site should they be moved. Such species include Black-footed Albatrosses (Bartholomew and Howell, 1964), Royal Terns (Buckley and Buckley, 1972), and Guillemots.

Failure to recognize eggs individually can occasionally result in birds incubating clutches containing eggs of two or more females. Such dump nests are notably frequent in ducks and pheasants, probably because in these species extra young do not require much extra care and selection for egg recognition is therefore weak. Egg discrimination is well marked in some species prone to parasitism by cuckoos, cowbirds, and other brood parasites. The marked resemblance of the parasite's egg to the host egg in the European Cuckoo, for example, could hardly have evolved in the absence of recognition of its own eggs by the host. A few species, such as the Catbird and American Robin, will normally eject Cowbird eggs from their nests and Cardinals will often desert parasitized nests, but other less regularly parasitized hosts do not discriminate the strange eggs.

Some ground-nesting species do not recognize eggs individually but will retrieve eggs lying a short distance from the nest. Retrieval consists in placing the bill beyond the egg and backing towards the nest whilst moving the bill from side to side to compensate for egg wobble. A wide variety of objects will elicit retrieval provided that they possess certain essential features (Tinbergen, 1973).

A few species have been proven to move their eggs from a disturbed nest site, most often by carrying the egg in the bill. Causes of disturbance eliciting such behaviour have included nest-site competition and drying out of the marsh around a Mallard's nest.

INCUBATION

The developing embryo does not produce enough heat to maintain the egg temperature at adequate levels and external heat is therefore an essential input to the energy balance of the egg. In most birds this is achieved by the parents applying their own bodies to the egg, but in the megapodes this heat is obtained by burial of the eggs in mounds of vegetation, in sand, or in volcanic material (the substratum depending on species). In grebes adult body temperature is rather low and parental incubation is assisted by the heat of decomposition of the decaying nest material (Bochenski, 1961). Where parental body heat is used, a brood patch – a defeathered and/or vascularized area or areas of the belly – is applied to the egg in most cases, but in some seabirds the webbed feet are used instead.

The behavioural aspects of incubation have been exhaustively reviewed by Skutch (1976) and the physiological factors have been reviewed by Drent (1970) and White and Kinney (1974). The primitive pattern of incubation is probably with both adults in attendance, either with both sharing incubation or with one incubating whilst the other brings food to the sitter. On either scheme the eggs are attended almost constantly. An alternative strategy is for a single parent, usually the female, to divide its time between attentiveness and foraging behaviour. This strategy usually occurs in association with a complex nest structure providing some degree of buffering against changes in environmental temperature whilst the adult is off the eggs.

Incubating birds do not adjust their heat production to regulate heat transfer to the egg. Instead, the amount of time spent in contact with the egg is adjusted in response to egg temperature. Egg temperatures are normally kept at about 35°C (see below). Below about 25°C attentiveness increases with falling ambient temperature. Above 25°C attentiveness decreases increasingly sharply until above 38°C, when the eggs are unattended. In open-nest species such temperatures often evoke special behaviours to cool the egg, such as the adult shading the eggs from direct sunlight or bringing water to the eggs.

Embryonic development does not begin to any significant extent until the egg is incubated to above about 35°C. Hence by deferral of incubation until clutch completion a bird ensures its clutch will hatch synchronously (or nearly so). Such delay is particularly important for birds with large clutches: a Blue Tit, for example, takes 12 days, occasionally more, to complete its clutch. Other species commence incubation with the laying of the first egg: the young hatch asynchronously as a result.

Incubation periods are themselves variable between species and are positively correlated with egg weight (Rahn and Ar, 1974). In general eggs laid in holes or burrows have longer incubation periods than have eggs of open-nest species, presumably because of their relative safety from predation. Within each species incubation periods tend to be shorter when the eggs are incubated at higher temperatures. In the Blackbird, the incubation period averaged 13.,7 days in March but only 12.7 days in June (when days were warmer than in March) (Snow, 1958). Such variations are significant in relation to egg mortality.

Incubation periods tend to be considerably shorter in cuckoos and other brood parasites than in non-parasitic species of comparable size. In the European Cuckoo this is in part due to the egg being partially incubated within the oviduct of the female before being laid (Lack, 1968). Such reduced incubation periods allow the cuckoo to hatch ahead of its host nest-mates, in time to eject them from the nest.

Incubation periods and nestling periods are frequently correlated within bird orders (Lack, 1968). Lack suggested that embryonic growth and nestling growth might be coupled together in such manner that the easiest (or even only) way to adjust nestling growth rate is to accept the related change in incubation period. But, as Case (1978) notes, no such correlation exists among mammals, as one would expect were such coupling physiological. Instead, Case offers the observation that both the eggs and the young of altricial species are tended in the same place (unlike mammals), so should be subject to parallel selective forces. That is, if it pays a species to accelerate the development rate of its nestlings, it also pays it to speed up the development rate of its eggs. Figure 3.4 shows that the mortalities of eggs and of nestlings

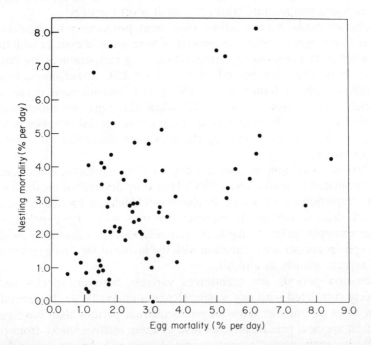

Figure 3.4 Daily mortality rates amongst the eggs and nestlings of altricial land-bird species, showing the general correlation between egg and nestling mortality. From data in Ricklefs (1969a).

within individual species are indeed correlated, as Case's idea demands.

Egg-shell thickness and porosity may provide the immediate mechanism for developmental adjustment. Species which must forage afar need eggs which do not dehydrate nor cool down too rapidly. Both are met by increase in shell thickness and by decrease in porosity, but the incubation period is consequently prolonged (Rahn and Ar, 1974).

Incubation temperatures

Most small birds probably incubate their eggs at 35 °C or higher. Spot measurements of temperatures of some 37 species of several orders averaged 34 °C during incubation (Baldwin and Kendeigh, 1972), but in most species they vary between 34 and 38 °C (Drent, 1973) and average still higher values during active incubation. In the Great Crested Grebe, egg temperature (measured directly under, rather than inside, the eggs) varied between 15.0 and 32.5 °C (Schiermann, 1927). There is thus significant variation between species in temperatures during incubation. In the domestic fowl no development took place when temperatures were below about 25 °C and no embryos survived continuous incubation below 35 °C or above 40.5 °C: optimal temperature for development was 37–38 °C (Lundy, 1969).

Departures from optimal temperatures have different effects at different stages of the incubation period. Unincubated eggs have considerable capacities for chilling resistance: Greenwood (1969) records the successful hatching of a Mallard egg frozen so deeply that the shell had cracked. The ability to resist chilling depends on the species. Eggs of the Peking domestic duck incubated for 24 hours suffered a 55 per cent mortality after 7 days at 4 °C whilst chicken eggs on the same regime all died (Hailman and Klopfer, 1962). Sensitivity to cooling increases with incubation in most species, both domesticated and wild (Hunter et al., 1976), and probably reflects the increasing development of the embryo and its temperature-dependent metabolism (p. 49). But chill resistance is well developed in embryos of all ages in certain species, particularly those in which the egg has frequently to be neglected by the adults hunting an erratic food supply during temporary shortages. European Swifts are occasionally hit by cold, wet, or windy spells, when food is difficult to locate, and the eggs may then be allowed to cool to temperatures lethal to less-adapted altricial species (Lack, 1956). Similar egg-neglect is common in the Procellariiformes, with eggs neglected for as long as 7 days subsequently hatching successfully (Pefaur, 1974). Pefaur suggests that such neglect extends the incubation period to avoid having the chicks hatch out at a time of poor feeding conditions, but such a tactic can work only if periods of food scarcity and abundance are relatively persistent.

Overheating of the eggs may be a problem for some species, particularly those nesting in sunny sites. The main physiological defence may be to apply the brood patch more closely to the egg so that heat flows from the egg into the adult's body, there to be eliminated by evaporative cooling (described in Chapter 8). Using model eggs whose temperature he could independently adjust, Franks (1967) found that closer application of the brood patch and increased panting (by gular flutter – see p. 131) were the principal responses of Ringed Turtle Doves to high egg temperatures.

The second major avenue of defence is behavioural. At moderate ambient temperatures the eggs can be shaded by the adult positioning itself between the eggs and the Sun, or the eggs may be covered over. In addition, water may be brought to the nest and sprinkled over the eggs to cool them. It is probably significant that belly-soaking – that is, going to a water source and immersing the underparts to soak up water in the belly plumage – has most frequently been recorded in the predominantly ground-nesting Charadriiformes, a group also the source of most egg-covering records (G. L. Maclean, 1974, 1975).

SUMMARY

Egg structure and composition are outside proximate control by the embryo but are subject to selection for their effects on survival. Eggs may be light in colour in hole-nests to let the adults see them, cryptic in ground-nesters to avoid predation, and individually variable in colonial species for individual recognition. Egg weights are greater in heavier species, but eggs are unusually small in brood parasites hosted by small species and are unusually big in species producing downy young. Large eggs have longer incubation periods but give rise to heavier hatchlings which may survive better in poor conditions. Within a clutch light eggs may be present wholly or in part as 'insurance' eggs. During development the yolk provides the main reserve of nutrients for the embryo and the air space provides a means of recycling metabolic water. Egg recognition is present mainly in colonial species and in those liable to parasitism. Embryonic development takes place at incubation temperatures above 35 °C, but certain species have well developed capacities to withstand chilling during periods of egg neglect.

CHAPTER 4

... To hatching

In the course of incubation the embryo changes from a totally dependent organism to one capable of a degree of independent life, producing a proportion of its own heat requirements and able in some species to communicate with parents and siblings even before hatching. This transition involves morphological and physiological development by the embryo. Altricial and precocial species alike must have sufficient behavioural competence to cut or break their way out of the egg at hatching, and the more precocial species must additionally be appropriately mature in behaviour and physiology to lead independent (or partly independent) lives on completion of their embryonic stage.

EMBRYO DEVELOPMENT

Embryonic development has been studied principally in domesticated species but also in a few wild species (Figure 4.1). Embryos are very small over the first 20 per cent or so of the incubation period but then enter a phase of rapid weight gain. A similar pattern is apparent in birds of all developmental modes.

Freeman and Vince (1974) provide a photographic record of the development of chicken embryos. Early embryos are small, light, and float on the yolk. On the second day the head region begins to bend into a cranial flexure at the level of the mid-brain, followed a day later by a further cervical flexure at the junction of trunk and head. Day 2 sees the start of differentiation of foregut, heart, brain, and inner ear, and, by day 3, of the stomach, lungs, and limbs. With the development of the central nervous system the embryo curls into the crescent shape apparent in most photographs. The albumen which initially surrounded the yolk loses water and sinks towards the small end of the egg, whilst the embryo becomes surrounded by the amnion and fluid begins to accumulate within the amniotic cavity. Eye pigmentation begins about this time, as do the first active movements of head and neck. By day 7 the egg-tooth has appeared, air sacs have differentiated, and the production of thyroid iodine has commenced; sex differentiation also becomes apparent at this stage. Bone mineralization starts on day 8. On day 10 the embryo is able to produce thryoxine and throid stimulating hormone (TSH), two thyroid hormones implicated in temperature regulation. On day 12 the embryo begins to absorb albumen and shell calcium and to consume the amniotic fluid. Two days later, the embryo has acquired control of neural, hormonal, and digestive function. At this stage various developments more directly linked to hatching considerations take place. The position of the embryo alters, still leaving the head just below the air space but re-aligning the body along the yolk, the yolk sac dwindles and the embryo increases rapidly in size. The yolk sac is finally withdrawn into the embryo about 20 hours before hatching.

This final phase of development is marked by the onset of substantial behavioural capability. Beak-clapping begins, providing a mechanism for communicating with the incubating hen (p. 46), and auditory sensitivity improves. On day 15 electrical activity in the optic lobes is detectable, thyroid activity increases, and growth hormone is secreted. By day 16 the embryo has cerebellar function, amino acid transport, and respiratory capacities. With the rapid growth of the embryo the head now pushes against the air sac. The position of the head alters to bring the beak across an amnion severely weakened by stretching under the size increase of the chick. By day 19 the hatching muscle has matured and clearance of fluid from the various sacs got under way. The following day the chick penetrates the air sac, can breathe and vocalize, and commences the uncapping of the egg.

Threlfall and his co-workers have recently begun to study embryonic development in wild birds (Maunder and Threlfall, 1972; Haycock and Threlfall, 1975; Mahoney and Threlfall, 1981). Figure 4.1 shows how the external morphology of a Kittiwake embryo develops. Head width and eye diameter have reached 50 per cent of final (embryonic) size by days 10 and 12 respectively; body length, head length, and bill length by day 13; tarsal length by day 18; and middle toe between days 18 and 20. Similar data for Herring Gull embryos reveal an almost identical pattern, except that tarsus length develops 2 days earlier. This is correlated with behavioural differences between the two species: the ground-nesting Herring Gull chicks run away when alarmed but the ledge-nesting Kittiwakes simply crouch and thus avoid the normally greater threat of falling from the nest (Cullen, 1957).

During development, embryos show evidence both of active and of passive movements which prevent adhesion of the embryo to the egg membranes. Passive movements of the embryo have three origins (Freeman and Vince, 1974): pulsations of the heart and large blood vessels between days 3 and 6, irregular myogenic contractions of the amnion (increasing in rate and amplitude from day 4 to day 9), and the changing geometry of the egg components as yolk sac and amniotic fluid are consumed and embryo size increases.

Active movements by embryos occur in three phases. Between day 3 and day 7 slight lateral flexions of the neck commence and head-lifting and head-bending also appear. These extend to the trunk region between day 4 and day 5 and become more regular by day 6. Limb movements are seen late

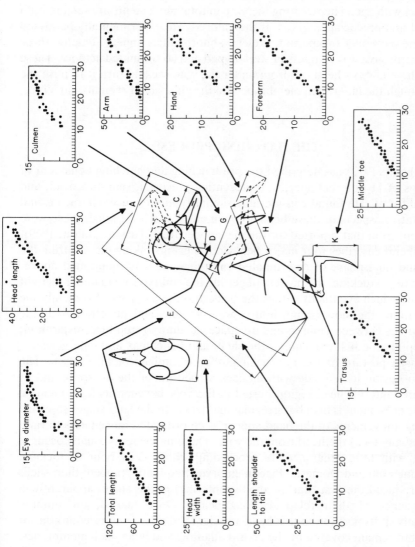

Figure 4.1 Growth in external dimensions of Kittiwake, *Rissa tridactyla*, embryos. Vertical axes all show sizes in millimetres; horizontal axes all show ages in days since laying. Modified from Maunder and Threlfall (1972).

in day 6, but their independent individual movement does not appear until day 7. Over the next week these rather regular movements become random and unco-ordinated. Nevertheless, they retain a periodicity, in that spells of activity alternate regularly with spells of stillness. The latter decrease in duration with age. The resulting increase in total movements are seen in both altricial and precocial embryos. If the embryos are experimentally paralysed for some days over this period, skeletal abnormalities appear. Finally, these movements give way to bouts of stereotyped and co-ordinated activity. These bring the embryo's head out from under the yolk sac and turn it to push the bill through the inner shell membrane into the air space (Freeman and Vince, 1974).

THE HATCHING PROCESS

Oppenheim (1972) has described the pre-hatching and hatching behaviour of embryos of 11 different species, representing precocial, semi-precocial, and altricial forms. He found considerable similarity in sequence, character, and quantitative aspects of these behaviours in all 11 species, though these were divergent from that reported for the megapode Brush Turkey (Baltin, 1969). Oppenheim describes four major behavioural events taking place during the third and final phase of co-ordinated embryonic activity mentioned above. These are (a) tucking, the process of getting the embryo's head under its right wing and which eventually brings the beak into the air space; (b) membrane penetration, the slow (many hours) wearing thin and penetration of the membranes between embryo and air space, leading to true lung respiration; (c) pipping, the first cracking of the shell; and (d) climax and emergence.

Embryo posture at the pre-tucking stage is similar for all species. The embryo lies on its left side with its neck arched under the air space and its beak and anterior head region buried in the yolk between its legs. General somatic movement is high but irregular and jerky. In the later stages so-called 'tucking' movements of the head appear as co-ordinated movements to raise (and usually turn) the head from the yolk. These movements usually occur in bursts, with little other activity proceeding within each bout of tucking. Laughing Gull and Domestic Pigeon embryos frequently 'pipped' their shells (i.e. produced the first 'star' of shell splintering) at this stage, apparently in consequence of their relatively long beaks. Once tucked, the embryo gradually shifts position so that beak and right shoulder penetrate into the air space but remain covered by the chorio-allantoic and inner shell membranes. In this so-called 'draped' position, head and bill movements wear a hole through the membranes. In most individuals this induces lung respiration (a few start shortly before penetration). Respiration is initially irregular but quickly strengthens. Oppenheim found that breathing in altricial species was generally weaker at this stage than was the case in precocial species. Altricial species also engaged in shorter bouts of bill-clapping than did precocial species at this time.

The hatching process is aided by two morphological adaptations of the full-term embryo. First, in most species except the megapodes (see below), a small hatching tooth is present on the upper mandible at hatching and is used to puncture and cut open the shell and its membranes (Clark, 1961; Parkes and Clark, 1964). Second, in many species a so-called hatching muscle at the back of the head and neck of the embryo provides the forces needed to penetrate the shell with this tooth (Fisher, 1962). The egg tooth is lost some time after the hatch and the hatching muscle also reduces in relative size.

Pipping (initial 'starring' of egg shells) takes place hours ahead of the hatching climax, the interval ranging from 15.4 hours in the House Wren to 40.9 hours in the Bobwhite Quail. Nice (1962) remarked the later pipping of eggs of the grebes which first 'star' just before hatching, presumably as an adaptation to the wet nests of this order (see Chapter 2). Pipping is achieved by forceful lifting or back-thrusting of the head and beak towards the shell. Such movements are significantly more frequent at this stage of the embryo's development and are accompanied by increased respiration rates. Vocalizations are first heard at about this stage.

Oppenheim found a rather sudden onset to hatching climax in embryos of all 11 species studied, followed by more or less sustained stereotyped hatching movements. These consisted of deep rapid exhalation (depressing the beak towards the chest), a strong extension of the tarsal joints deeper into the narrow (non-airspace) end, and a vigorous upthrust of head and beak towards the shell. Each sequence lasted 1–3 seconds and was repeated every 11–30 seconds. All but the earliest included a rotation of the embryo's body. This turning was always counter-clockwise from the site of the original pipping and resulted in steady chipping of the shell circumference. The shell cap fell off after a characteristic rotation, ranging from 65° in the Laughing Gull to 528° in the Bobwhite Quail. The value for the quail was due to the very thick membranes of this species: the shell frequently cracked after a 360° rotation but remained in place as the membranes had not torn.

Some species do not cut a shell cap systematically. Vince reports that Puffins move around the egg only 5° before the very thin shell breaks up longitudinally (Freeman and Vince, 1974). Hatching is similar in the Black-tailed Godwit (Lind, 1961). Brunning (in Freeman and Vince, 1974) found that in the Common Rhea the embryo enlarges the pip-crack and simply straightens its massive legs, shattering the large end of the egg. Such use of the legs differs between species, for when Oppenheim (1973) cut holes in the small end of eggs of chicken and of Bobwhite Quail and unfolded the legs of the embryos through the hole to dangle there, the quails hatched but the chickens did not.

Megapode eggs hatch at the base of a fermentation mound (Chapter 2) and the chicks have to dig their own way to the surface, obviously needing air around the beak during the ascent. Hatching position is consequentially modified in this species, with the head remaining between the legs at emergence and during the climb to the surface (Freeman and Vince, 1974).

VOCALIZATION BY EMBRYOS

Embryos of precocial and semi-precocial species begin to vocalize with lung ventilation but altricial species are silent until hatched. These vocalizations begin 1–3 days before hatching and in several species are markedly correlated with changes in parental behaviour by the sitting adult: rising, egg shifting, and calling each increase with pre-natal calling by the embryos (Impekoven, 1976). The link has been shown experimentally in Laughing Gulls, by playing recorded calls from a hatching chick into small loudspeakers placed below the nest. These calls appear to suppress aggressive behaviour by the adults: 15 of 26 Laughing Gulls sitting on unpipped eggs at the normal time of hatch (their own eggs had been experimentally replaced with younger ones) pecked a newly-hatched chick introduced into the nest although none of 31 birds sitting on a pipping egg did so (Impekoven, 1976).

PHYSIOLOGY OF THE EMBRYO

Oxygen

Oxygen requirements of the developing egg are met initially by simple diffusion. A blood vascular system then quickly develops, with haemoglobin synthesis detectable at 35 hours and a circulatory system detectable after 48 hours in the domestic fowl (Freeman and Vince, 1974). Erythrocyte concentration in the blood increases linearly with age until hatching, most rapidly in precocial species. Heart rate rises from about 200 beats/min on day 4 to 280 beats/min on day 10. It then decreases to about 260 beats/min in conjunction with an increased cardiac output (4.8 ml/min at 12 days but 6.3 ml/min at 17 days). In addition, the oxygen affinity of blood rises between days 14 and 17. Thus a variety of changes improve the chick's ability to acquire oxygen for metabolism.

Lipids

Lipids form the major energy source for the developing embryo, with about 80 per cent of the consumption taking place after day 14 in the domestic fowl. Eggs contain relatively little carbohydrate – less than 500 mg in chicken eggs – and this is consumed during early development. Once the liver differentiates (about day 6 or 7, perhaps earlier), it can synthesize and store glycogen, but the yolk sac membrane is an even more significant store. Both sites increase to a sharp peak in contents at day 18.

Water

Embryonic metabolism is closely coupled with the availability of water since no external source of water is available until hatching. Water is lost from the egg during incubation, about 8 g of the 40 g in a chicken egg being lost in this way. This loss is regulated in part by characteristics of the shell and incubation regime (Rahn and Ar, 1974) and in part by embryo physiology. During complete metabolism each gram of fat oxidized yields about 1.07 g of water, so the 2.5–3.0 g of lipid consumed in a chicken egg during incubation can provide up to 40 per cent of the water lost. Not until the final week does water balance come under neurohypophyseal control (Freeman and Vince, 1974). The developing excretory system of the embryo is consequently adapted for the conservation of water (see also the 'cold nose' effect of the air space described in Chapter 3). Ammonia is present from the start of development but its concentration decreases after the first week as the concentration of urea rises. Synthesis of uric acid can probably begin as early as day 5 and it becomes the major excretory product by about day 10. These products are initially stored in a diverticulum of the hind-gut, the allantoic sac, which increases in volume to a maximum of 7 ml about day 13. At this point water is re-absorbed, urates are precipitated, and sac volume declines (Freeman and Vince, 1974).

Hormonal control

Hormonal control of growth and thermoregulation has been demonstrated in respect of the thyroid (Freeman and Vince, 1974; Hall, 1979). In domestic fowl the thyroid develops slowly from its first appearance as a diverticulum of the pharynx until it achieves adult appearance on day 14. Figure 4.2 shows the pattern of its growth in Japanese Quail. Secretion of the two thyroid hormones thyroxine and triiodothyronine (T3) is regulated by the neurohypophyseal hormone thyrotrophin (also called thyroid stimulating hormone or TSH). Growth rates and metabolic rates can be depressed (and hatching retarded) by use of drugs antagonistic to thyroid secretions, but injections of the hormones do not accelerate embryonic growth. However, the uptake of amino acids by developing bones has been shown to be stimulated by T3, and treatment of the embryo with T3 or with thyroxine induces a small (less than 1 hour) advance in the onset of pulmonary respiration and a larger (4–5 hours) advance in the hatching time itself (Freeman and Vince, 1974). McNabb and McNabb (1977) have compared thyroid activity in an altricial and a precocial species (Ringed Turtle Dove and Japanese Quail respectively) and found that thyroid activity in the former was lower at hatching than it was in the latter. They relate this difference to corresponding differences in thermoregulatory ability (Chapter 8). Thyroid hormone secretion rates are probably low through most of development in Japanese Quail (Figure 4.2), but are activated shortly before hatching in preparation for independent thermoregulation by the hatching (McNabb et al., 1972).

Developing domestic fowl embryos synthesize at least two cortical hormones, corticosterone and aldosterone. The adrenal glands develop slowly until day 11, then increase linearly with age to hatching, and adrenaline content increases sharply over the latter period. Corticosteroids have been implicated in bone development but their functions are poorly understood.

48

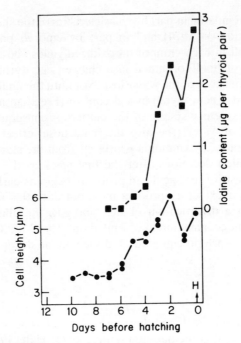

Figure 4.2   Mean cell height (●) and total iodine content of the thyroid (■) in Japanese Quail, *Coturnix coturnix*, embryos in relation to hatching time. Modified with permission from McNabb *et al.* (1972).

Pancreatic hormone functions are similarly not fully understood but injected insulin increases glycogen concentration in the yolk sac membrane. In the developing chicken embryo both insulin secretion and glycogen storage increase in the final week (Freeman and Vince, 1974). The pituitary gland has likewise been poorly studied but is known to secrete growth hormone in domestic fowl from day 15. Removal of this gland results in such embryos as survive being dwarfs (Freeman and Vince, 1974).

Calcium needs

Chicken embryos draw about 100 mg of calcium from the egg shell between day 13 and hatching on day 21, thus accounting for the change from about 22 mg calcium in the unincubated egg to about 125 mg calcium in the newly-hatched chick. Labelling with calcium isotopes shows that calcium withdrawal is slight prior to day 13. Bone mineralization occurs earlier, from about day 8, but the calcium required is drawn almost exclusively from the yolk stores. Once shell calcium withdrawal begins the shell membranes weaken in adhesion to the shell, a point of relevance to later hatching (Freeman and Vince, 1974).

EMBRYONIC HEAT PRODUCTION

Oxygen consumption

Embryonic oxygen consumption is very largely an allometric function of yolk-free embryo mass, plotting linearly on a double logarithmic plot (Figure 4.3). Within individual species the exponents vary, from 0.79 in Zebra Finch and 0.81 in Japanese Quail to 1.23 in Domestic Pigeon, but the significance of these exponents, calculated as they are over a small range of embryo weights, is not clear. Vleck $et~al.$ (1979) found that oxygen consumption increased exponentially with incubation time in each of five (three altricial, two precocial) species, with yolk-free embryo weights varying in very similar fashion. But whilst in altricial species the exponential rise continues to hatching, in precocial species it ceases at a plateau value 60–70 per cent of the way through the incubation period. In both groups oxygen consumption rises again when the eggs pip and active hatching starts. Using the values prevailing immediately before pipping as a measure of embryonic metabolism, Hoyt $et~al.$ (1978) found for 18 species the relationship

$$c = 25.2~W^{0.730}~, \tag{4.1}$$

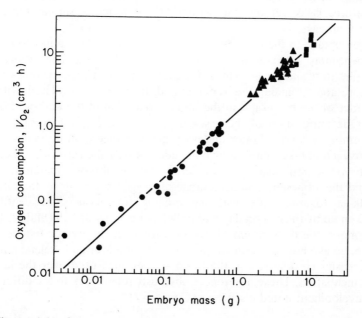

Figure 4.3 Oxygen consumption in relation to yolk-free weight in embryos of various ages. The regression line shown $V_{O_2} = 1.61~m^{0.92}$) accounts for 98% of the variance in oxygen consumption. Key: ●, Zebra Finch; ▲, $Coturnix$ Quail; ■, Pigeon. Redrawn from Vleck $et~al.$ (1979), $Physiol.~Zool.$, **52** : 363–377.

where c is the 'pre-internal pipping' rate of oxygen consumption (in cubic centimetres per day) and W is fresh egg weight (in grams). Converted to heat production the equation gives heat production m (in kilocalories per day) as

$$m = 18.5 \ W^{0.730} \ . \tag{4.2}$$

This equation thus describes a regression line parallel to but below the Lasiewski–Dawson (1967) regression for adult metabolism on body weight (p. 123). Thus, full-term embryos have metabolic intensities (oxygen consumption per unit body weight) about one-third of those in adults of the same weight. The difference in level is not in itself surprising, for the embryos are yet dependent upon the chorio-allantois for their oxygen supply and both blood pressure and heart rate are lower than in hatchlings (Freeman and Vince, 1974), but the parallelism is unexpected. As we shall see later (p. 75), post-natal growth rates in birds are similarly linked to body weight with such an exponent, so the results suggest that avian development rates are fundamentally coupled to weight-dependent processes.

Oxygen consumption is not constant throughout the day but rises through the afternoon from a morning low, to peak values at around 19.00 hours (Freeman and Vince, 1974). The significance of this rhythm, which persists on controlling such environmental variables as illumination, pressure, and temperature, is as yet unclear. Drent (1970) found a similar pattern in Herring Gull eggs.

Oxygen uptake in the chicken differs between the sexes, males consuming 74 per cent more between days 10 and 19 than do females. This difference is in part due to the faster growth of the male embryo, which averages 55 per cent more than female; the remaining difference is correlated with the appearance of sex hormones in the blood circulation of the embryo (Shilov, 1973). One implication of these sex differences in respiratory rate is that female embryos survive better under oxygen shortage than do males.

The oxygen uptake figures above include not only the respiratory needs of the embryo itself but also the uptake of the extra-embryonic membranes, the yolk, and the albumen. The percentage of oxygen taken up by the allantois and yolk sac together varies with age, ranging from about 35 per cent of the total taken up by the egg on day 6 to only 5 per cent by day 19 (Shilov, 1973). Furthermore, the distribution of oxygen within the embryonic tissues varies with age, uptake rate increasing in, for example, the liver and skeletal muscle – but decreasing in others, e.g. the brain – when measured over the last few days of incubation. These differences obviously relate back to the differential body development noted elsewhere (Chapter 7).

Energy balance

Khaskin (1961) determined the main avenues of heat loss from domestic duck eggs (Figure 4.4). During the initial stages of development heat was lost from

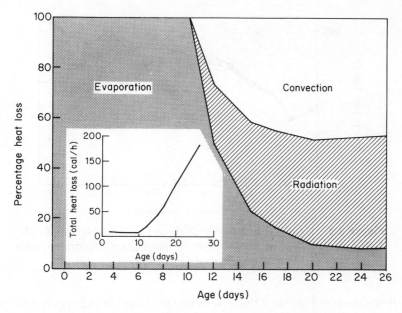

Figure 4.4 The principal channels of heat loss with (inset) total heat emission (in calories per hour) in relation to incubation stage in domestic duck eggs. From data in Khaskin (1961).

the egg principally by water loss across the shell but in the second half of the incubation period this proportion declined, to less than 10 per cent towards hatching. Instead, both convective losses (the movement of warmed air away from the egg to be replaced by colder air) and radiative losses (the emission of energy as electromagnetic radiation, a property of all hot objects) increased in importance, from about 25 per cent each on day 12 to about 45 per cent each on day 26. The need to minimize such convection and radiation losses thus provide good biophysical grounds for the enormous thickness of down found surrounding the eggs in a duck nest and for the duck's habit of covering over the eggs with down when she leaves the nest unattended.

Drent (1970) found a more or less steady increase in embryonic metabolism in the Herring Gull as incubation progresses, due largely to the increasing size of the embryo. Figure 4.5 shows that, as the Herring Gull embryo nears full term, its share in the total budget of the egg rises to about 75 per cent, the remainder being supplied by parental incubation. This is about the same percentage as is contributed by the full-term House Wren embryo. Such thermoregulation involves an energy cost. In the case of an embryo the energy is obtained by the metabolism of fat reserves within the egg. Table 4.1 shows estimates of the proportion of egg reserves actually used by the embryos of various species in this way: the balance is available to the

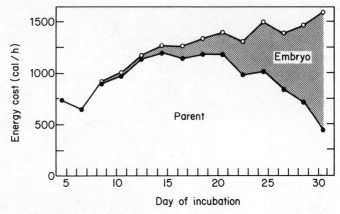

Figure 4.5 Heat production by adult and embryo Herring Gulls, *Larus argentatus*, in the course of incubation. Redrawn with permission from Drent (1970).

newly-hatched bird to tide it through its first hours out of the egg. For the two precocial species mentioned above, more than half the original reserves were still available at hatching, thus allowing an extended period without food should this be necessary, in travelling from nest site to feeding grounds, for example. In the altricial House Wren, on the other hand, the original fat reserves of the egg were just sufficient to meet the embryo's energy expenditure within the egg and no reserves were available to the newly-

Table 4.1 Energy expenditure of avian embryos during incubation. From Siegfried *et al.* (1976)

Species	Energy content of fresh egg (kcal)	Energy expended by embryo (kcal)	Percentage of available energy expended (%)	Source
House Wren, *Troglodytes aedon*	1.6	0.9	55	Kendeigh (1963)
Herring Gull, *Larus argentatus*	144.4	57.9	40	Drent (1970)
Domestic fowl, *Gallus domesticus*	90.9	25.3	28	Romijn and Lokhorst (1960)
Domestic duck, *Anas platyrhynchos*	138.7	39.7	29	Khaskin (1961)
Maccoa Duck, *Oxyura maccoa*	202.2	48.5	24	Siegfried *et al.* (1976)

hatched young. The Herring Gull is intermediate in this respect, with about 40 per cent of its reserves available to a newly-hatched chick.

Metabolism during egg neglect

Certain species regularly neglect their eggs during incubation, usually when the adults have difficulty in feeding themselves (p. 11). Embryonic metabolism in such species shows specialized adaptations to cope with such neglect. Fork-tailed Storm Petrels are one such species, where eggs at times are neglected for up to 28 days. Vleck and Kenagy (1980) found that metabolism decreased steeply as the egg cooled, so that even 28 days of neglect at typical burrow temperatures cost only about 4 per cent more than without neglect. Embryos were able to continue to develop as temperature fell to as low as 30 °C, but cooling to 10 °C for a period simply prolonged the incubation period by the corresponding period. The habit of laying eggs in a cool burrow or crevice may have allowed the evolution of ability to develop at what, for other species, would be suboptimal temperatures. Even so, burrow temperatures must both be cool enough to ensure energy savings to the embryo if it goes torpid through egg neglect, and yet be warm enough for the embryo to survive the chilling.

Hatching synchrony

In many precocial species the eggs of a clutch hatch synchronously. When eggs of Bobwhite Quail are incubated in contact with each other in an incubator they similarly hatch within an hour or two of each other. If they are incubated out of contact with each other, however, they hatch over a longer time-span, though with about the same mean time as the eggs incubated as clutches (Vince, 1968). Such a finding requires two effects: eggs that would have hatched late if incubated in isolation must be accelerated in some way on incubation in clutches, and eggs that would have hatched early in isolation must be correspondingly retarded when in a clutch. Figure 4.6 shows experimental evidence on these points. The diagram also suggests that advanced eggs were less frequently retarded (or were retarded by smaller amounts) than were the less incubated eggs advanced. Vince and Cheng (1970) found the same with Japanese Quail: advanced eggs of this species had to be in contact with at least three retarded eggs for hatching synchronization to occur. Vince (1966, 1972) found that the synchrony is the outcome of mutual stimulation by the embryos in the nest. She recorded audible sounds and low frequency vibrations emanating from the eggs in the last few days before hatching. Recordings of 'egg clicking' (Driver, 1965) – a special form of breathing – proved effective in accelerating quail embryos (Freeman and Vince, 1974). This clicking commences a few hours after embryos have commenced silent breathing and its onset is loosely synchronized (maximum range 10 hours) within Bobwhite clutches (Vince, 1969). This applies even

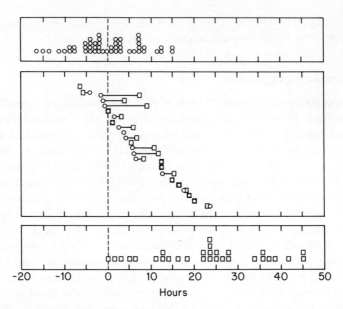

Figure 4.6 Hatching time distribution of two batches of Bob-white Quail, *Colinus virginianus*, eggs starting incubation early (○) and 24 hours later (□), when incubated in isolation (top and bottom) or as pairs in contact with each other (middle). Note the synchrony within each pair, brought about by retardation of early-incubated and acceleration of late-incubated eggs when in contact. 0 hours is the mean hatching time of the early isolates. Reproduced with permission from Freeman and Vince (1974).

though the eggs concerned have pipped up to 48 hours apart. Regular clicking at rates of 3–6 per second were most effective, though embryos were accelerated in development with clicks rates of from 1.5 to 60 per second. Hatching times were subsequently even more tightly synchronized than the onset of clicking.

Pre-hatch vocalizations are unknown in altricial species, for which such extreme synchrony of hatch is not likely to be of adaptive value.

SUMMARY

Embryos of most species commence a phase of intensive growth about 20 per cent of the way through their incubation, though many organs are already identifiable by then. Development is not passive, the embryo instead receiving much stimulation, initially from myogenic contractions and changing geometry of egg constituents but later from rhythmic, self-generated muscular activity. Hatching follows broadly similar lines in most species, but precocial species may vocalize some time before hatching and are responsive to external stimulation. Within a clutch the hatching of advanced embryos is retarded and the hatching of younger embryos advanced by exposure to 'egg-clicking' respiratory noises from siblings, thus synchronizing brood hatch. Embyronic metabolism develops with increased oxygen-carrying by the blood and with the secretion of thyroid hormones controlling thermogenesis, and is fuelled by lipid reserves in the egg yolk. Metabolism increases exponentially with incubation time in both altricial and precocial species, but for the latter this increases about two-thirds of the way through the incubation period. Across a wide range of species weights, full-term embryos function more or less uniformly at one-third the rate of an adult of the same weight. Full-term embryos provide about three-quarters of the heat budget of their egg, the remainder being supplied by the incubating parent.

CHAPTER 5

Senses and behaviour

Growing young are on their way to becoming adults with the full repertoire of adult behaviours, but individual components of that repertoire appear only gradually with age. Some components first appear in incomplete form and improve with practice or with the continuing physical development of the young. Other components appear in a complete and stereotyped form all at once, though their subsequent integration with other elements may still depend on nervous or on hormonal developments. We begin by considering the gross patterns of development in behaviour, then consider individual items in detail.

PATTERNS OF BEHAVIOURAL DEVELOPMENT

Nice (1962) recognizes five stages of post-embryonic development in altricial species. There are outlined in Table 5.1 for the Song Sparrow, with comparable stages for a precocial species, the Spotted Sandpiper, also indicated. There are considerable resemblances between the two patterns if the newly hatched altricial nestling is regarded as approximately equivalent to the 12–13 day old embryo of the precocial species (Figure 5.1). The initial stage is devoted entirely to post-natal growth in the altricial species but occurs as pre-natal development in the precocial species. This is *the* feature of the altricial state, to use Nice's description, a period of parental brooding and feeding in which the sole task of the nestling is to ingest the food presented to it and convert it into its own biomass with high efficiency. During the equivalent stage of precocial development the only direct input which the embryo receives is the heat of incubation, its food requirements being served by the yolk and albumen laid down at egg laying. The adult has, as it were, performed its food collection duties in advance.

Both species have a Stage II phase in which elementary comfort movements first appear but in the precocial species this is compressed into some hours (in a few precocials it may extend into the second day). The earlier occurrence (Table 5.1) of head-scratching and of pecking in the precocial group is obviously of value in the light of the chicks' independence. In this and the following stage, the young begin to behave as individuals for the first time and to develop relationships with their parents and siblings. The greatly

56

Table 5.1   Major features of development stages in the altricial Song Sparrow and in the precocial Spotted Sandpiper:
+, present; L, occurs in following stage. Modified from Nice (1943)

Development stage	Principal features	Song Sparrow	Spotted Sandpiper
I. *Post- or Late Embryonic* Concerned mainly with nutrition	Hatching	+	L
	Gaping	+	—
	Size increase	+	Embryonic
	Start of feathering	+	Embryonic (natal down)
II. *Preliminary* Comfort movements begin	Main weight increase	+	L
	Rapid feather growth	+	L
	Temperature control	+	+
	Eyes open	+	+
	Standing	+	+
	Scratch head	L	+
	Preening	+	+
III. *Transitional* Maturation of comfort movements	Stretching	+	+
	Crouching	+	+
	Exploratory pecking	L	+
IV. *Locomotory* Nest departure and fledgling life	Leave nest	+	+
	Fleeing	+	+
	Self-feeding	L	+
V. *Socialization* Social interactions	Flight	+	+
	Bathing	+	+
	Play-fleeing	+	+
	Aggression	+	+

shortened duration of this phase in precocial species is accompanied by the imprinting phenomenon, a process of greatly accelerated recognition of their parents by the young (Chapter 10). Altricial species show less pronounced imprinting since they have several days to recognize the characteristics of the birds feeding them. Parental recognition is essential once the nest is left, at least for those young dependent on adults for subsequent support.

Precocial species and many open-nesting altricial species leave the nest before the young can fly but hole-nesting species, normally subject to much lower risks of nest predation, usually remain until flight capabilities are fairly well developed.

58

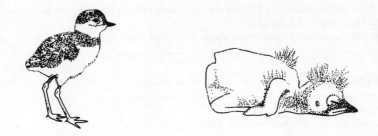

Figure 5.1 The contrasting morphology and behaviour of (left) a young precocial Lapwing, *Vanellus vanellus*, and (right) an altricial nestling Robin, *Erithacus rubecula*. After illustrations in Harrison (1975).

There have been surprisingly few detailed studies of the timing of behavioural development of altricial species. Figure 5.2 summarizes the results of a study of the Brewer's Blackbird (Balph, 1975). The various categories of behaviour of the altricial young are ordered here by time of appearance and reflect the bird's transition through the stages outlined by Nice. Individual species differ in absolute age at which a particular component first appears, but these differences are probably more a reflection of size and of variation in the period of dependency than of any absolute significance.

In some species individual behaviour patterns occur early in development, in obvious correlation with nesting behaviour. Young Least Bitterns in their

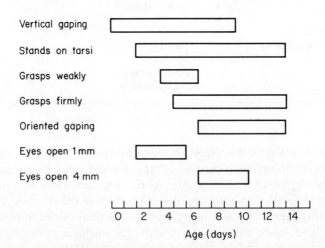

Figure 5.2 The transition from vertical to oriented gaping in Brewer's Blackbird, *Euphagus cyanocephalus*, nestlings in relation to postural and visual capabilities. Data from Balph (1975).

waterside nests, for example, show innate responses at hatching which include weak but persistent grabbing with the feet. The young bittern can nearly hold itself on a branch when newly hatched and 1 day old birds can keep themselves upright on a branch or other vegetation by this means, even though it is several days before they can firmly grasp twigs.

SENSE ORGANS

Birds rely more on vision than on other senses and the development of sight is therefore of major importance to them. Many species open their eyes either whilst in the egg or within a short while of hatching, and this is in fact one of the characteristics used by Nice (1962) in her classification of maturity at hatching. Apart from the typically altricial orders the principal groups to hatch with their eyes closed are the owls, the tropicbirds, penguins, and some but not all procellariforms.

The timing of eye-opening has not been extensively studied but it is probably linked to each bird's overall development rate. In the Common Grackle the eyes are closed at hatching, the lids begin to part at about 60 hours, are half-way open by 150 hours and fully open by 280–300 hours. The eyes are at first opened only when gaping, but from about 180 hours on they gradually remain open for longer (Schaller and Emlen, 1961). A similar pattern is apparent in most passerines. During the period of closure, morphological changes take place: in the House Sparrow the ciliary muscles are not functional until about 6 days of age (Slonaker, 1921) and in Starlings myelinization of the optic nerve is incomplete until the eighth day (Schifferli, 1948). In the Grackle both the pupillary reflex and an optomotor response appear rather suddenly between 150 and 163 hours, about half-way through the period of eye-opening (Schaller and Emlen, 1961).

Prior to eye-opening nestlings are responsive mainly to mechanical jarring of the nest and to tactile stimulation of their beaks to elicit begging. Thereafter the range of acceptable stimuli is reduced down to those coupled to the arrival of the adult with food. In hole-nesters, for example, a shadow darkening the nest entrance will evoke intense begging in hungry young. The onset of visual capability also results in the modification of behaviours previously present. Thus in Brewer's Blackbird initial begging is directed vertically at first but at an angle once the eyes are open and the position of the parent is visible (Figure 5.2). In nestling thrushes the begging is oriented towards the 'head' end of simplified models of the adult once sight has been attained (Tinbergen and Kuenen, 1939). Other activities may also acquire a directional component once eye function starts: Eastman (1969) relates how nestling Kingfishers regularly defaecated towards her camera lights during filming, the floodlights being the brightest light source in the burrow: in an undisturbed nest the burrow entrance would have been the light source.

Visual inputs are also essential before nestlings are able to balance properly. Several altricial species show strong grasping abilities prior to their eyes opening but, until they can see, all are unable to balance properly when perched on a test twig (Holcomb, 1966a,b).

Several studies have shown that nestlings and chicks possess colour vision (e.g. Tinbergen and Perdeck, 1951; Hess, 1956a). These birds are able to discriminate both the colour intensity and the colour saturation of objects presented to them (Kear, 1966a; Delius and Thompson, 1970).

Nestling Song Sparrows are able to hear from birth (Nice, 1962) and embryos of ducks and quail can communicate successfully from some days before hatching (Vince, 1968). Hearing is first detected in the chick embryo on day 12 of incubation, in the form of electrical activity in the cochlea when experimentally stimulated with low frequency tones. Sensitivity and frequency range improve with age and in duck embryos adult-type responses are present by day 22 (hatching is on day 26) (Impekoven, 1976). Newly-hatched Wood Ducks can recognize by voice alone the mother they have never seen, an obvious advantage for a hole-nesting species. Precocial species respond to parental warning calls from within a few minutes of hatching, but altricial species show overt response only after some days in the nest. Several studies have shown that young birds learn to recognize their parents as individuals partly as a result of arrival calls heard during feeding visits (e.g. Evans and Mattson, 1972).

Touch receptors are well developed by the time of hatching. In altricial species the oral flanges are particularly sensitive to touch and nestlings can often be induced to gape for the food they previously ignored by the adult's bill touching the flange.

Bobwhite Quail have the capacity to discriminate between sweet, sour, salt, and bitter solutions at 10 days of age (Brindley and Prior, 1968). They show a preference for sweet (10 per cent glucose) or slightly sour (0.05 per cent hydrochloric acid) solutions and reject bitter or salt solutions, and this preference proved to be independent of age. Electrophysiological studies show that taste bud receptors in newly-hatched pigeons and chickens can respond to bitter, sour, and salty substances, but not to sugars (Kitchell *et al.*, 1959). Thus, taste is functional in neonates but not to adult standards. The difference is associated with a disparity in the number of taste buds present, for histological examination shows the newly-hatched chicken to have about eight taste buds in its mouth whilst adults have about 24 (Lindenmaier and Kare, 1959).

MATURATION

Some behaviour patterns alter with age simply because the bird is better able to perform the behaviour as a result of independent changes in its physical (or, indeed, physiological) structure. Laughing Gull chicks show an improvement with age in accuracy of pecking at their parents' bills (or models thereof) for food. Hailman (1967) compared normally developing young with those force-fed and reared in the dark (so that they had no prior practice at pecking). Even without practice the proportion of hits increased in older chicks with longer tarsi, indicating that bone and muscle growth contribute to body stability when pecking, but practice contributed substantially. Bar-headed Geese also increase their pecking frequency and accuracy over the first 48 hours (Wurdinger, 1974), during which time the choice of pecking target also improves. Over the first 24 hours pecking is directed at dark contrast patches; at 24 hours at the gosling's own feathers; at 36 hours at the feathers of its siblings; and at food particles late in the third day. Until this time dark objects are preferred but the preference thereafter shifts to green. Similar improvements in feeding efficiency as a result of continued muscle growth have been recorded in young Chaffinches (Kear, 1962), in young Reed Warblers (Davies and Green, 1976), and in domestic chicks (Hess, 1956b).

A rather different example of maturation underlying developmental improvement in behaviour is provided by the diving behaviour of young White-headed Ducks (Figure 5.3). Diving was first observed after only as little as 7 minutes on the water, with major improvements over the first 3 days. Newly-hatched young had obvious difficulty in plunging their downy bodies through the water and needed a distinct leap up and over when

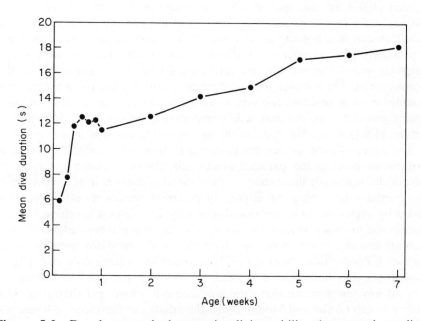

Figure 5.3 Developmental changes in diving ability (measured as dive duration) by young White-headed Ducks, *Oxyura leucocephala*. Based on data in Matthews and Evans (1975).

immersing. This leap, and the twin jets of water kicked up, were noticeable into the fourth week, by which time the young were spending some 40 per cent of their time diving. As their feathers developed in the fifth week their performance improved noticeably (Figure 5.3) and by the seventh week their performance approached adult levels. Over this period their time spent underwater also decreased, to 23 per cent of the daylight hours. Thus changes in body size and proportions and in the development of feathers governed the development of diving ability independently of any improvement due to practice (Matthews and Evans, 1975).

In at least some species motivation to investigate new objects is provided by age-dependent curiosity. When Vince (1960) presented young Great Tits with small brightly coloured objects eliciting pecking, she found that the birds' interest was highest between 10 and 15 weeks after fledging. This period roughly coincides with the autumnal switch towards winter feeding patterns, when most young birds must learn to take new foods or perish. In still older juveniles the readiness to respond decreased gradually, reaching the levels found in adults about 30 weeks after fledging.

FEAR AND THREAT

Although small nestlings typically gape at any stimulus, older birds show a crouching reaction to stimuli other than those of their parents. Schaller and Emlen (1961) list the ages at which nestlings of a variety of species first displayed the crouching reaction. These ages are broadly those at which the nestlings can first hope to survive outside the nest (Nice, 1943). Nestlings, particularly those in open nests, disturbed much later than this are liable to 'explode' out of the nest into the surrounding vegetation. The timing of this behaviour has been shown to depend on increasing experience of the parents, coupled with a negative reaction to all non-familiar objects. By removing young grackles from the nest or by temporarily blinding them, Schaller and Emlen (1961) were able to delay the age at which fear reactions occurred.

The young of most species, but particularly those of precocial birds, tend to crouch on hearing the parental alarm call. As such young are typically cryptically coloured, this enhances their survival chances in the presence of the predator disturbing the adults. In precocial species its effectiveness is aided by aggressive or distraction displays by the adults. Crouching in these species occurs long before the young are thermally independent of their parents and the chicks may cool dramatically if a predator remains in the vicinity for some time. Norton (1973) recorded an average drop of 5.7 °C in sandpipers thus immobilized in Alaska. Willow Ptarmigan can become so cold at low temperature that they are unable to move, yet they retain the ability to call (Aulie and Moen, 1975): the adults can therefore re-locate the chicks and brood them as soon as it is safe to do so.

Large young, such as those of herons and birds of prey, can defend themselves vigorously and the period of crouching responses is of short duration. The threat display usually involves plumage erection and wing waving, both contributing to an increase in apparent size, and an auditory component, often bill snapping. Closer disturbance usually evokes striking actions with bill or feet but in nestlings of a few species it involves body secretions: from the day of hatching Fulmar chicks can regurgitate and spit oil over an intruder and coucal and hoopoe nestlings produce foul-smelling excreta when disturbed. In some groups (coucals, titmice) nestlings hiss when disturbed, possibly mimicking a snake. Reactions may, however, be chosen according to the identity of the intruder. Adult Least Bitterns defend their nests by threat against small intruders but 'freeze' on the nest when faced with larger predators such as Great Blue Herons. The young first adopt this posture when about 3 days old but at 5–7 days they take up an aggressive posture with the neck down, head ready to strike, and wings held open with dorsal surfaces forward. In this posture they sway back and forth from side to side (Weller, 1961), a behaviour common in other bitterns and possibly of camouflage value as marsh vegetation sways in the wind.

Adult gulls have elaborate displays for communicating the fears and threats associated with territorial and other social interactions. The majority of these calls and display notes can be traced back to the changing vocalizations of the chicks as they develop (Figure 5.4). Newly-hatched Franklin's Gull chicks have two types of call. One call, catalogued by Moynihan (1959) as the 'high intensity distress note', is a loud, harsh, hoarse, and rather quavering call

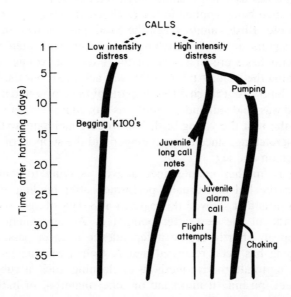

Figure 5.4 The course of development of calls in Fanklin's Gull, *Larus pipixcan*, chicks, showing approximate relationships between calls. Redrawn with permission from Moynihan (1959).

given by chicks in extreme discomfort – when too cold or too hot, after being pecked by an adult, etc. The other calls are cheeping notes (Moynihan's 'low intensity distress note'), whistle-like and monosyllabic, and are given in conditions of slight discomfort. In response to either note the parents adjust their behaviour overtly, to cool or brood the chick and so on, with greatest alacrity to the high intensity calls. From these two basic calls the other patterns develop (Figure 5.4).

Young Ring-billed Gulls develop in a very similar way. The process is one of elaboration of the initial distress calls (which serve to acquaint the parents with the existence of chick discomfort) into variants which become increasingly segregated and confined to a particular situation calling for attack, escape, or appeasement. Thus, upon fledging the young gulls have an array of social signals with which to conduct their relationships with conspecifics. However, although equipped to convey messages of fear or threat, the fledglings have not yet acquired a fully adult repertoire (see p. 260).

MAINTENANCE BEHAVIOUR

The ability to attend to their own plumage and body care develops only gradually in young birds. Ficken (1962) describes the pattern for hand-reared American Redstarts from 6 days onwards. Preening movements were already present by that age, when feathers were still largely in their sheaths. Preening helps the fracture of these sheaths and the removal of the broken bits. The uropygial gland was used in preening only from day 12 (fledging is on day 9), the oil so obtained being applied to the feathers. Bathing was first noticed in birds 18 days old. Birds cannot reach the head feathers with the bill and to preen them depend on head-scratching movements with their feet. Incomplete head scratches were first seen in the 1 day old young, the foot not actually reaching the head. At 7 days this was achieved, but the birds moved their heads to let the foot reach different parts of the crown and nape: in older birds the head was held still and the foot was moved instead. For several days after leaving the nest the young birds had trouble maintaining their balance during head-scratching, since they no longer had the support of the nest and had to balance on one leg.

Bill-wiping is another maintenance activity, serving to remove foreign material from the bill and usually performed after feeding and drinking. Nestling Redstarts first showed this pattern at 8 days of age, wiping the bill against the side of the nest after being fed. As one might expect, the behaviour appears shortly before independence in most passerines (Table 5.2). Bill-wiping has been recorded mainly from passerine families, other species resorting to alternative methods of cleaning, such as rubbing the bill against the feet, pushing it into sand or other material, or bathing (Clark, 1970).

Various forms of stretching movements are observable in young birds. In passerine adults three stretching movements are regularly observed (Nice,

Table 5.2 Earliest appearance of bill-wiping in some passerine species (after Clark, 1970)

Species	Nestling age	Nestling period
Orange-fronted Parakeet, *Cyanocitta cristata*	15	*c.* 42
Cactus Wren, *Campylorhynchus brunneica-pillus*	24	21
Curve-billed Thrasher, *Toxostoma curvirostre*	16	18
Loggerhead Shrike, *Lanius ludovicianus*	33	24
American Redstart, *Setophaga ruticilla*	8	9
Brown Cowbird, *Molothrus ater*	14	—
Rose-breasted Grosbeak, *Pheucticus ludovi-cianus*	7–11	12
Song Sparrow, *Melospiza melodia*	11	9–10
Serin, *Serinus canarius*	11	13–17

1943; Ficken, 1962). A movement designated *wing-and-leg-sideways stretch* involves a stretching sideways of one wing and leg and fanning of half the rectrices, all to the same side. The second, the *both-wings-up stretch*, extends the unfolded wings above the body. The third, the *both-legs stretch*, involves only a simultaneous stretching of the two legs. Young Redstarts (and young of several other passerine species) also have a fourth stretch movement, the *both-wings-down stretch*, in which both wings are stretched down but not outwards. Six day old Redstarts lack the wing-and-leg-sideways stretch but show the other three. Apparently the both-wings-down stretch replaces the wing-and-leg-sideways stretch in young birds since the former is transitory (lasting only 3 days in Song Sparrows). The wing-and-leg-sideways stretch develops gradually in Redstarts, first as a wing only stretch at 8 days, later more fully; the both-wings-down stretch declines in frequency as the new movement appears (Ficken, 1962). What is interesting here is the restriction of the sideways movements to the very end of the nestling period (9 days): until then the young are effectively constrained for space. In the cramped cavity nest of the Cactus Wren wing stretching is even more modified, consisting of a lowering and posterior movement of the wing wrist, so that the primaries are spread over the back and thus require no additional space (Ricklefs, 1966).

In the American Redstart and in the *Emberiza* buntings studied by Andrew (1956), exercises were performed mainly after a period of rest. Apparently the muscles become cramped during inactivity, particularly within the nest, and the stretchings occur in response to proprioceptive feedback from the muscles. But in birds requiring competent flight on fledging, such as the hirundines and various birds of prey, wing exercising is a regular feature of the later stages of the nestling period. The young flap vigorously whilst holding the nest tightly: older nestlings may jump up and down on the nest using the lift provided by their flapping. These exercises presumably strengthen the flight muscles and their co-ordination in preparation for flight. They are not absolutely essential for flight: when Grohmann (1939) confined some nestling pigeons in tubes to prevent this flapping, he found that flight movements and co-ordination when released on completion of growth were as good as in normal fledglings. Nevertheless, the muscles of some young had deteriorated, so sustained flight suffered. Nestling Swifts also show such pre-flight exercises when well nourished but not when short on food, presumably to conserve energy for essential processes (Lack, 1956). In this hole-nesting species the form of the exercises has been modified to accommodate the lack of space for flapping. The young undertake a form of 'press-ups' in which the partly-opened wings are pressed down on the floor and the body raised off the ground. These exercises start at about 30 days and continue through the remaining 2 weeks or so of the nestling period.

Precocial species are dependent upon their legs for their early departure from the nest and must perform any preliminary exercises for their use whilst still within the egg. In domestic fowl eggs the embryo commences a series of co-ordinated flexures of its legs and toes from about day 17, and these apear to provide a measure of practice in preparation for nest departure. When eggs are artificially stimulated to hatch unusually early the period of flexures is prematurely terminated and the chicks concerned are less competent at standing and walking than are normally hatched chicks (Vince and Chinn, 1972).

ECOLOGICAL CORRELATES OF BEHAVIOURAL DEVELOPMENT

Development mode is broadly correlated with species ecology. Precocial development of flightless young is an adequate strategy for terrestrial or aquatic insectivores but is not a strategy open to aerial hunters. Similarly obvious considerations limit the relative merits of producing down-covered rather than naked hatchlings, or of self-feeding rather than parentally-fed young. Within these broad limitations, though, comparisons of related species have often shown adaptive variations in the development of particular behaviours. Two such cases – the avoidance of cliff edges by ground-nesting and by cliff-nesting birds, and the incidence of climbing behaviour in cavity-nesting and in open-nesting ducks – are discussed as detailed examples below.

Visual cliff responses

In various species that regularly nest at a height above ground and where a fall could be fatal, the young show well-developed edge avoidance behaviour from an early age. In investigating whether such avoidance is innate or whether it is learned in some way, 'visual cliff' experiments of the type

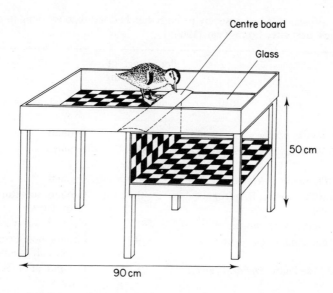

Figure 5.5 A typical visual cliff apparatus used in testing cliff-edge avoidance by young birds. Modified from Kear (1966b).

illustrated in Figure 5.5 have been particularly valuable. In such experiments, a bird placed on the centre board sees on the left a chequered board at its own level and on the right a similar sheet of material some 20 cm below it. The bird is protected from falling, should it step off at the 'deep' side, by a sheet of glass laid across both halves of the apparatus. The drop is therefore only visual and the bird's responses can be studied without it learning to avoid cliffs as a result of falling over the edge.

Kear (1966b) used this apparatus to test the reactions of visually naïve chicks of 12 species (Table 5.3). The results are strikingly correlated with the normal position of the nest of each species. Ground-nesting species such as Pheasant and Partridge showed a strong disinclination to move to the deep side, whilst tree-nesting species such as the Wood Duck showed a more random response. Species which sometimes use trees, such as the Mallard, were intermediate. Observations of the behviour of the young showed that the birds going on the deep side did so purposefully: they walked to the edge and jumped off, as they would have done in the wild on a real drop. These behaviours have obvious survival value: flightless birds need to avoid sharp drops in nature but tree-nesting species must have young prepared to jump from the nest cavity when called down by the female. Similar adaptations are shown by cliff-nesting species: young Guillemots seek tactile contact, warmth, and darkness coupled with an aversion to visual cliff edges and thus tend to stay under their parent where the risk of falling from the nesting ledge is minimal (Wehrlin, 1977). Comparative studies of Kittiwake and Herring Gull similarly show that cliff avoidance is pronounced in the former (cliff-

Table 5.3   Visual cliff responses by recently-hatched nidifugous young in relation to their species' nest site. From Kear (1966b)

Species	Percentage of tests favouring shallow side of 'cliff' (%)	Normal nest site
Partridge, *Perdix perdix*	94	Ground
Australian White-eye, *Aythya australis*	87	Ground
Pheasant, *Phasianus colchicus*	86	Ground
White-faced Tree Duck, *Dendrocygna viduata*	81	Ground, sometimes in hollow trees
Moorhen, *Gallinula chloropus*	81	Ground, rarely in trees
Marbled Teal, *Anas angustirostris*	79	Ground holes, sometimes holes off ground
Red-billed Tree Duck, *Dendrocygna autumnalis*	78	Ground, at times in hollow trees
Mallard, *Anas platyrhynchos*	77	Ground, 8–10 per cent in trees
Muscovy Duck, *Cairina moschata*	64	Always in holes, ground level or above
Mandarin Duck, *Aix galericulata*	54	Tree holes, etc.
Comb Duck, *Sarkidiornis melanotus*	51	Tree holes, etc.
Wood Duck, *Aix sponsa*	46	Tree holes, etc.

nesting) species, the behaviour being reinforced by several other behavioural features modified from the typical gull pattern in an adaptive fashion (Cullen, 1957; McLannahan, 1973).

Climbing ability in ducklings

A second example of how a simple ecological factor may influence development is provided by the morphological and behavioural adaptations for climbing present in young tree-ducks. Newly hatched young of these species face the problem of getting out of their lofty nest cavity and down to the ground where food can be found. Cavity nests of tree-ducks usually have the exit hole near the top of the cavity. The ducklings have thus to climb a near-vertical smooth surface to make their exit, usually on their second day. Siegfried (1974) tested five species for their ability to leave an artificial chimney in this way and related his findings to the ducklings' morphology (Table 5.4). The method of climbing in Wood Duck is as follows: starting from a stationary position, the bird pulls the body close to the wall, then upwards whilst braced by the tail; at the top of the lifting sequence the feet are released and the wings swing outwards, upwards, then inwards, all the

Table 5.4 Climbing ability of 2 day old ducklings in relation to their morphology. After Siegfried (1974)

Species	Percentage successful[a] (%)	Leg length (mm)	Claw curvature (degrees of arc)	Caudal down Length (mm)	Width (mm)	Hook on wing
Wood Duck	75	60	89	18.2	23	Yes
Goldeneye	71	71	110	17.7	20	Yes
Mandarin	63	62	102	17.9	17	Yes
Hooded Merganser	33	74	107	17.4	18	No
Mallard	0	65	94	9.1	15	No

[a] The ducklings had to climb out of a brick chimney whose walls sloped 10° from the vertical.

while supported by the tail; the wall is re-grasped by the feet and the tail drawn up into the starting posture. The Wood Duck was the most successful species at climbing (Table 5.4), in which it was aided by short legs, sharp curved claws, and a long and broad unusually stiffened block of tail down. The ducklings possess small wing hooks but Siegfried noted that they made no use of these when climbing. The other species with well-developed climbing abilities also shared these features. Other features of hole-nesting species noted were a greater jumping rate in attempting to leave the cavity: Wood Duck young averaged between seven and nine jumps per minute against only 4.1 jumps per minute by Mallard young; the latter also showed less well-oriented jumping and a greater tendency to give up their attempts to get out.

SUMMARY

Individual behavioural patterns first appear in incomplete form and improve with practice or maturation or both. The sequence in which different elements of the adult behaviour repertoire appear is very similar in altricial and in precocial species, though the onset of scratching and pecking behaviour is earlier in precocial young, as is appropriate to their needs. Some behaviour patterns improve with practice but others depend on the continuing physical development of the young. Touch and hearing abilities may be present in hatchlings of all groups, but altricial species do not open their eyes for some days; during this time morphological development of the eyes continues. Taste discrimination in newly-hatched birds is inferior to that of adults, again because of incomplete morphological development of the appropriate organs. Young of most species initially respond behaviourally to a broad range of environmental stimuli and later restrict their responses to the appropriate stimuli, but in certain species ecological constraints have resulted in very selective types of behaviours being apparent from the moment of hatching.

CHAPTER 6

Growth and development

Quite substantial differences of growth rate may exist even between closely related species of similar size. There are two schools of thought as to why this variability is present (Case, 1978). Some workers, especially Lack (1968), have equated development rate with nestling period and related them to clutch size and adequacy of nestling food within each species. Such analysis assumes that growth rates are correlated more or less linearly with nestling periods. Extension of such analysis to take account of non-linearity in weight change and ill-defined fledging ages in precocial species has generated an alternative school of thought (Ricklefs, 1968a, 1973, 1976). This regards interspecific variation in growth rate as the outcome of internal physiological constraints, with each species growing as fast as it can within the design constraints on its life-style (Ricklefs, 1973). Here we review these ideas in relation to the various determinants of growth. Later we will see that single factor explanations are inadequate, development patterns being most properly viewed as only one component in total life histories (Ricklefs, 1977b; O'Connor, 1978a; Dunn, 1980).

GROWTH CURVES

Growth in birds has most frequently been described in terms of weight versus age curves for nestlings or chicks. Little attention has been given to the internal development of the bird or to the energetic consequences of parallel metabolic development with age. Figure 6.1 presents examples of growth curves for a variety of species, in each case with weight expressed as a percentage of adult weight.

The curves reveal several points of general interest. First, most birds show some form of sigmoidal growth, in which weight increases are initially small, then build up, and eventually level off towards some final weight. In some species this weight is at or close to adult weight, as in the Starling. In others, the weight at which growth levels off is substantially below adult weight (Wood Pigeon, American Robin). In others again, nestling weights increase to some maximum above adult weight and then lose weight before finally fledging (Swift, Manx Shearwater). Finally, some species leave the nest whilst still increasing in weight, completing their progress along their growth curve

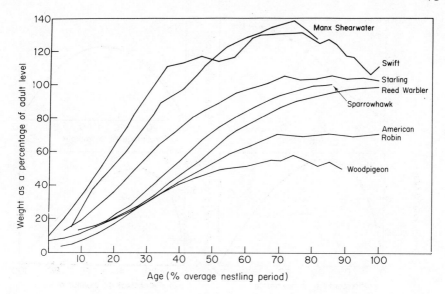

Figure 6.1 Postnatal growth curves for a variety of species, with weight at each age expressed as a percentage of adult weight.

outside the nest (e.g. Reed Warbler). Second, there exists substantial interspecific variation in the amplitude and time scale of these growth curves. This variation with adult size and nestling period, and its causes form the subject of much of this chapter.

GROWTH EQUATIONS

Analysis of growth rates is greatly facilitated by the use of fitted growth equations (Ricklefs, 1967a, 1968a). By summarizing the information in graphs of weight versus age, such as those of Figure 6.1, into a small number of constants with biological meaning, it is possible to undertake interspecific comparisons of growth. The variables of most interest are the form of the growth curve, its final magnitude or asymptote, and the rate at which it is traversed.

The equation most frequently found to fit avian growth data is the logistic equation

$$W = A/[1 + \exp(-k(t-t_0))], \qquad (6.1)$$

where W is the weight at time t, A is an asymptotic weight towards which the young grow, and K and t_0 are constants. If this equation is differentiated with respect to t we get an expression for the instantaneous growth rate at time t:

$$\frac{dW}{dt} = KW(1 - W/A). \qquad (6.2)$$

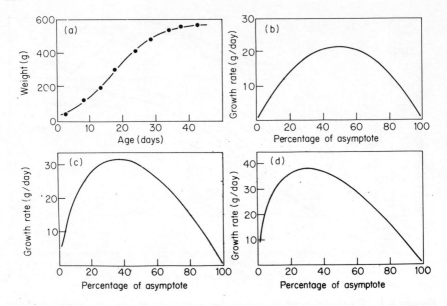

Figure 6.2 Examples of different forms of growth curve, with fixed asymptote and rate constant: (a) logistic growth equation fitted to growth data for Barn Owl, *Tyto alba* (asymptote, 570 g; rate constant, 0.152); (b) instantaneous growth rate (in grams per day) as a function of growth completed for this logistic equation; (c) the same parameters but with a Gompertz growth equation, showing the earlier peaking of growth rate; (d) the same parameters in a von Bertalanffy growth equation, showing the still slower completion of body growth. Data and equations from Ricklefs (1968a).

As an example, growth data for the Barn Owl are plotted in Figure 6.2, showing both the logistic equation and its derivative for instantaneous growth rate. The figure shows the equation to mimic the initially slow growth of the nestlings, the acceleration of weight gain in mid-period, and the subsequent tailing off of growth. The asymptote A is obviously a good approximation to the final weight of the bird. The parameter t_0 simply defines the absolute position of the growth curve in time, a point of little significance for the biology of growth. The growth constant K serves as a standard measure of growth rate. It is proportional to the fraction of asymptotic weight achieved at the moment of fastest weight gain. In the example of Figure 6.2, the fitted value of K is 0.152; this corresponds to the young growing at a rate of 3.8 per cent of their asymptote ($A = 570$ g) at the point of inflection, i.e. at 7.6 per cent of their then weight.

Not all species have weight–age curves that can be well fitted by a logistic equation; that is, their growth form is different from the symmetrical logistic curve (Figure 6.2). However, two other growth equations have been found to cover the majority of non-logistic avian growth. One of these is the Gompertz equation:

$$W = A \exp [- \exp (-K_G (t - t_0))], \qquad (6.3)$$

where K_G is subscripted to denote the nature of the growth equation. This equation reaches its inflection point slightly earlier than does a logistic curve with the same numerical value of K (Figure 6.2), thus describing species accelerating in weight gain rapidly at first but then tailing off.

A third growth form appropriate to a few species is that of the von Bertalanffy equation:

$$W = A [1 - \exp (-K_B (t - t_0))]^3, \qquad (6.4)$$

with notation as before. Figure 6.2 shows that this equation describes species with rather rapid early growth and a very slow completion phase.

Strictly speaking, rate constants from different growth equations should not be compared. It is, however, possible to calculate the relationships between K, K_G and K_B and to use these relationships to standardize otherwise heterogenous growth curves (Ricklefs, 1973). The technique is an adequate approximation for gross comparisons of growth rates over orders of magnitude ranges but is otherwise imprecise.

FACTORS AFFECTING GROWTH ASYMPTOTES

Most species grow along a logistic weight curve towards some asymptotic weight which they may or may not eventually achieve but whose values are nevertheless worth examining in view of the generality of logistic fits. Of particular interest is the relationship of these asymptotes to the normal adult body weights of the species. More than half the species considered by Ricklefs (1968a) had some additional growth to complete as juveniles. Nearly 19 per cent of all species have nestlings growing towards asymptotic weights 10 per cent or more above their adult weight. This tendency differs strikingly between passerines and non-passerines: two-thirds of all passerines have lower asymptotic weights than they will attain as adults but only 28 per cent of non-passerine species fall into this category.

In several species, but particularly amongst aerial insectivores, the excess nestling weight has been shown to be due to the high water content of growing tissues, much of it being lost as the tissues mature (p. 106). In these species lipid levels do not exceed the levels present in adults, but in others deposition of fat may contribute to the peak in chick weight.

Above-adult weights and subsequent weight recession occur primarily in nestlings of oceanic species and in swifts and hirundines (Ricklefs, 1968b). These species generally spend much time in flight to forage and their young are in many cases capable of feeding themselves on fledging. The generally secure nesting places of these species allow them accept the long nestling periods needed for the flight muscles to mature fully enough. Perhaps for the same reason several hole-nesting species also display above-adult weights whilst in the nest, though to a slighter extent than in hirundines (O'Connor, 1977).

This correlation of high nestling asymptote with flight ability also extends to less specialized fliers (Ricklefs, 1968a). Figure 6.3 shows that 78 per cent of the passerine species with asymptotes more than 10 per cent below adult weight were ground feeders such as finches, thrushes, and icterids. Similarly, species with intermediate ratios tended to forage amongst foliage (tanagers, vireos, warblers) or from perches (flycatchers) (Figure 6.3). These trends reflect the very large proportion of a bird's biomass formed by the flight muscles, particularly in the aerial species. Young swallows could not hope to survive outside the nest without considerable flying abilities and therefore need large pectoral mass, but a young thrush can both feed and escape predators if its legs are well enough developed.

The possibility that weight recession in nestlings is due to starvation has been suggested for several species but the empirical evidence is that weight loss occurs even where the chicks are fed (Harris, 1978).

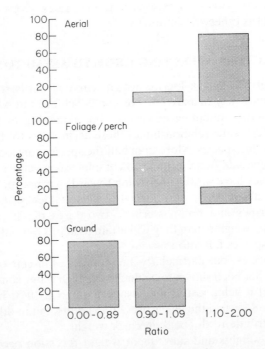

Figure 6.3 Ratio of nestling growth asymptote to adult weight in relation to adult foraging behaviour among small landbirds. Data from Ricklefs (1968a).

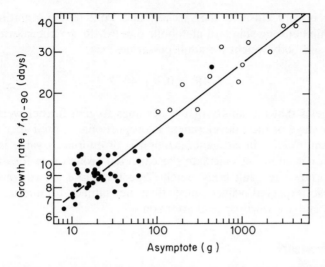

Figure 6.4 Growth rate (as time taken to grow from 10 per cent to 90 per cent of final weight) as a function of body size amongst passerines (●) and raptors (○). The regression line slope is 0.278. Redrawn with permission of The British Ornithologists' Union from Ricklefs (1968a).

FACTORS AFFECTING GROWTH RATES

Adult body weight

A major determinant of growth rate in birds is that of body weight. Figure 6.4 shows the relationship for Temperate Zone passerines and raptors, selected for the uniformity of their breeding ecology and behaviour as altricial land-birds (Ricklefs, 1968a). Some 89 per cent of the variation in growth rate is attributable to an exponential dependence of growth rate upon asymptote described by the formula

$$t_{10-90} = 3.94 \, A^{0.278} \, . \tag{6.5}$$

Since this equation is for species growing, at least to a first approximation, along a logistic weight curve, it is possible to re-cast the relationship to give the following general expression for the growth rate expressed in gram per day:

$$\frac{dW}{dt} = CA^{0.722} \, , \tag{6.6}$$

where C is a parameter constant for the part of the growth curve at which the absolute rate of growth (which varies continuously during the growth period – see Figure 6.2) peaks. The interesting point about this equation is that the exponent is very close to the Aschoff and Pohl (1970) equation for the relationship between standard metabolic rate (SMR) (in kilocalories per bird per day) and body weight for adult passerines, viz.

$$SMR = 0.761 \, A^{0.726} \, . \tag{6.7}$$

This suggests that the ability of the nestlings to gain further weight at any particular stage of their development is proportional to their species-specific metabolism. Such a broad generalization is of enormous value in studying growth, for it allows us calculate the expected growth rate for any species once we know the adult body weight. Species whose growth rates deviate from these expected values must then be subject to unusual selective pressures and are worth special attention.

Nestling mortality

Lack (1968) considered that development rate was a compromise between mortality and food supply. Mortality, especially that due to predators, can be reduced by shortening the duration of the vulnerable nestling stage. Such reduction is countered by the greater energy demands of young growing at faster rate, since the parents can then rear fewer young. There is indeed a general negative correlation between clutch size and mortality, Lack's most significant examples being hole-nesting species and seabirds. The former escape much predation by virtue of their nest site and generally produce more young in a brood than do related birds with accessible nests. The latter breed on inaccessible cliffs and stacks or on remote islands and consequently suffer little predation. Both groups have extended incubation and nestling periods, as required by Lack's explanation (Lack, 1954, 1968).

Lack's evidence requires the assumption that differences in nestling period are proportional to differences in energy demand. With non-linear growth curves (Figure 6.2) this cannot be the case unless the energy density of tissues and the maintenance energy of young vary in complementary fashion to the weight changes. Ricklefs (1969a) therefore re-tested Lack's (1968) hypothesis that growth rates were optimized by the mortality patterns of each species. He collected mortality rates for a variety of species from the literature and plotted these values against similarly derived growth rate data.

An absence of clear correlation between the two variables seems to argue against mortality-driven growth rates but the validity of this analysis depends substantially on the extent to which the growth data and mortality data are both representative for their species (Case, 1978). Growth rate variation within a species is negligible in comparison to that between species (Ricklefs, 1968a) but the mortality data are less variable between species (range 0.6–3.6 per cent per day) than within species. For example, nestling death rates over the 14 years of one Polish study have a coefficient of variation of 63.4 per cent for House Sparrows and 77.4 for European Tree Sparrows (Dyer *et al.*, 1977), whilst the corresponding coefficients of variation in growth constants are only 15.4 per cent and 21.0 per cent respectively (calculated from data in Pinowski and Myrcha, 1977). Consequently, Ricklefs' evidence is not as strong as it first appears to be. On the other hand, plotting nestling losses against growth constants for a species, where both variables are for the same area and time provides no stronger correlation than is obtained interspecifically (e.g. House Sparrow, $r = 0.179$, not significant; Tree Sparrow, $r = 0.469$, not significant; data from Pinowski and Kendeigh, 1977). Hence even after controlling for Case's (1978) objection a correlation between growth rate and mortality is not apparent.

However, even if Ricklefs' findings are confirmed, Lack's conclusions are not entirely rebutted, for it is possible for nestling periods to shorten under predation pressure whilst growth rates do not. Lack's data refer principally to nestling periods, though his interpretations were in terms of energy limitations. This area of analysis is one which currently remains open to further research.

Energetics of growth

Energy models

Ricklefs (1969b) developed a model of energy expenditure in relation to nestling mortality. This model predicted that altricial species do best if they grow always at the maximum physiologically possible rate, rather than at some limit set by the availability of food for growth. In this way, growth rates would be maximized rather than optimized, driven to this extreme by the reduction of nestling mortality which results (Lack, 1968). Case (1978) has revised this model to take account of factors neglected in the original formulation, with rather different conclusions.

Lack's (1968) argument was that rapid growth reduces the period over which the young are vulnerable to predators. But there are two ways in which a predator may discover a nest or young. The first is by random search: if a predator repeatedly searches some defined area in this way, any nest there will be discovered sooner or later. However, the chances that it contains eggs or young at that moment will depend on the length of the nest cycle. For such predation, shorter development times reduce mortality. But the second category of predation is not so independent of growth rate. Suppose now the predator locates nests not by random search but by watching for adult birds carrying food to the nest, as Skutch (1949) suggests. Species with rapid growth require more frequent feeding of nestlings and these species are now more, not less, vulnerable to predation than are slow-growing species. The outcome – whether total mortality increases or decreases with alteration of growth rate – depends in this situation on the balance of the two types of predation intensity. Similar ideas hold in relation to other mortality: for instance, faster growth requires more foraging by parents, and the reduction in brooding then risks greater mortality to adverse weather. When such effects are taken into account in models of growth energetics, it turns out that there are indeed circumstances in which very fast growth is less productive in fledged young than is some lower rate (Case, 1978). However, the equations describing such optimal growth rates are mathematically intractable and amenable only to computer simulation.

Ricklefs' (1969b) model is also deficient in some assumptions other than those about mortality (O'Connor, 1980). We know now that metabolism in nestlings varies as a power function of nestling weight rather than remaining constant (Mertens, 1977a), that nestlings can conserve heat by huddling together (Chapter 8), and that the energy density (in kilocalories per gram of body weight) of nestling tissue is age-dependent in some species (Westerterp, 1973; Ricklefs, 1974). Each factor can contribute to peak productivity within a re-defined model of the type originated by Ricklefs (1969b) and can alter the relative speed with which number of broods and brood size vary with growth rate. In addition, the energetics of incubating several eggs at once (rather than as a series of one-egg clutches) do not obey the assumptions of Ricklefs' model. Equations for productivity taking these points into account allow the most productive brood size to exceed unity but, as before, are too complex to be useful except within computer simulations (O'Connor, 1980). One special way in which some birds by-pass these limitations is for male and female to incubate and tend separate nests. In such cases, slow growth and the security of the nest site permit reduction of the daily energy requirement of the adult and brood below the level sustainable by one adult (Hilden, 1975; Haftorn, 1978): the double breeding attempts more than offset any reduction in productivity due to the slowed growth.

Figure 6.5 shows one way in which energy considerations can constrain growth rates. If the adults have some fixed capacity for gathering food and this capacity is less than twice the peak energy demand of a single nestling growing at the optimal rate, the adults will be under-employed. If, however, the growth rate of the nestling is reduced as shown, then two nestlings can be accommodated at the time of peak demand. Similarly, if another species can rear two young through its particular energy bottleneck at optimal rates, a third young could be reared on the surplus capacity of the adults if growth rate is reduced slightly. The process can be repeated with larger clutches, with the adjustment of growth rates serving as a 'fine-tuning' of productivity between the integer changes possible by clutch adjustment. Ricklefs (1968c) used the standard deviations of incubation period (usually well correlated with nestling period) to test the idea quantitatively and found values different from, but closely correlated with, those expected. Fretwell et al. (1974) show how inclusion of nestling maintenance costs in the argument will tend further to increase the ratio values.

Energy requirements may markedly limit growth rates in those seabirds which lay a one-egg clutch. If the growth of young is limited by the ability of the adults to feed their offspring, clutch size must be adjusted to this limit.

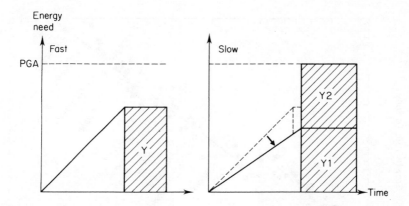

Figure 6.5 A simple model showing how slowed growth can increase nesting productivity. Left: a nestling growing at physiologically optimal rate plateaus at some level of energy need above 0.5 times the optimal parental food gathering ability (PGA); only one young can be sustained at this rate. Right: slower growth to a lower plateau level permits a second young to be accommodated within the parental feeding limit but the period of growth must be correspondingly prolonged.

Once clutch size is reduced to one egg, however, further adjustment can take place only through an alteration of the form of growth, to reduce the rate at which energy is required of the adults. The chick must therefore either pack more growth into the earlier stages of the nestling period, thus exploiting the additional foraging capacity of its parents before the bottleneck is reached, or must reduce its growth rate in the later part of the period. Of these options the former increases nestling biomass (and therefore maintenance costs) earlier, thus imposing heavier maintenance charges on the chick on reaching the energy limit. Conversely, a reduction in growth rate late in the nestling period reduces the relative expenditure on maintenance and should therefore be more advantageous. A logistic curve retarded selectively in this way would resemble a Gompertz or, in extreme form, a von Bertalanffy curve. These are in fact the forms most commonly found amongst oceanic species, the most likely candidates for energy limitation (Ricklefs, 1968a).

On the basis of the model above, one might also expect to find a difference in growth rate between parentally-fed precocial species and parentally-fed altricial species, since the former are functionally mature in respect of thermoregulation and locomotion. Certain seabirds do not match this expectation, however. Altricial young in the families Diomedeidae and Fregatidae grow more slowly than the precocial or semi-precocial young of the families Stercorariidae, Laridae, and Alcidae. On the other hand, Drent and Daan (1980) point out that, whilst growth rate has some maximum set by

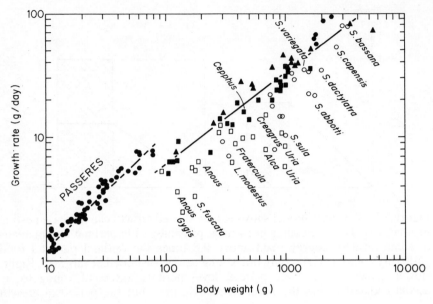

Figure 6.6 Absolute growth rate (in grams per day) at the mid-point of development in relation to adult body weight in passerines (●, left) and in pelicaniforms (○, right), charadriforms (■, □) and falconiforms (▲). Nidicolous species only. Open symbols indicate reduced brood sizes excluded from the regression line (slope 0.74). Redrawn with permission from Drent and Daan 1980).

physiological constraints, the ecological question of most interest is whether growth rates can be reduced if the daily cost of the maximum rate cannot be met by the parents. As summarized above, adjustment of clutch size provides a primary channel of adjustment of parental reproductive effort, with variation of growth rate only a fine-tuner of effort to the environmental conditions prevailing. However, in species with broods of one young, growth rate depression is the only means of adjusting to poor food supply. Figure 6.6 summarizes this point, showing that most seabirds laying one-egg clutches (and those laying two eggs, with one an insurance against egg loss) have depressed growth rates for their body size. The list includes most pan-tropical boobies, frigatebirds, auks, and tropical gulls and terns. Perhaps even more interesting are the exceptions within these groups. The Peruvian Booby has a richer and more predictable food supply than its tropical relatives and often raises two or three chicks at normal growth rates. The Atlantic Gannet has the growth rate expected for its size and can raise experimentally introduced twins, whilst the related Cape Gannet has slower growth and fails to rear twins. Similarly, Pigeon Guillemots can rear two chicks at normal rates where most auks rear only one at slow rates. Drent and Daan (1980) detail other examples.

Patterns of energy use

How do the energy models of optimal growth relate in practice to the energetics of avian growth: Figure 6.7 shows the pattern of energy use during growth by an altricial and by a precocial species. Gross energy intake (GEI) is

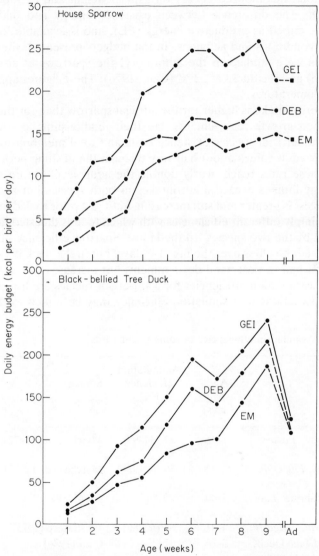

Figure 6.7 Energy utilization during growth in an altricial species (top: House Sparrow, *Passer domesticus*) and a precocial species (bottom: Black-bellied Tree Duck, *Dendrocygna autumnalis*). *Key*: GEI, gross energy intake; DEB, daily energy budget; EM, existence metabolism. Data from Kendeigh *et al.* (1977), from studies by Blem (1975) and Cain (1976).

the energy consumed by the growing birds, the daily energy budget (DEB) or total metabolism defines the energy actually metabolized by the bird during each day, and existence metabolism (EM) is the rate at which energy is used by birds maintaining a constant weight. With growing young this latter cannot be measured directly, but it can be estimated within an age group by regressing total metabolized energy on gain in weight and extrapolating to zero growth. The difference between existence energy and daily energy budget is described as productive energy (PE), and is available for energy-demanding conditions and activities. In the budgets presented, the young of both species were studied in the laboratory, the sparrows at about 36 °C (Blem, 1975) and the ducks at 32 °C (Cain, 1976). These figures approximate brooding temperatures.

Gross energy intake is higher for the altricial sparrow than for the duck, in line with their growth requirements, but their relationships to adult values differ. In the altricial species existence costs and total metabolism level off sharply near adult values about half-way through the nestling period. In the ducklings these rates reach nearly double the adult level before declining. Gross energy intakes exceed adult intakes in both species but again in the duck the excess is greater and still increasing until just prior to fledging. These differences imply different efficiencies with which food is digested, absorbed, and utilized by the two species. In the House Sparrow efficiency of food use increases to adult values over the first 4–5 days, but in the duck it varies little over the first half of post-natal development, but increases thereafter.

Average assimilation efficiencies for a variety of species are listed in Table 6.1. These variations in assimilation efficiency may be linked to diet, meal

Table 6.1   Assimilation efficiencies in some young birds

Species	Age span (days)	Assimilation efficiency (%)	Source
House Sparrow, *Passer domesticus*	1–15	82 (34–95)	Myrcha *et al.* (1972)
Tree Sparrow, *Passer montanus*	1–15	76 (62–91)	Myrcha *et al.* (1972)
Red-backed Shrike, *Lanius collurio*	3–14	70	Diehl (1971)
Starling, *Sturnus vulgaris*	4–16	64 (60–80)	Westerterp (1973)
Black-bellied Tree Duck, *Dendrocygna autumnalis*	3–63	77	Cain (1976)
Black Duck, *Anas rubripes*			Penney and Bailey (1970)
Double-crested Cormorant, *Phalacrocorax auritus*	11–21	80–88	Dunn (1975b)
Dunlin, *Calidris alpina*		57	Norton (1973)

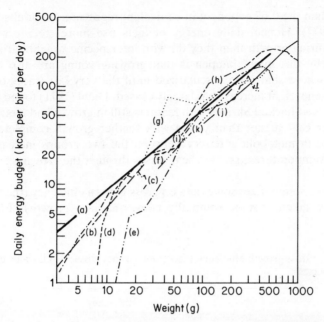

Figure 6.8 Changes in daily energy budget with growth in various species: (a) regression line for House Sparrow, *Passer domesticus*, and Black-bellied Tree Duck, *Dendrocygna autumnalis* (see text); (b) House Sparrow, *Passer domesticus*; (c) Red-backed Shrike, *Lanius collurio*; (d) Dunlin, *Erolia alpina*; (e) Rook, *Corvus frugilegus*; (f) Starling, *Sturnus vulgaris*; (g) Coot, *Fulica americana*; (h) Black Duck, *Anas rubripes*; (i) Lesser Scaup, *Nyroca affinis*; (j) Double-crested Cormorant, *Phalacrocorax auritus*; (k) Blue Grouse, *Dendropagus obscurus*. Redrawn with the permission of Cambridge University Press from Kendeigh *et al.* (1977).

size, season, brood size, and other variables. Amongst different species with equal daily energy budgets, differences in assimilation efficiency determine the additional energy intake needed in the diet. In energetic terms, variation in the size of DEB needed with different species is of interest. Figure 6.8 shows that DEBs are strongly weight-dependent within species. Some of the budgets used in this diagram reflect retarded growth in laboratory-fed birds but Kendeigh *et al.* (1977) provide a best estimate of

$$DEB = 1.353W^{0.814} , \qquad (6.8)$$

where DEB is measured in kilocalories per bird per day and W is body weight in grams. The regression was obtained only from the House Sparrow and Tree Duck data of Figure 6.8, to avoid the effects of retarded growth. The same line fits both altricial and precocial species. The exponent 0.814 is greater than the 0.67 exponent in equivalent regression for adults (Kendeigh *et al.*, 1977). Hence, daily energy budgets rise more steeply with weight change during growth than they do with interspecific weight variation. One reason why this might happen is that growing young are also developing feathers whose shafts are vascularized until the very last stages of development, when weight increase has largely ceased. Diehl (1971) found that young fledgling Red-backed Shrikes with feathers still in growth had metabolic rates 20–30 per cent higher than young with feather growth completed. As the difference in metabolic intensity between the two groups increased at low ambient temperatures, greater heat losses through the growing feathers are indicated.

Table 6.2 summarizes some data on gross growth efficiency (the proportion of energy taken that is eventually converted into growth). Efficiency is

Table 6.2   Gross growth efficiency (kilocalories in new tissue ÷ kilocalories eaten) for various species

Species	Efficiency (%)	Source
Wood Stork, *Mycteria americana*	24	Kahl (1962)
Herring Gull, *Larus argentatus*	26	Brisbin (1965)
Chicken, *Gallus domesticus*	16	Davidson *et al.* (1968)
Pigeon Guillemot, *Cepphus columba*	34	Koelink (1972)
Starling, *Sturnus vulgaris*	14	Westerterp (1973)
House Sparrow, *Passer domesticus*	14	Kendeigh *et al.* (1977)
Black-bellied Tree Duck, *Dendrocygna autumnalis*	14	Kendeigh *et al.* (1977)

obviously not correlated with developmental mode since duck, chicken, sparrow, and Starling are equally efficient (or profligate) in their use of food. The three species with high efficiency are all fish eaters but this may be an artefact of the small number of species considered.

Table 6.3 summarizes the relative use of energy for growth and maintenance in some other species. The youngest nestlings tend to devote relatively more of their metabolized energy to growth than do older birds. This was also apparent in a multiple regression analysis undertaken by Blem (1973). He found that metabolized energy (in kilocalories per bird-day) was positively correlated with both nestling weight and daily weight change but with the former three times more important a predictor than the latter, i.e. increasing size lowered the proportion of energy going into growth. Over their entire growth period House Sparrows used 19 per cent of their energy supply for new tissue, 57 per cent for respiration and maintenance (including the cost of

85

Table 6.3 Percentage of metabolized energy expended on growth by passerines nestlings. Data from Blem (1973)

Species	Age span (days)	Percentage to growth (%)
Carolina Wren, *Thryothorus ludovicianus*		
(a)	4–6	23.7
(b)	6–8	20.7
Brown Thrasher, *Toxostoma rufum*	1–3	61.6
Mockingbird, *Mimus polyglottos*	1–3	43.8
American Robin, *Turdus migratorius*	1–3	73.0
Starling, *Sturnus vulgaris*		
(a)	4–8	16.3
(b)	8–10	20.6
(c)	10–12	12.8
Common Grackle, *Quiscalus quiscula*	8–12	15.8

biosynthetic activity), and 24 per cent for activity (Blem, 1975). Starlings used 22 per cent for tissue production, 61 per cent for maintenance and other metabolism, and 17 per cent for activity. Comparing precocial and altricial species, over the period from hatching to independence the House Sparrow used about 89 per cent of its DEB for existence and 11 per cent for growth, whilst for the Tree Duck the figures were 75 per cent and 25 per cent respectively (Kendeigh et al., 1977).

Physiological constraints on growth

In complete contrast to the idea that energy supply to young birds is limiting to their growth is the idea that growth is regulated by what might be termed the physiological design of the young bird. Two types of internal constraints to growth rates have been proposed (Ricklefs, 1969b, 1973). The first is based on the rates at which energy and nutrients are distributed about the growing bird and assumes there are internal limits to the distribution of energy or nutrients about the body. Since species differ in the extent to which the young are active and responsible for their own thermoregulation, the diversion of energy to these functions might limit the proportion of energy ingested that can be made available for growth. The second explanation is based on the idea that biochemical and molecular constraints may limit the extent to which cells in a tissue can both differentiate functionally and continue proliferation and growth. The first type constrain the nutrition of the organism as a whole, whilst the second operate at tissue level and affect each body organ differently.

Organismal constraints

The nutritional status of a young bird fed regularly by its parents is set finally by its ability to metabolize and assimilate the food supplied. These processes are influenced by the energy and nutrient composition of the diet. Dietary deficiencies are well known to retard growth in many animals but amongst birds growth rates are largely independent of dietary quality. A few tropical frugivores have unusually low growth rates (p. 143) but there is no clear-cut relationship between diet and growth rate (Table 6.4). On the other hand, Perrins (1976) has suggested that the slow growth of late broods of Great Tits may be due to poor quality diet. At that time of year, the oak leaves, on which the tits' caterpillar prey feed, have accumulated a high phenolic

Table 6.4 Growth rates[a] of neotropical birds in relation to nestling diet. Data from Ricklefs (1976)

Diet	Growth constant		
	Range	Mean ± s.d.	(*n*)
Fruit	0.098–0.460	0.262 ± 0.151	(5)
Mixed fruit–insect, mostly fruit	0.280–0.464	0.375 ± 0.076	(6)
Mixed fruit–insect, mostly insect	0.199–0.536	0.379 ± 0.102	(9)
Insect	0.236–0.524	0.357 ± 0.078	(13)
Nectar–insect	0.256–0.362	0.317 ± 0.055	(3)
Seed	0.472–0.520	0.496 ± 0.034	(2)

[a] Rate constants of logistic equations fitted to the growth data.

content, chiefly of condensed tannins. These chemical compounds form part of the oak's defences against grazing and impair the growth and survival of caterpillars consuming them. Perrins conducted some simple laboratory experiments to see if these effects in the insects affected tit growth. Broods of Blue Tit nestlings were fed mealworms either with or without oak-leaf tannin added. Weight increments were 9 per cent higher in the tannin-free birds after 2 days of such feeding, and 21 per cent higher after a further day than in tannin-contaminated birds. Uncontaminated birds also begged more and fed more readily than did experimental birds. Perrin's experiments thus suggest that the ability of nestlings to cope with diets of different types affects their growth rate. Tannin contamination must be a regular feature of nestling diets in the wild and one might expect evolved adaptations to the problem. Perrins speculates on this: for example, Great Tits dealing with a large caterpillar usually behead it, then draw out the gut from the body and discard it, thus reducing the toxic effects of the gut contents of the caterpillar!

Another group of species in which problems of nutrient allocation may arise is the procellariiforms. The remote foraging of the adults has led to the use of regurgitated 'oil' as the food of the young, presumably to reduce the mass brought back to the nest. Analysis of oil fed to Leach's Petrel showed it to contain 33 per cent water, 63 per cent lipid, and only 4 per cent non-lipid dry matter (Ricklefs et al., 1980a). This amounts to about 130 kcal of metabolizable energy per gram of protein present, a loading in excess of even the 91 kcals/g of tropical fruits. The chicks thus have special problems in handling this imbalance of energy and protein and presumably sink some of the excess calories into the enormous fat deposits they develop.

Despite these individual cases, Ricklefs (1979a) suggests that growth rates are rarely limited at organism level. Two lines of evidence support his argument. First, there is much less variation in growth rate between species with very different foraging energetics than one would expect were such limitation the case. Second, there are no abrupt changes in growth rate at hatching, despite gross changes in diet and in the degree of self-sufficiency of the chick. Constraints at the nestling level are all of a type open to parental assistance: an excessive thermoregulatory burden may be met by brooding; too difficult a food location task may be met by the parents locating the food; and too specialist a prey-capture technique may be met by parental feeding. Hence one would expect parental aid to evolve to match any organismal-level limits to growth. It is significant in this respect that it is those groups with high skills in foraging – terns, herons, raptors, etc. – that display the greatest prolongation of parental feeding.

Precocity constraints

The precocity hypothesis advanced by Ricklefs (1973) suggests that growth rates in birds are limited by the extent to which a bird's tissues are called upon to function from an early stage, mature function being inconsistent with continued cellular development. Why such a trade-off exists at cell level is unknown, though presumably it originates in the nature of DNA replication and protein synthesis (Ricklefs, 1979a).

Amongst birds of similar body size precocial chicks gain weight more slowly than do altricial species (Figure 6.9). Deviations of growth rates from the common regression line of growth rate on asymptote clearly fall into two groups, precocial and altricial. The difference between modes corresponds to altricial growth rates 3.0–3.5 times greater than among precocial species.

Precocial functions absent from very young altricial birds are temperature regulation, ground locomotion, and flight (Ricklefs, 1973). The three functions develop at different rates within individual species: Table 6.5 sets out the relative speeds of acquisition of each in different groups. Any one of the three functions could be the source of the constraint on growth rate, or some factor common to the three might be involved. Both flight and temperature regulation require plumage development, the former of primaries and

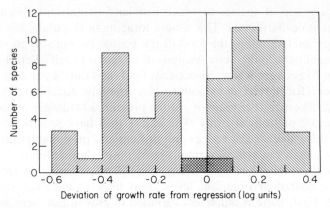

Figure 6.9 Distributions of growth rate constants for precocial (□) and for altricial (□) species, with growth constants corrected for adult body size by reference to a regression of growth constant on size. Redrawn with permission from Ricklefs (1979a).

rectrices, the latter of contour feathers (Chapter 7), and the two are rarely synchronized in development. All three do require the development of muscle mass, though. Muscles provide for heat production for temperature balance (Chapter 8) and for locomotory power in walking and flight. Final muscle size is generally large in birds, the flight muscles accounting for 15–39 per cent of adult body mass in most groups and up to 44 per cent in some

Table 6.5 Relative timing of temperature regulation, ground locomotion and flying in various groups. Modified from Ricklefs (1973)

Group	Mode of development	Development of Thermoregulation	Walking	Flight
Corvidae	Altricial	Late	Late	Late
Columbidae	Altricial	Late	Late	Late
Falconiformes	Semi-altricial	Early	Late	Late
Strigiformes	Semi-altricial	Early	Late	Late
Phalacrocoracidae	Semi-altricial	Late	Late	Late
Pelicanidae	Semi-altricial	Late	Late	Late
Laridae	Semi-precocial	Intermediate	Early	Late
Charadriidae, Scolopacidae	Precocial	Early	Early	Intermediate
Anatidae	Precocial	Varied	Early	Late
Rallidae	Precocial	Varied	Early	Late
Galliformes	Precocial	Early	Early	Early

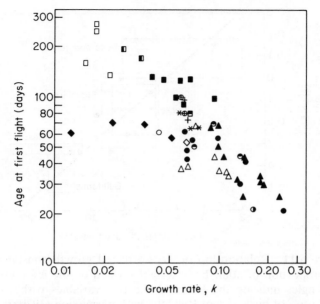

Figure 6.10 Age of first flight (= maturity) in relation to speed of growth in various families, *Key*: *, Spheniscidae, Procellariidae and Hyrobatidae; ⊕, Phaetontidae; ■, Sulidae; ◨, Fregatidae; □, Diomedeidae; ▬, Pelecanidae; ○, Phalarocoracidae; ◑, Ardeidae; ○, Ciconiidae; ◆, Anatidae; ●, Laridae; △, Alcidae; ▲, raptors and Corvidae. After Murton and Westwood (1977) and Ricklefs (1973).

pigeons and doves. Even in a group with weak flight, such as the Rallidae, these muscles constitute 12–16 per cent of body weight and here the leg muscles are correspondingly large, at 11–23 per cent of total weight (Ricklefs, 1973). The extent to which muscles assume mature function thus provides a possible constraint on growth rates common to all three precocial functions.

Examining this further, development of flight generally appears to be the dominant factor limiting growth amongst birds (Figure 6.10). On this basis, those species achieving flight late in development have proportionately slower overall growth rate because they need large functionally mature muscles before being able to fly. Excluding the precocial species, the relationship between growth rate and attainment of flight is good. The most important exceptions are the larids and alcids. In both groups, flight is attained relatively early and growth is rather slow for their body size.

The Anatidae are a notable exception to the trend, flying unusually early for their growth rate. However, growth rates amongst ducks are both unusually variable and rather rapid for their body size. Accessibility of food for young may be of greater importance within this group, for growth rates decrease as the difficulty of foraging increases from field species through

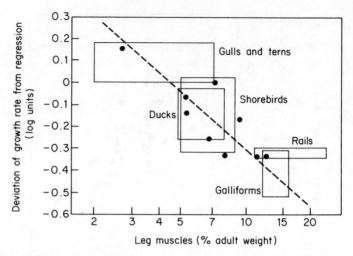

Figure 6.11 Relationship of size-corrected growth constants to the proportion of adult body weight formed by the leg muscles. Rectangles indicate the limits to the two variables within each group of birds. From Ricklefs, with permission (1979a).

dabbling ducks to sea-ducks (Ricklefs, 1973). Age of flight also differs between diving ducks and dabbling ducks, being about 10 days later in the former (Weller, 1957). Flight ability has also been implicated in comparisons of growth patterns in domesticated and wild ducks: the weight versus age curves of wild birds show two marked checks to weight gain, the first at 20–25 days when contour feathers are growing rapidly, the second at about 60 days when active flight (absent in the domesticated strain) commences (Mahelka, 1973). The same study showed that the growth of internal organs was most modified in the cases of flight muscle and wing loading. As a result the non-flying Peking Duck reaches adult weight between 120 and 140 days, ages at which the flying Mallard has achieved only 85 per cent of its final weight.

Since flight is so uniformly linked to growth rates, it is possible to test the influence of muscle function on growth rate by considering the differential effects of different proportions of tissue functioning in leg muscles. Figure 6.11 shows that size-corrected growth rates decline as the proportion of adult weight devoted to the leg muscles increases: in gulls and terns, with legs forming a few per cent of adult weight, growth rates approach those of altricial species, but in rails and galliform birds the large legs reduce growth rates by a factor of 3 or 4. These findings dispose of one of the more serious reservations raised by Case (1978) about the maturity of tissue function hypothesis. Case had noted, for example, that members of the Charadrii and of the Scolopacidae grow faster than do individuals of the Phasanidae and Meleagridae and that young Anatidae grow faster than do young Rallidae of equal weight, trends not easily accounted for on a simple scale of 'general precocity' but readily accounted for in relation to mature muscle mass (Ricklefs, 1979a).

Physiological experiments are also consistent with the idea that cellular processes will not support both growth and mature function in skeletal muscle. In chick embryos, for example, cells in skeletal muscle, in the lens of the eye, and in neural tissue cannot proliferate once differentiated; erythrocytes, feather pulp, cartilage, cornea, heart, and liver, on the other hand, all retain a capacity to proliferate after differentiation. Skeletal muscles increase in size mostly by cell generation until day 19 of incubation, but increase in cell size is involved thereafter. In Starlings and Japanese Quail the rate of uptake of tritiated thymidine – a measure of the mitotic activity of cells in the tissue – is closely related to the instantaneous percentage growth of various organs, thus linking organ growth to cell proliferation. In the Starling, but not in the quail, cell size also increases from day 14 to day 27 (measured from incubation) in all tissues examined (Ricklefs and Weremiuk, 1977). Uncertainties as to the length of the mitotic cycle in the muscles prevent unambiguous interpretation of these experiments but their results are at least consistent with the postulated trade-off between embryonic and mature function.

INTRASPECIFIC VARIATION

Intraspecific variation in growth rates does not appear to be at all correlated with variation in nestling asymptote, a situation in total contrast to the interspecific situation (cf. Figure 6.4). Ross (1980) notes that most studies of passerines fail to show distinct relationships between growth rate and asymptote. One likely reason for this is that growth rate and asymptote may be independent parameters of individual bird's development, coupled only by external factors such as nutritional level. Figure 6.12 presents empirical data for this being the case (O'Connor, 1978b). Here principal component analysis was used to search for coupling between the daily weights of individual House Sparrow nestlings as they grew from hatching to fledging weights. Were the weights at each age strongly constrained by some single developmental process during nestling life, the loadings on each variable would have been approximately equal within a major component. In practice the sparrow weights were coupled together in two groups, the largest (accounting for 48 per cent of the variance in growth patterns) involving weights over the first 6 days, the second (32 per cent variance carried) involving mainly those weights recorded from the asymptotic region of the curve. Principal components are orthogonal to each other, so these structures for growth phase and asymptote reflect independent aspects of House Sparrow growth. There thus exists scope for separate regulation of growth and of body weight either by genotype or by environment. In a similar analysis of Blue Tit growth a like segregation of growth and asymptote loadings was found and the asymptote component was a function of brood size, decreasing slightly in larger-than-average broods despite a prolongation of their growth phase (O'Connor, 1978b). Savannah

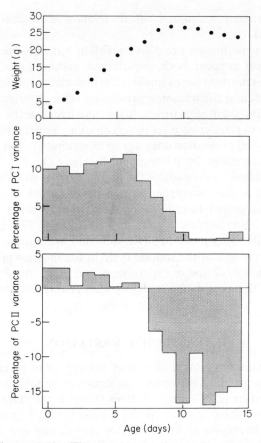

Figure 6.12   The structure of growth pattern in the House Sparrow, *Passer domesticus*. Daily weights of nestlings during the main phase of weight increase (middle and top) are mutually correlated but uncoupled from those during weight recession (bottom and top). Vertical axes of histograms are the percentage loadings on the daily weights in first and second principal components. Redrawn from O'Connor (1978b) by permission of The Zoological Society of London.

Sparrows have also been shown to reduce asymptote variation by prolonging the time taken to reach plateau weight (Ross, 1980).

An extended growth period allowing achievement of normal or near-normal fledging weight under adverse conditions is likely whenever post-fledging success is weight-dependent, as in titmice (Perrins, 1965; Garnett, 1981), or where the adversity is likely to be of short duration, as in aerial insectivores (Koskimies, 1950; Bryant, 1975). In some circumstances, howev-

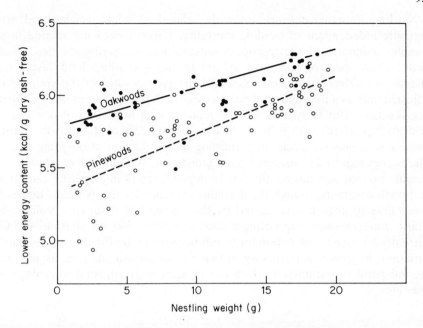

Figure 6.13 Growth economy in Great Tit, *Parus major*, nestlings in food-poor pinewoods, showing the lower energy content of body tissues of nestlings of any given weight relative to that of young reared in oakwoods. Redrawn with permission from Drent and Daan (1980).

er, extension of the growth period may be avoided by economizing on energy reserves normally maintained as insurance against temporary food shortages. Figure 6.13 shows how Great Tits nesting in sub-optimal pinewood reduced their total energy cost of growth by decreasing the size of fat depots maintained through the nestling period, most sharply in early life (Mertens, 1979). Such energy savings (about 5 per cent) are feasible if feeding conditions are relatively stable relative to the duration of the nestling phase, but not otherwise (O'Connor, 1978a).

SUMMARY

Growth in birds can be described in the form of weight versus age curves and summarized by the parameters of growth equations which approximate the form, magnitude, and rate of growth. For most species growth in weight is asymptotic and fastest when half-way towards the weight asymptote. For passerines, especially ground-feeding species, the asymptotic weights are often below adult levels, but in aerial insectivores and in pelagic seabirds nestling asymptotes tend to exceed adult weights as a result of fat deposition or of muscle development.

Nestling growth rates are inversely related to adult weights and are arguably independent of nestling mortality. Growth rates are related in a complex manner to the energetics of parental food-gathering abilities. Most birds adjust their reproductive efforts to the prevailing food levels by increasing or decreasing clutch size and the growth rates of the young are adjusted only as a fine-tuning of productivity to the integer steps of the clutch size function. But in species already reduced to a maximum of one egg – principally seabirds with difficult foods – the slowing of the growth of the single young may be crucial in permitting the production of offspring at all. The energy required to sustain a given growth rate is a function of adult body weight and may account for the weight dependence of interspecific variation in growth constants, though the available evidence for this is weak. Internal constraints to growth rate caused by the inability of tissues to proliferate whilst simultaneously supporting mature function offer an alternative and currently favoured explanation for trends in growth rate. Finally, intraspecific variations in growth pattern suggest that rate of growth and final asymptote are independent parameters which can be separately adjusted to ecological pressures.

CHAPTER 7

Plumage, moult, and other differential development

The use of weight versus age curves for the study of postnatal growth is widespread but tends to under-emphasize the importance of the differential development of body organs and structures. Young birds do not develop with all organs increasing at the same rate. Instead, growing birds develop by elaborating tissues differentially to form discrete body organs, such that at any stage a young bird is appropriately equipped to meet the functional demands made on it at the time. Indeed, the young bird may even at times possess organs or structures not present in the adult, egg teeth and hatching muscle being extreme examples. Growing birds also pass through one or more non-adult plumages before reaching maturity. This chapter reviews differential development in young birds, considering first the plumages adopted during progress to adulthood, and then the patterns of relative growth in body organs and constituents as birds grow.

DOWN

Down is the earliest postnatal plumage of developing birds and is characteristic of precocial species. However, although many passerines look naked when they hatch, natal down becomes apparent once the amniotic fluids have dried off. Wetherbee (1957) estimated an average of 130 neossoptiles (the elements constituting down) in the species he examined (an average rising to 151 if the genuinely naked species are excluded). Genuine loss of all natal plumage has occurred independently in members of seven families: Hirundinidae, Corvidae, Chamaeidae, Bombycillidae, Vireonidae, Parulidae, and Ploecidae – and may have occurred in four other families. Figure 7.1 shows the variation in typical distribution of natal feather tracts or *pterylae* in a variety of families. Although there exists substantial variation between species, five of the major tracts – the mid-dorsal, coronal, occipital, scapular, and femoral – are always represented but some others, notably the ventral, crural, rectrix, and alar tracts, have often been lost in evolution (Wetherbee, 1957).

The presence or absence of downy feathers is one of the characteristics used by Nice (1962) to classify development types in birds. To a large extent

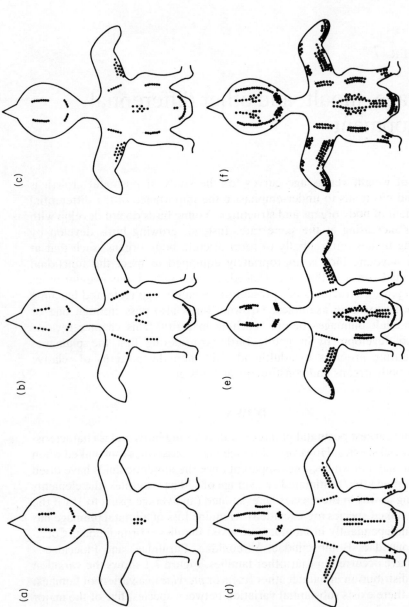

Figure 7.1 Basic neossoptile pterylosis of some small land-bird groups, showing the location of natal feather origins: (a) Troglodytidae, wren family; (b) Hirundidae, swallows; (c) Parulidae, American wood warblers; (d) Mimidae, mockingbirds; (e) Sturnidae, starlings; (f) Tyrannidae, tyrant flycatchers. Redrawn from Wetherbee (1957), courtesy of the Library Services Department, American

the presence of down is determined by the development pattern of the young. Precocial chicks have the most developed down but little research has been directed to the morphology of these plumages. Natal down consists essentially of a short proximal quill with a spray of slender, simple barbs with many barbules, each ending in a fine hair-like tip (Fjeldstra, 1977). Most quills lack a marked shaft, except among waterfowl and in a few odd groups such as the ratites, tinamous, and megopodes. Fjeldstra was unable to identify any clear taxonomic correlates of down type, for even within a group such as the terns (Sternidae) great variety of structure is present: thick soft down with long, silky terminal filaments in Black Terns; spiny spiky down with clumped sheathed filaments in Sandwich Terns; and matted woolly plumage of variable thickness in Little Terns.

A few species possess down apparently specialized to significant ecological demands on the young. In young Sandgrouse, for example, the feather barbules are directly controlled by skin muscles. When depressed they form a dense closed insulating coat but when elevated allow circulation of air through to the skin, thus permitting heat shedding by this desert species (Fjeldstra, 1977). In the Cape Barren Goose the down is relatively long compared with other Anseriformes (26.1 mm neoptiles ventrally, 29.8 mm dorsally), an elongation presumably associated with the winter nesting of this species: the species nearest in neoptile length is the Andean Goose which breeds at high altitudes and has neoptiles averaging 26.3 and 36.2 mm on ventrum and dorsum respectively (Veselovsky, 1973).

PLUMAGE SEQUENCES AND MOULT

Down, where present, is only one in a series of plumages which the bird goes through in its lifetime, though the length and complexity of the series varies with species. Hatchlings of a few groups such as kingfishers, woodpeckers, and some passerines are completely naked. Others have a few traces of down in areas (termed prepennae) from which feathers will later emerge. Others hatch with substantial coats of down, as in most ducklings, young gallinaceous birds and many seabirds. In some slow-growing species with severe thermal demands on them – penguins, petrels, owls – the natal coat of down is followed by a second coat. These downy feathers are eventually lost as the growing contour feathers beneath emerge. In the crane family, Gruidae, for example, two generations of down are produced by the young and a third coat is developed as an under-down (present also in many other species) to the contour feathers. Some young birds display specialized downy tracts, the prominent cover of the young Skylark being a striking example. Amongst altricial nestlings these downy patches are frequently found on the exposed back and head surfaces. These may perform a thermoregulatory function by insulating growing contour feathers. These are particularly concentrated on the dorsal surface of nestlings and their blood-laden quills must tend to lose heat from this surface. In waterfowl the definitive or adult down first develops on belly and flanks, the most exposed parts of these species.

The first coat of contour feathers is termed the juvenile plumage and in many species lasts only until the first autumn moult. Generally looser and softer in texture than adult plumage, this plumage may additionally be distinctively different in colour. The occurrence of such differences has in some cases been attributed to specific ecological or behavioural demands. In the autumn most young undergo either a partial post-juvenile moult or a complete moult into a first winter plumage. A post-juvenile moult typically includes the main body or contour feathers as well as a variable number of greater coverts, and often includes the renewal of part of the tail. A complete moult involving the wing and tail feathers is less common in first-autumn birds, though present in such species as Long-tailed Tit and European Tree Sparrow. The course adopted is usually species-specific but in some species a degree of individual variation is apparent. Selander and Giller (1960) found that only 28 per cent of the female Red-winged Blackbirds they examined underwent a complete post-juvenile moult. Ligon and White (1974) found that the extent of post-juvenile moult in Piñon Jays in New Mexico was dependent on their season of birth. Birds hatched in spring (February–June) replaced more feathers, including most of the greater (secondary) coverts and some secondaries and alar feathers, whilst birds hatched in autumn (September) did not renew any secondaries and few or none of the greater coverts and alar feathers. Spring-hatched birds also began their moult at a later age (60 days) than did autumn-hatched birds (45 days). These two features – moult of fewer feathers by late hatched broods and faster moult by late birds – have been observed in other passerines. Factors possibly responsible for such differences include differences in photoperiod during moult, in ambient temperatures, in food availability, and in the need to complete moult before departure on migration or before winter stress sets in. In the case of Piñon Jays early-hatched young fed on insects whilst moulting but late broods had to feed in piñon seeds; the former might be expected to be a better protein source for the feather synthesis required.

The main function of the autumn moult (which is usually undergone by adults as well as by juveniles) appears to be to renew the bird's plumage ahead of winter stress. Where only a partial moult is involved, it is the insulating contour feathers of the body that are renewed, the flight feathers being retained. Most first winter plumages are duller than are adult ones, to some extent because the bright breeding plumage colouration is hidden by dull-coloured tips to each feather. In spring the tips wear away through abrasion and the breeding plumage becomes apparent. In other species the breeding plumage is acquired by a partial moult in the spring.

As a general rule, the longer-lived a species is, the more complex its plumage sequence. Although most small songbirds attain breeding plumage in their first spring, many of the long-lived non-passerines attain only one of a series of immature plumages at that time. In most of these species each age-class has a distinctive patterning of plumage.

In most passerines the primaries are shed at regular intervals throughout a well-defined moult season and the moult, migration, and breeding seasons are mutually exclusive. In non-passerines slower moult cycles are not infrequent and may lead to overlapping of moult and breeding. Immature Shags may in this way retain some of their primaries for 26 months after birth, apparently because the reduced physiological stress of a slow moult during the summer more than offsets the reduced efficiency of aged feathers (Potts, 1971). Moult is suspended during the winter, when the risks of adverse conditions being met are high. European Swifts similarly stagger their primary moult, losing one or two primaries at a time over a total moult period of 6 months or more but meanwhile retaining essential flight efficiency.

Juveniles generally differ from adults as to the timing of autumn moult. In multi-brooded species early brood juveniles may already be starting moult whilst adults are still tied to late broods. In single-brooded species, however, the adults can often complete their moult faster than can juveniles. In long-lived species the non-breeding immatures can undertake autumn moult at times or in places prohibited to adults by breeding commitments.

Many of the larger non-passerines have unfavourable wing-loadings which preclude the further loss of flight efficiency inherent in a staggered moult (gradual replacement of flight feathers, a few at a time) and instead moult in a flightless condition in some secluded area. Amongst waterfowl, in particular, this has led to the evolution of moult migrations to special moulting grounds on which thousands of birds may be present in flightless condition. Thus up to 100000 Shelduck from all over Europe may be present in late summer on the tidal flats at Knechtsand on the German Waddensee coast (Salomonsen, 1968). The extent to which immature birds participate in these migrations varies between tribes of waterfowl (Table 7.1), in part because of parental duties. In *Cygnus* swans, for example, immatures can become flightless on safe feeding grounds but breeding birds could be separated from their flying

Table 7.1 Participation of immatures of different waterfowl tribes in a moult migration[a]. From Salomonsen (1968)

| | Adults | | |
Tribe	Males	Females	Immatures
Anserini	—	—	××
Tadornini	××	××	××
Anatini	××	—	—
Aythyini	××	×	—
Somateriini	××	×	××
Mergini	××	×	××

[a] ×× indicates that all individuals of the group participate; × indicates that only some individuals participate.

offspring if they underwent wing-moult after the cygnets fledged. Instead, the adults moult their flight-feathers during the breeding period, with the female completing hers before the male moults to ensure that at least one flying adult can defend the nest or young (Salomonsen, 1968).

ADAPTIVE VALUE OF NATAL AND JUVENILE PLUMAGES

A variety of adaptive advantages have been attributed to sub-adult plumages different from adult forms, including concealment by colour and form, heat retention (Chapter 8), protection from the Sun's rays (Chapter 8), and signal function. Cryptic colouration is particularly common on the part of ground-reared young such as ducks, gallinaceous birds, Woodcock, and Nightjar. The young of these species crouch on hearing parental alarm calls and the cryptic effect of their plumage match to their habitat is increased by the distracting effect of parental activities, whether in distraction display or in mobbing behaviour. Chicks of a few species enhance their camouflage by specialized behaviours. Young Black Skimmers as young as 1 day in age use their feet to excavate depressions on their nesting beach so that they settle flush with the beach. The excavation process additionally results in some sand falling on their backs, providing further camouflage against the danger (Hays, 1970).

In some species chick plumage seems adapted to a signalling function (Fjeldstra, 1977). Young grebes, for example, have head and neck distinctly striped with black and white (or tawny) stripes, with naked lores and, in many species, a bare patch on the crown. If attacked by an adult, the young display a complex appeasement behaviour, with alteration of exposure and hiding of the facial pattern and the bare skin flushing at intervals from pink to red. If begging for food, on the other hand, the bare skin is quite pale.

Other evidence for a signal function to chick plumage may be inferred from a correlation between the conspicuousness of chicks of various Rallidae and the degree of dependence on parental care shown by each species (Fjeldstra, 1977). Corncrakes receive least attention, being fed only over the first couple of days, and the chicks hatch with a dark blackish-brown plumage and pink bill, the latter darkening within a couple of days. Other rail chicks are fed for longer, until about half-grown, and hatch with lustrous black plumage and white bill, the latter augmented with yellow, red, or black markings. Sparse down in the head region reveals the red or blue skin. Finally, three other species, Moorhen, Purple Gallinule, and Coot, each feed their young until full-grown or beyond. Their young have almost naked blue or red skin on the crown, augmented with specialized down bristles tipped with white or yellow. These colours serve as feeding and appeasement signals until individual recognition of parent and chick develops. This gradient towards increased signal function on the part of the plumage presupposes a corresponding absence of or reduction in predation risk, presumably brought about through the protection against many predators afforded by the aquatic habitat of these birds.

DIFFERENTIAL DEVELOPMENT OF BODY STRUCTURE

As with the nestling and juvenile plumages just discussed, nestling body structure does not develop uniformly throughout the nesting period. Instead, those structures used most by nestlings whilst in the nest develop rapidly and those not functional until the birds fledge develop more slowly (Holcomb and Twiest, 1968).

Bill structure

The mouth is one of the first structures used by a young bird. In Red-winged Blackbirds gape width and ramus length increase most rapidly between day 0 and day 3 and almost double in size over the first 5 days. Bill lengths in Blue Tits and House Sparrows increase at a rather constant rate with age, but in House Martins initial growth is faster than is later growth. In the last two species, the gape also initially widens with age. Amongst older nestlings the gape often narrows a little as the fleshy rictal parts of the mouth harden (O'Connor, 1977). These changes in mouth size and structure undoubtedly serve to allow an increase in meal size with age and increased energy needs (Chapter 9). They also serve to alter the shape of the mouth, which initially has to serve as a feeding target for parents seeking to place food into the beak of a relatively helpless nestling but later has to serve the fledgling as a foraging tool. In newly-hatched young the bill is often enhanced as a target by conspicuous colouration of the beak cushion, with this fading later. Presumably these older nestlings have great enough co-ordination and co-operation with the parent to permit the bill shape to develop into the adult structure.

Precocial species show similar differential development of the bill in relation to adult care. Young flamingos lack the dense lamellar filter in their bills until their third week, when bill form alters markedly. After this the young are capable of feeding themselves. In other species such as the Curlew the young first feed themselves by picking insects from amongst vegetation, use of the adult method of probing waiting on the elongation and curving of the bill.

Locomotory organs

In passerine nestlings the legs and their associated musculature undergo an early increase in relative size (Austin and Ricklefs, 1977; O'Connor, 1977). Early development of these limbs is a necessary concomitant of improvement in stability and orientation of begging behaviour. In the American Goldfinch the development of grasping and balancing abilities has been shown to develop in parallel with these morphological changes (Holcomb, 1966a). The nestling that extends its head and neck nearest the parent may be fed preferentially (Löhrl, 1968), so early leg development is selected for by competition within the brood. This may explain why passerines faced with sibling competition hatch with disproportionately large legs which do not increase greatly in relative size later (O'Connor, 1978a).

Among semi-precocial species, leg size is rather constant in proportion to body weight at first but decreases late in nestling life. In Herring Gulls this is partly an artefact, for although relative dry weight actually increases over the first 3 weeks the water content of each muscle decreases by an average of 40 per cent in the same period (Hall, 1979). Similarly, Common Terns lose substantial proportions of muscle water during development but leg size relative to non-lipid dry weight nevertheless decreases (Ricklefs, 1979b). More precocial species decrease in relative leg mass from hatching. Reinecke (1979) reports that tarsus growth in the Mallard is related to body weight with only a 0.30 coefficient of allometry.

Rapid development of the leg muscles may also be linked to the onset of thermoregulation. Most heat production in birds is by shivering (West, 1965) and in young birds is obtained from leg musculature. In Willow Ptarmigan the legs carry out this task until the pectoral muscles have developed sufficiently to do so (Aulie, 1976). Even more striking an example is the young Leach's Storm Petrel, which receives no brooding after the first 5 days. Leg size in this species increases disproportionately rapidly with age, growing exponentially at 10 per cent per day (or nearly twice the rate of 6 per cent per day seen in Common Terns and Japanese Quail) and thus provides the musculature needed to produce heat whilst the parents are absent (Ricklefs et al., 1980b).

The structure of the leg bones also changes with age. In newly-hatched European Blackbirds and Song Thrushes primary bone has hardly begun to form in the tibia, which is mostly cartilage, but at fledging, when the bone has about trebled in length, the marrow cavity is well developed and is surrounded by a thin wall of cortical bone (Bilby and Widdowson, 1971). Similarly, the femur progressively calcifies during the nestling period, containing only about 1.1 per cent calcium at hatching but an average of 6.2 per cent at fledging. Similar changes occur in precocial species, though the young are better developed at hatching: a neonatal domestic chick femur contains about 2.9 per cent calcium but takes about 17 days to reach the state of development attained by the thrushes at the end of their 12–13 day nestling period.

In the Little Blue Heron the development pattern of nestlings is dominated by the rapid early growth of the feet and by the rapid onset of ambulatory skills (Werschkul, 1979). This acceleration is driven by predation pressure, for at about 260 g the adults are too small to defend young against large predators. Young have therefore to fend for themselves if attacked and the older young do so by leaving the nest and scrambling away among the branches. This ability to escape is aided by the considerably accelerated growth rate of the feet (which have already reached 85 per cent of their final size by 11 days), by rapid growth in body size (29 per cent higher than expected for a heron of its size), and by early independence in thermoregulation (achieved 2 days earlier than expected for its size (Dunn, 1975a)).

Finally, associated behavioural development – the diminution of visual cliff avoidance (Chapter 13) and increase in readiness to jump from a perch on an approach by predators – also progress in step.

Other species with similar habitat refuges have disproportionately large legs and feet at an early age. Young Reed Warblers a week old can leave the nest and escape through the reedbeds by clambering from stem to stem with their well-developed legs and feet, despite a barely half-grown flight plumage. Young Mallard have tarsi 50 per cent of adult size at a time when their body weight is barely 3 per cent of the adult level (Reinecke, 1979). The acceleration this ratio affords the ducklings can take them running across a water surface at near-flying speeds, if threatened by predators. Such precocious growth of leg muscles contrasts with the early but adaptive development of flight in terrestrial game birds (Aulie, 1976; Ricklefs, 1979b).

Kushlan (1977) found leg growth of White Ibis chicks well advanced at fledging, even though body weight was then only 79 per cent of adult weight. Kushlan considered the reduced body size to be a means of reducing total food needs below adult levels, useful if feeding abilities of fledglings were inferior to those of adults. Large legs were needed early to allow nestlings to perch and climb and walk about the nest tree and later to forage. Since post-fledging foraging was on foot, full flight ability might not be necessary as in more aerial birds.

Late development of wings, and still later development of the pectoral muscles, suggests that a principle of economy operates within growth patterns (O'Connor, 1977). If the available food either is or is ever likely to be limited, what food is obtained is always better devoted to urgent needs – legs to stand on, feeding structures, alimentary organs, etc. But due regard for future needs is also necessary. The bone structure of the wings is needed to support the flight muscles and the flight feathers they drive, so each must develop in optimal sequence. This appears to be the case amongst passerines, even to interspecific differences in the relative timing of different feather tracts (O'Connor, 1977). In some tropical and Temperate Zone species with slow growth, the availability of unusually rich food can accelerate feather growth in captive young to the point where the wing muscles are unable to support the weight (Kear, 1973). Ironically, this condition is commonest in the heaviest members of the broods and in females (who put on weight faster relative to their wing length), but illustrates the need for differential development to proceed in proper sequence.

Growth of digestive organs

In most species the gizzard, alimentary tract, and liver at first grow at a greater rate than the body as a whole (Figure 7.2). In young Rooks and Jackdaws the digestive organs have already reached or exceeded adult size by their second week (Voronov, 1974). The ability of the young to process food for growth thus increases very rapidly with age. The peak proportions of total

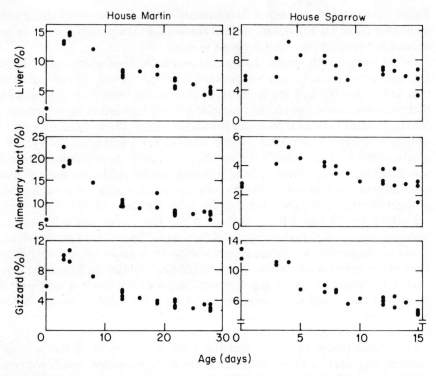

Figure 7.2 Relative size of liver (top), alimentary tract (middle), and gizzard (bottom) in House Martin, *Delichon urbica*, and House Sparrow, *Passer domesticus*, nestlings. Size is given as lean dry weight of organ expressed as a percentage of lean dry body weight. Redrawn with permission of The British Ornithologists' Union from O'Connor (1977).

body weight reached by individual organs vary between species (and also with whether wet weights, dry weights, or lean weights are used for the comparison). The intestines account for about 20 per cent of lean dry weight in the Blue Tit, 33 per cent in the House Martin and 16–17 per cent in the House Sparrow (Figure 7.2). In Herring Gulls the peak proportion is about 8 per cent (Hall, 1979) and in Double-crested Cormorants it is about 8 per cent of wet weight (Dunn, 1975b). In the same species the corresponding figures for liver weights are respectively 20, 15, 10, 6–7 and 6–7 per cent. In the Cactus Wren the liver peaks at 7–9 per cent of lean wet weight, according to figures presented by Ricklefs (1975).

Precocial species appear not to develop their digestive systems in quite the same way. Liver size increases slightly both in absolute size and in relative size around the time of hatching, largely because of the transfer of nutrients from the residual yolk sac (Kear, 1970b). In Black Duck the liver remains at

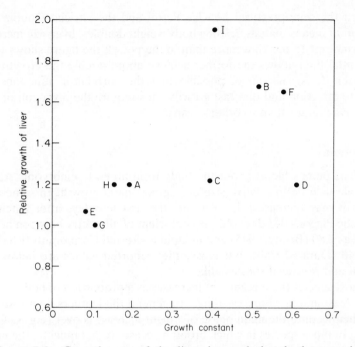

Figure 7.3 Growth rate of the liver in newly-hatched young in relation to their subsequent growth rate constants. Liver growth is given as coefficient of allometry with body weight. Species are as follows: A, Double-crested Cormorant, *Phalacrocorax auritus*; B, Red-winged Blackbird, *Agelaius phoeniceus*; C, Cactus Wren, *Campylorhynchus brunneicapillus*; D, Vesper Sparrow, *Pooecetes gramineus*; E, Black Duck, *Anas rubripes*; F, Rufous-winged Sparrow, *Aimophila carpalis*; G, Leach's Storm Petrel, *Oceanodroma leucorhoa*; H, Herring Gull, *Larus argentatus*; I, Starling, *Sturnus vulgaris*.

this level – about 5 per cent of body weight – for the next 4 weeks, but then slows in growth relative to other organs and by 8 weeks forms only 3.7 per cent of body mass. The gizzard, however, grows almost exactly in proportion to body mass, forming a steady 3.9 per cent of the duckling's weight (Reinecke, 1979).

The relative growth of the liver in neonatal young is correlated with the growth rate achieved in total body weight (Figure 7.3), even though the liver constitutes only 4–6 per cent of adult weight. In this figure the relative growth rate of the liver was assessed as coefficient of allometry during the first few days of growth. The relationship holds across altricial and precocial species, with the precocials in the low growth quadrant and small passerines to the upper right (Figure 7.3). The 20 per cent difference in allometry coefficients between the semi-precocial Leach's Petrel and the altricial Double-crested Cormorant means that, as nestling body weight doubles, liver size increases in the cormorant 15 per cent more than in the petrel; the figure shows that this differential then allows the former achieve about double the growth rate of the latter. It is, of course, possible that the correlation functions in the opposite direction and that fast growth rates require the diversion of growth to the liver rather than to other organs.

Other organs

The heart pumps blood about the body from an early embryonic stage and hence varies little in relative size during post-natal growth. In some altricial species it may increase a little during the first few days after hatching. In Domestic Pigeons 0–3 days old the coefficient of allometry between heart and body was 1.80 (Brody, 1945) and in Double-crested Cormorants 0–8 days old was 1.40 (Dunn, 1975b), but most other reported values are below 1.1, in altricial and precocial species alike.

In most species the integument increases as a proportion of body mass with age, at least among young nestlings. In general the feathers do not split from their sheaths until the main phase of weight increase is over (e.g. O'Connor, 1975c). In those species in which brooding ceases rather quickly, the nestlings usually develop a coat of down through which the contour feathers eventually appear. In precocial species this down is present from hatching. These points are reflected in the proportional weight of the integument in different species. In young brooded altricials it forms 9–10 per cent of lean dry weight but is about 14 per cent in House Martins (O'Connor, 1977). In Ricklefs' studies Starling neonates had integuments forming 9 per cent of lean wet weight, Common Tern 13.5 per cent, Japanese Quail 14.3 per cent, and Leach's Storm Petrel (which are brooded only over the first 5 days) 15.4 per cent.

BODY COMPOSITION AND GROWTH

Water content

Variation in body constituents with age differs substantially between species. In the Cactus Wren water content increases from hatching to age 13–14 days (to 21–22 g) and then remains constant or decreases slightly (Ricklefs, 1975). In Rufous-winged Sparrows a similar increase to 7.0–7.5 g takes place over the first 6–7 days but continues to rise (to about 9 g) at least to days 13–15. In Blue Tits the rise is to about 7.5 g by day 10 and level thereafter whilst in House Sparrows water content peaks at about 17 g by day 11 (O'Connor, 1977). In House Martins water content is maximal (16 g) by day 20 but decreases steeply (to about 11 g) over the next 8 days. However, most of this variation relates to nestling size and relative development. Figure 7.4 shows

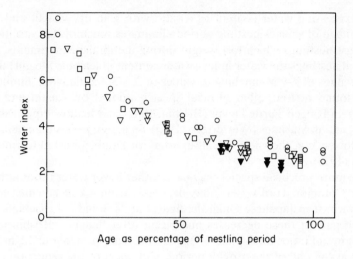

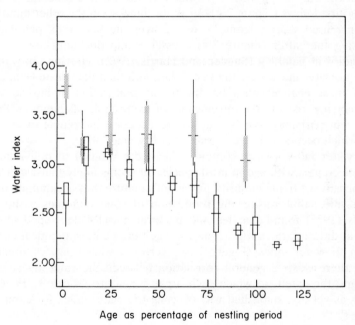

Figure 7.4 Top: Changes with age in water index (water content of nestling ÷ lean dry weight of nestling) in three altricial species. *Key*: ▽, Blue Tit; □, House Martin; ○, House Sparrow. Bottom: Changes in water index in two precocial species – Wood Duck (□) and Japanese Quail (▨). Redrawn with modifications from O'Connor (1977), Brisbin and Tally (1973), and Clay *et al.* (1979).

that expressing water content as a ratio with lean dry weight and age as a percentage of species nestling period eliminates much of this variation, even amongst nestlings which lost weight during metabolic experiments. In these altricial nestlings the water index (water content ÷ lean dry weight) decreases from values of 7–9 at hatching to values of 2.5–3.0 at fledging. Similar values were found in four other altricial species studied by Austin and Ricklefs (1977), in Ringed Turtle Doves (Brisbin, 1969), in Meadow Pipits (Skar *et al.*, 1972), and in Starlings (Westerterp, 1973). In a few species the final ratios are even lower, e.g. the value of 2.0 reported for Double-crested Cormorants by Dunn (1975b).

The more precocial species may have rather lower water ratios at hatching. In the Common Tern water index decreases from 4.1 at hatching to 1.76 in adults whilst in Japanese Quail the figures are 4.3 and 2.4 (Ricklefs, 1979b), though part of these decreases may occur after fledging. Brisbin and Tally (1973) found ratios for Japanese Quail to decrease from about 3.8 at hatching to 3.1 at the end of the growth period, with most of the reduction occurring over the first 8 days (Figure 7.4). In Wood Ducks, on the other hand, water index increased sharply from 2.7 to 3.2 over the first week post-hatching, decreasing thereafter (Figure 7.4). Lesser Scaup ducklings also show low water ratios at hatching (Sugden and Harris, 1972). Herring Gulls increase slightly in water index over the first 4 days and then dehydrate (Hall, 1979). Part of these changes must be due to the maturation of muscle with its accompanying rise in the proportion of contractile elements within the tissues, but part may also be due to the decrease in volume of circulating blood which occurs with age (Kostelecka-Myrcha, 1976).

The water index (water content ÷ lean dry weight) of individual body organs decreases with age in most species. It seems likely that water indices are a measure of tissue maturity, for as cells mature they accumulate proteins and enzymes, thus raising the proportion of solid materials within them. Ricklefs (1975) found that leg and pectoral muscles developed in Cactus Wrens at different rates with respect to age but in almost identical fashion in relation to water index (Figure 7.5). Across a wide range of organs and species there exists a general correlation between the value of each water index and the extent to which each organ is mature (Ricklefs, 1979b), but there is as yet no established way of using the water index for comparative purposes.

Fat content

The extent of fat reserve varies between species and with age within species (Table 7.2). Lipid indices – fat content ÷ lean dry weight – correct for the changing body size of the nestlings and thus provide an estimate of the amount of fat available to support metabolism if food runs short. Minimum lipid levels vary little between species, though it is worth remarking the high values for the House Martin and the European Swift, species heavily

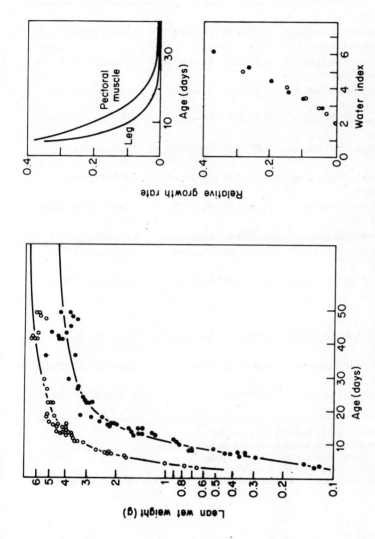

Figure 7.5 Growth of legs (○) and pectoral muscles (●) in the Cactus Wren, *Campylorhynchus brunneicapillus*. Left: Lean wet weight in relation to age. Right: the relative growth rates (percentage rate of increase) of these organs in relation to age (top) and water index (bottom). Redrawn with permission from Ricklefs (1975).

Table 7.2   Lipid index values and age trends during nestling growth in various species

Species	Range in lipid index[a]	Trend with age	Source
Altricial species			
Double-crested Cormorant, *Phalacrocorax auritus*	0.02–0.30	Increase[b,c]	Dunn (1975b)
Cactus Wren, *Campylorhynchus brunneicapillus*	0.03–0.35	Decreases	Ricklefs (1975)
Red-winged Blackbird, *Agelaius phoeniceus*	0.05–0.30	Constant	Ricklefs (1967b)
House Sparrow, *Passer domesticus*	0.05–0.55	Increase	O'Connor (1977)
Barn Swallow, *Hirundo rustica*	0.05–0.60	Increase	Ricklefs (1967b)
Ringed Turtle Dove, *Streptopelia risoria*	0.08–0.35	Increase[b]	Brisbin (1969)
Rufous-winged Sparrow, *Aimophila carpalis*	0.10–0.20	Increase	Austin and Ricklefs (1977)
Blue Tit, *Parus caeruleus*	0.20–0.40	Constant	O'Connor (1977)
House Martin, *Delichon urbica*	0.30–0.80	Increase	O'Connor (1977)
Red-backed Shrike, *Lanius collurio*	0.33–0.46[d]	Increase	Diehl *et al.* (1972)
Other species			
Herring Gull, *Larus argentatus*	0.08–1.18	Increase	Dunn and Brisbin (1981)
Japanese Quail, *Coturnix coturnix*	0.00–1.20	Increase	Brisbin and Tally (1973)
Wood Duck, *Aix sponsa*	0.05–0.60[e]		Clay *et al.* (1979)

[a] Lipid content divided by lean dry body weight.
[b] Except immediately post-hatching, when decreases.
[c] Hatching fat levels are due largely to yolk lipids.
[d] Considerable variability over the first 4 days.
[e] Decreases sharply in first week, then peaks about 75 per cent of the way through the growth period.

dependent on aeroplankton. In most species fat levels increase during nestling growth and the highest levels achieved are more variable between species but aerial species again have especially high values, as have the two precocial species. Ricklefs (1979b) gives values of 0.14, 0.18, and 0.34 for lipid indices of adult Starlings, Common Terns, and Japanese Quail respectively. These adult values are low compared to the upper limits attained by young of most species in the table, a difference which presumably reflects the greater need of young birds for energy reserve during their transition to independence. A second factor making for high fat content in growing birds may be a need for an 'energy sink' in disposing of excessive calories in low protein diets. In growing broilers an increase in the energy to protein ratio of the diet causes increased fat deposition and the degree of saturation of these fats increases in parallel (Bartov and Bornstein, 1976). Some natural diets may be so low in protein content that similar energy sinking is necessary, e.g. in Leach's Storm Petrel nestlings (Ricklefs et al., 1980a).

In ducklings size at hatching has been shown to be related to the neonate's need for food reserves in this way. Energy demands in young are higher (1) in northern latitudes, (2) when nests are remote from feeding grounds, and (3) when feeding skills are hard to learn. Evidence for the first of these effects comes from comparative studies of the Aythya genus (Kear, 1970a). Five species in the genus – Tufted Duck, Lesser Scaup, Canvasback, Scaup, and Pochard – breed northwards to the Arctic Circle or beyond to 70° N and produce ducklings weighing 3.9–4.2 per cent of female weight at hatching. Three other species – Ferruginous Duck, Redhead, and Baer's Pochard – do not lay further north than 50° N and have young weighing 3.6–3.7 per cent of female weight. Finally, the Australian White-eye stays below 42° S and has young only 3.4 per cent of the mother's weight. Thus relative hatching weights are greatest in the colder high latitudes.

Figure 7.6 shows yolk and liver size in Tufted Ducks and Mallard around the time of hatching, showing how food materials in the yolk are transferred to the liver or consumed in fuelling the hatching process. As they approach hatching Tufted Duck embryos have more yolk than have Mallard embryos, both absolutely (11.5 g versus 9.1 g) and relatively (25.0 per cent versus 20.3 per cent), but the two species have similar levels 4 days later. Liver sizes are, however, rather similar in the two species throughout development. Young Tufted Ducks usually spend their first 24 hours as ducklings within the nest, whereas Mallard leave at 12 hours, and Tufted Ducks also undertake longer journeys, of up to 4 km, to their feeding grounds (Kear, 1970b).

The greater yolk reserves of the young Tufted Duck are also correlated with differences of behaviour. Young Tufteds do not begin to feed as early as do Mallard, perhaps not until their third day, and they are more strongly dependent on live insect food. The ducklings are unusually vigorous, leaping and dashing after insects and already diving to the bottom on their third day. This greater activity necessitates greater metabolic reserves (Kear, 1970b).

Fat is not distributed uniformly around the body but is concentrated into distinct depots (Table 7.3). In very young nestlings most fat is derived from the residual yolk sac and is therefore present at the abdomen. As this is consumed total fat concentration may drop, but fat deposition increases in the cervical and subscapular depots and in the oldest nestlings studied these may be almost as important a store as is the abdominal depot.

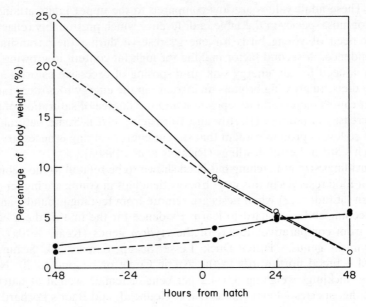

Figure 7.6   Yolk (○) and liver (●) sizes in Tufted Ducks, *Aythya fuligula*, (solid line) and Mallard, *Anas platyrhynchos* (dotted line) in relation to time of hatching. Data from Kear (1970b).

Table 7.3   Distribution of body fat between various depots in nestling Red-winged Blackbirds, *Agelaius phoeniceus*. From Brenner (1964)

Age (days)	Abdominal fat (mg)	Cervical fat (mg)	Subscapular fat (mg)	Nestling mean weight[a] (g)
1	118	—	—	4.2
3	87	4	18	10.3
5	86	50	74	16.3
7	89	63	77	22.8
9	261	223	217	29.2

[a] Not confined to birds measured for fat content.

Energy accumulation

Growing birds increase in energy density during growth, largely because of the decrease in water content with age (Myrcha *et al.*, 1972) but in part because fat content increases with age. The change in body constituents can be expressed in energy terms by multiplying ash-free lean dry weight (almost

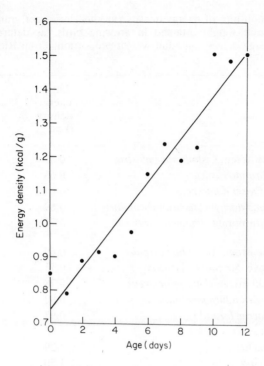

Figure 7.7 Energy density (based on wet weight) in relation to age in House Sparrow, *Passer domesticus*, nestlings between hatching and 12 days in age. From data in Myrcha *et al.* (1972).

all protein) by the energy equivalent of protein (5.7 kcal/g) and lipid weight by its energy equivalent (9.5 kcal/g). Carbohydrate is usually considered negligible in these calculations. The resulting caloric density or energy density (in kilocalories per gram wet weight) of the bird varies more or less linearly with nestling weight (Ricklefs, 1974). Figure 7.7 shows the pattern involved. The intercepts of such plots describe the initial energy loading of young of different species at a hypothetical zero weight (Table 7.4). For most young intercept values lie below 1 kcal/g, due to their high water content at hatching in the immature state. Precocial species have rather higher energy loading as they carry less water in their more mature tissues. Adult energy densities do not vary between precocial and altricial species, so species with high intercepts have correspondingly low rates of energy gain in moving towards adult weight. An exception to this inverse relationship of intercept and slope is Leach's Storm Petrel, in which fat is accumulated extremely rapidly to bring nestling peak weight substantially above the adult level (Ricklefs *et al.*, 1980a).

Table 7.4 Relationships of energy density (in kilocalories per gram wet weight) to proportion of adult weight attained in growing birds, as determined by linear regression of energy density on adult weight proportion. From Ricklefs (1974) and Myrcha *et al.* (1972)

	Initial energy density (kcal/g)	Increase in energy density during growth (kcal/g)
Long-billed Marsh Wren, *Cistothorus palustris*	0.51	1.32
Barn Swallow, *Hirundo rustica*	0.63	0.99
House Sparrow, *Passer domesticus*	0.64	0.78
Cactus Wren, *Campylorhynchus brunneicapillus*	0.69	1.11
Double-crested Cormorant, *Phalacrocorax auritus*	0.71	1.53
Rufous-winged Sparrow, *Aimophila carpalis*	0.69	1.31
Ringed Turtle Dove, *Streptopelia risoria*	0.75	1.61
Red-winged Blackbird, *Agelaius phoeniceus*	0.78	1.11
Mourning Dove, *Zenaidura macroura*	0.92	1.05
Common Tern, *Sterna hirundo*	0.92	0.96
Japanese Quail, *Coturnix coturnix*	1.05	1.33
Sooty Tern, *Sterna fuscata*	1.29	0.68
Dunlin, *Erolia alpina*	1.50	0.30
Domestic Fowl, *Gallus domesticus*	1.58	0.86
Leach's Storm Petrel, *Oceanodroma leucorhoa*	1.87	1.50

Chemical composition of growth

The requirements of young birds for minerals and other chemical elements during growth have hardly been studied, except in respect of the calcium needed for bone formation. Table 7.5 compares chemical concentrations in newly-hatched and fledgling European Blackbirds with those in the precocial hatchling of the domestic fowl. Most of the elements decrease in concentration during the nestling period towards the values shown by precocial chicks, the exception being calcium. Calcium content in domestic chick embryos in fact increases almost exponentially with age until 4–5 days from hatching, when its accumulation rate decreases slightly. In altricial nestlings calcium accumulation is sigmoid over time, initially slower but later faster than body growth as a whole. These high requirements for calcium intake – Scott (1973) estimates 20–30 mg per gram of dietary protein – are potentially limiting to growth rate and in some species have led to the habit of feeding the young on shells, mammal bones, and other sources of extraneous calcium. Bilby and Widdowson (1971) showed that nestling Blackbirds and Song Thrushes derive their calcium requirement not primarily from the tissues of the earthworms and caterpillars on which they are fed but from the gut contents of these

Table 7.5 Concentration (in milligrams per 100 mg lean dry weight) of various elements in newly-hatched and fledgling European Blackbirds, *Turdus merula*, and in newly-hatched domestic chicks, *Gallus domesticus*

Element	European Blackbird[a]		Domestic chick,[b] newly-hatched
	Newly-hatched	Fledgling	
Calcium	0.80	2.44	2.00
Phosphorus	2.04	1.86	1.81
Magnesium	0.148	0.133	0.013
Sodium	3.07	0.72	0.64
Potassium	1.43	1.01	0.59

[a] Calculated from data in Bilby and Widdowson (1971).
[b] From Romanoff (1967) for dry weight values.

invertebrates. Calcium is one of the main excretory products of insects and the green leaves favoured by caterpillars are rich in calcium. Earthworms similarly concentrate calcium into their guts, which may be more than seven times richer in calcium content than are earthworm tissues themselves.

Sulphur is a key element in the chemical compounds that go to make feathers and the availability of sulphur-bearing amino acids may be limiting to feather synthesis during moult (Newton, 1968). In domestic chick embryos its accumulation is similar to that with calcium, though with a lower rate constant. The extent of this accumulation is a correlate of the extent of down cover in the neonate and during embryonic development has to be met from stored materials within the egg. As noted in Chapter 3, there exists a marked correlation between relative egg size and the occurrence of down in different orders of birds.

SUMMARY

Few species are truly naked at hatching, though in most altricial species natal down is very sparse. Most birds moult in their first autumn, to renew their plumage for the stresses of winter, but the extent of this moult may be modified by ecological factors, especially those associated with time of fledging. Chick plumages may serve for insulation, for crypsis, or for signalling. Body organs develop differentially in the course of postnatal growth, with feeding organs disproportionately large in very young nestlings and locomotory organs large in young precocials. In non-flying young the large leg muscles are probably functional in thermogenesis. These general trends are modified to meet ecological constraints in various species. Water content decreases (per unit weight) as young mature and the resulting water index probably indexes general tissue maturity. Fat contents in young birds increase with age and are generally higher than in adults, particularly in species such as aerial insectivores and pelagic seabirds which have uncertain food supplies. As a result the young accumulate energy during growth. Mineral concentration also increases with age, especially in the case of calcium, and the necessary elements may require special dietary supplements to the young by their parents.

CHAPTER 8

Maintenance of body temperature

Birds and mammals are the only two classes of animal able to produce heat for the maintenance of their own body temperature. The embryos of birds are, however, poikilothermic, their temperature varying directly with that of their environment. They are therefore largely dependent on their parents for the maintenance of this environment at a suitable temperature. Thus, the adults warm the eggs by brooding them at low ambient temperatures and cool them by shading or other means at high ambients. At some stage in their development, therefore, young birds must take over the maintenance of their own body temperature from their parents.

Precocial species are largely (though not entirely) thermally independent of their parents when they hatch. Altricial species, on the other hand, continue as poikilotherms into the middle part of the nestling period, or beyond. The two conditions are, however, merely the extremes of a continuum of thermoregulatory patterns. Ducks, particularly the sea-ducks, are well able to respond metabolically to changes in ambient temperatures within a few hours of hatching (Koskimies and Lahti, 1964), gulls can do so within 1–2 days (Hall, 1979), galliforms take a few days (Freeman and Vince, 1974), and altricial species take at least several days, the exact time depending on their growth rate and length of nestling period (Dunn, 1975a).

DEVELOPMENT OF TEMPERATURE STABILITY

Studies in the nest contribute rather little to the understanding of the thermoregulatory physiology of the young. Firstly, parental behaviour normally maintains the nestlings in near homeothermic conditions, so measurements in natural conditions can be taken only immediately after a brooding spell; non-regulating young simply cool by amounts related to the length of parental absence (see, for example, Yarbrough, 1970). Secondly, there may be well-marked diurnal variation in both nest and body temperatures. For example, Cactus Wren nestlings of all ages undergo such a cycle in body temperature in response to the daily cycle of air temperatures in the Sonoran desert (Ricklefs and Hainsworth, 1968).

The standard procedure for the analysis of temperature stability is to expose the young bird in a constant temperature chamber for some fixed

Table 8.1   Nestling body temperatures in Great Tits, *Parus major*, exposed to various ambient temperatures for a 30 minute period. Data from Shilov (1973)

Age (days)	Mean body temperature (°C) at			Range (°C)[a]
	29–32 °C	20–30 °C	9–12 °C	
6	30.9	27.9	16.0	15.2
8	28.9	27.8	19.8	10.6
10	34.9	32.0	25.3	9.6
12	36.1	33.6	30.7	5.4
14	36.6	37.1	35.2	2.5
16	35.7	37.2	36.0	4.2
18	38.8	37.2	36.4	2.4

[a] Observed body temperature range over all ambient temperatures tested.

period and to measure the resulting drop in body temperature. Results for Great Tit nestlings (Table 8.1) show the features found in many experiments of this type (Shilov, 1973). Firstly, at any given temperature older nestlings had generally higher body temperatures than had younger nestlings. This indicates that nestlings are better able to counter or compensate heat loss as they get older. Secondly, the more severe the cold stress the lower were the nestling body temperatures. Third, ability to resist the heat loss increased with age: 6 day old birds at 30 °C had body temperatures averaging 30.9 °C but at 10 °C they averaged 16.0 °C, only 14.9 °C (and not 20 °C) lower. Older birds showed even greater resistance to increases in cold stress. The final column of the table shows the range of variation in body temperature encountered over all experiments at each age. The range dropped from 15.2 °C for 6 day old nestlings to only 2.4 °C in 18 day old birds (a day or two before fledging). Thus as body temperatures rise with age they also come under increasingly better control. Note, though, that even the 18 day old chicks showed variation in body temperature with ambient temperature, indicating that full temperature stability or *homeothermy* had yet to be achieved.

A revealing way of analysing data such as these is by the construction of a thermoregulatory index. This consists essentially of expressing nestling temperature as a function of the temperature of the adult bird when exposed to the same environment. The index used is usually

$$I = 100 \ (T_n - T_e)/(T_a - T_e), \tag{8.1}$$

in which T_n and T_a are the body temperatures of nestling and adult respectively and T_e is the environmental temperature to which both are exposed. Figure 8.1 shows the results of such an analysis for House Martins

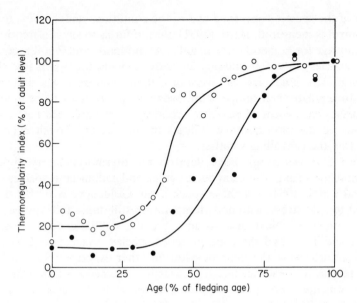

Figure 8.1 Developmental change in thermoregulatory ability in House Martin, *Delichon urbica*, nestlings (○) and in House Sparrow, *Passer domesticus*, nestlings (●). After O'Connor (1975c).

and House Sparrows (O'Connor, 1975c). For the sparrows thermoregulatory abilities were low over the first week of nestling life, then improved gradually until shortly before fledging. In the martins thermoregulatory ability increased significantly over the first 2 weeks and further increased sharply as the plumage developed insulating ability. This difference is probably because the young martins need to be able to survive long parental absences during temporary shortages of the aerial insects on which they feed (p. 11).

Analyses of temperature data in this form are particularly useful because they permit comparison of data collected under diverse environmental conditions. Statistical analysis of a large set of data (22 species) shows that about 75 per cent of the variation in the timing of altricial homeothermy is associated with the growth rate of the species concerned. Fast-growing species attain internal control of body temperature (*endothermy*) sooner than slow-growing species. But if one considers species with a given growth rate, then those with the longer period in the nest tend to achieve homeothermy slightly later (Dunn, 1975a). Thus, although homeothermic development is not exclusively due to the physical development of the nestling, it is fairly tightly bound to the overall rate of development, as one would expect intuitively.

Very few precocial species can fully regulate their body temperature in the immediate post-hatching period. However, the exact pattern is very much a feature of the species concerned and of the conditions under which temperature control is measured. Barth (1951) goes so far as to say it is impossible to define any age for homeothermy in gulls. At moderate ambient temperatures, several species show fair stability in body temperature when less than 24 hours old, if the cold stress is severe or if their plumage is wet, but they are unable to regulate their temperature even when 1 week or more in age. For such species the extent of parental brooding is probably the best available guide to the thermoregulatory abilities of the young, though even this is affected by the prevailing weather.

Table 8.2 shows comparative data on the thermoregulatory abilities of newly-hatched young of several ducks, gulls, and gallinaceous species (Koskimies and Lahti, 1964). Of these species, the Goldeneye is by far the most resistant to cold stress, with body temperature still under full control after 20 minutes at 10 °C. Next in resistance are the dabbling ducks tested, the Mallard and Teal, and three gallinaceous species, Blackcock, Capercaillie, and Pheasant (whose thermal behaviour in other experiments at 20 °C were similar to that of domestic chicks). A fourth gallinaceous species, the Willow Grouse, was much poorer in thermoregulation at 10 °C. Herring and Lesser Blackbacked Gulls dropped about 6 °C in body temperature, and the Black-headed Gull about 10 °C, but even these figures are better than the nearly 16 °C temperature drop of 1 day old Domestic Pigeons exposed to these same conditions. We see here, then, a graded series of species groups, from essentially thermal independence in the sea-ducks through dabbling ducks, gallinaceous species, and gulls to the altricial pattern of the domestic pigeon. The series thus parallels the trend in parental care.

Table 8.2 Temperature regulation in recently-hatched young of various species when experimentally kept in ambient temperatures of 10 °C for 10 minutes[a]

Species	Drop in body temperature (°C)
Goldeneye, *Bucephala clangula*	0.2
Mallard, *Anas platyrhynchos*	1.0
Teal, *Anas crecca*	1.3
Blackcock, *Lyurus tetrix*	3.8
Capercaillie, *Tetrao urogallus*	4.0
Pheasant, *Phasianus colchicus*	4.4
Willow Ptarmigan, *Lagopus lagopus*	8.0
Domestic Pigeon, *Columbus livia*	15.4

[a] Data were estimated from the graphical presentation of Koskimies and Lahti (1964).

Auks developed thermoregulatory control in a manner intermediate to those of altricial and precocial species. In both Common and Brunnich's Guillemot held at 2 °C the chicks develop homeothermy between 5 and 9 days of age (Johnson and West, 1975). At 10–12 °C Common Guillemots showed stable body temperatures from about 3 days of age, Razorbills from 3–4 days, and Puffins from 6–7 days; Black Guillemots showed stable body temperatures at 10–12 °C from their fourth day but brooding in this last species in fact continues to 5–6 days, thus cautioning against too rigid an interpretation of experimental findings (Drent, 1965). Drent also found that although Pigeon Guillemots on Mandarte Island, British Columbia, could stabilize their body temperature when 1 day old, they were brooded by the adults nearly continuously through their first 3 days and partially to at least day 5. The young did not maintain adult temperatures in the wild until 15 days old.

Hardly anything is known of temperature regulation in waders. Norton (1973) found that young of several sandpipers at Barrow, Alaska, had body temperatures of 34–39 °C (average 37.2 °C) when brooded, 30–38 °C (average 34.3 °C) when actively feeding, and 23–31 °C (average 28.6 °C) when crouching in response to parental warning of the observer's approach. These figures imply rather poorly developed thermoregulatory capacities in this group.

HEAT PRODUCTION AND THERMOREGULATION

Changes in thermoregulatory ability can be brought about either (1) by developments in the ability of the young to produce heat or (2) by age-changes in their ability to retain heat. Of these, heat production is more important in practice. Many young birds first maintain steady body temperatures by functioning at high levels of metabolism, later reducing these levels to near-adult rates as their insulation improves.

Since oxygen is required for practically all metabolic processes in birds, oxygen consumption rates provide an index of heat production. Figure 8.2 shows results of oxygen measurements on Blue Tit nestlings of different ages (O'Connor, 1975c), results typical of many altricial young. In the youngest age-groups metabolic rate varies directly with ambient temperature, a typical poikilothermic response. By 7 days the nestlings are able to increase their heat production in response to lowered ambients over the range 30–40 °C, a response typical of the homeothermic adults. At lower ambients they are unable to maintain this response and their metabolism falls. Older nestlings (9–10 days) can compensate for heat loss over a wider range of ambients. By 11–12 days the fully inverse relationship of metabolism and ambient temperature found in adults is already apparent. Full temperature stability at ambients of 15–25 °C first appears in this species about day 10, but rather earlier at warmer ambients. It is therefore also noteworthy that metabolic intensity (heat production per gram body weight) at 30–40 °C falls with increasing age, once thermoregulation is established. Maintaining temperature stability is thus energetically cheaper in the older nestlings.

122

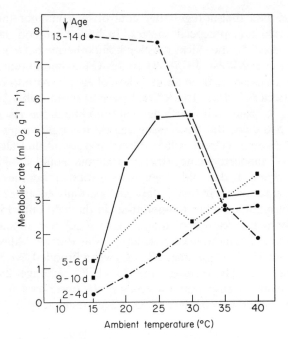

Figure 8.2 Age changes in the metabolic rates of Blue Tit, *Parus caeruleus*, nestlings exposed to various ambient temperatures, showing the progressive improvement in heat production ability. After O'Connor (1975c) by permission of The Zoological Society of London.

The heat production of most altricial species, then, is initially that of a poikilotherm, followed by a gradual transition to the homeothermic pattern. The reduction in metabolism in very young nestlings under cold stress is not pathological: when they are warmed up again, they resume their normal rates of heat production. In this they differ from older nestlings, for whom such a cold exposure would, in general, prove lethal.

The transition to metabolic or chemical thermoregulation takes place in different species at different ages. Table 8.3 presents some representative data for small passerines. In general, heat production appears in a regulated fashion about 8–10 days after hatching. It also appears slightly earlier in open-nest species than in those with more secure nests. This is potentially adaptive since open nests are more prone to predation: when so disturbed, a brood able to thermoregulate can scatter into the surrounding vegetation and can subsequently be fed there by the adults.

Table 8.3 Onset of chemical thermoregulation (i.e. by physiological heat production) in relation to nestling period in various altricial species. Data from Shilov (1973) and O'Connor (1975c)

Species	Nesting period (days)	Onset of chemical thermoregulation (days)
Field Sparrow, *Spizella pusilla*	8	4–6
Chipping Sparrow, *Spizella passerina*	10	6–7
Garden Warbler, *Sylvia borin*	9–10	6–9
Blackcap, *Sylvia atricapilla*	10–12	8–12
Redwing, *Turdus musicus*	11–12	4–9
Fieldfare, *Turdus pilaris*	12–13	5–10
Pied Flycatcher, *Muscicapa hypoleuca*	14	8–10
House Sparrow, *Parus domesticus*	14	7–11
Tree Sparrow, *Parus montanus*	15–16	10–12
Coal Tit, *Parus ater*	16–18	9–12
Blue Tit, *Parus caeruleus*	19	7–11
Great Tit, *Parus major*	18	8–17
Starling, *Sturnus vulgaris*	20–21	8–12
House Martin, *Delichon urbica*	29–35	10–19

Metabolic rate and body weight

Basal metabolic rate among adult passerines varies with body weight to an 0.724 power, according to the Lasiewski–Dawson (1967) equation:

$$M = 6.25 \, W^{0.724} , \qquad (8.2)$$

where M is the metabolic rate (in watts) and W is the body weight (in kilograms). Similarly, the equation predicting metabolic rate for non-passerines is

$$M = 3.79 \, W^{0.792} . \qquad (8.3)$$

Figure 8.3 shows these equations in relation to similar data for nestlings and chicks of several species. In altricial passerines metabolic output increases more steeply than body weight demands, thus indicating that developmental improvement in metabolic capacity is taking place. In Great Tits metabolic rates are directly proportional to body weight in 6 day old nestlings ranging in weight from 5 g (retarded growth) to 11 g (normal growth). This suggests that body weight is a better predictor of metabolic rate than is age (Mertens,

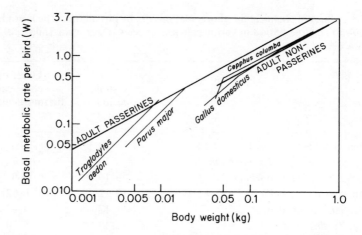

Figure 8.3 Basal metabolic rates during growth of nestlings of various species, in relation to adult metabolism. Redrawn with the permission of Springer Verlag. After Mertens (1977a).

1977a). This is also true for Blue Tit and House Martin at low ambient temperatures but does not appear to hold for House Sparrows (O'Connor, 1975c).

Non-passerines reach the Lasiewski–Dawson levels of weight-specific metabolism fairly soon after hatching. Thereafter they increase their output in line with the Lasiewski–Dawson equation (Figure 8.3). Their thermoregulatory processes thus seem to be more mature shortly after hatching than in passerines of the same weight, in keeping with their general precocity of development. Mertens (1977a) notes that basal metabolic rates (per unit weight) of passerine embryos are about one-quarter those of adults of the same weight, whilst in non-passerine embryos they are about one-half the adult levels. This difference is reflected in the slower growth of the non-passerines.

PHYSIOLOGICAL CORRELATES OF THERMOREGULATION

Air sac development

Adult respiration in birds involves the complete air sac system rather than just the lungs alone. Full heat production capacities must therefore wait on air sac development. In young birds the air sacs do not reach a functional state of development until some days into the nestling period, e.g. between 5 and 9 days in the House Wren (Kendeigh and Baldwin, 1928).

Table 8.4 Heart size in relation to body weight in newly-hatched birds. Data compiled from various authors

Development mode	Species	Heart weight (% nestling weight)
Altricial	House Martin, *Delichon urbica*	0.95
	Starling, *Sturnus vulgaris*	1.0[a]
	Double-crested Cormorant, *Phalacrocorax auritus*	1.0[b]
	Cactus Wren, *Campylorhynchus brunneicapillus*	1.1[b]
	Blue Tit, *Parus caeruleus*	1.2
	House Sparrow, *Passer domesticus*	1.6
Semi-altricial	Herring Gull, *Larus argentatus*	1.1[b]
	Common Tern, *Sterna hirundo*	1.3[a]
	Sooty Tern, *Sterna fuscata*	1.4[a]
Precocial	Japanese Quail, *Coturnix coturnix*	1.5

[a] As a percentage of fat-free weight.
[b] Estimated from graph in original source.

Oxygen transport

The oxygen taken up at the lungs by the blood has to be transported to the sites of active tissue respiration. In many altricial species the heart is initially small (Table 8.4) but grows faster than the body as a whole over the next few days. It levels off in relative size as thermoregulation commences. Further, both the number of erythrocytes per unit volume of blood and the amount of haemoglobin each increase steadily throughout the poikilothermic phase and level out thereafter (Figure 8.4). Functionally both cell number and haemog-

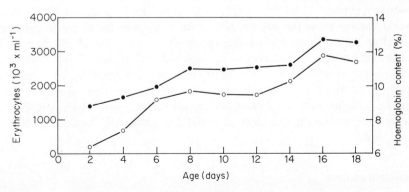

Figure 8.4 Erythrocyte concentration (●) and haemoglobin content (○) in relation to age in Wren, *Troglodytes troglodytes*, nestlings. From data in Shilov (1973).

Table 8.5   Blood sugar and liver glycogen in nestling Pied
Flycatchers, *Ficedula hypoleuca*, at various ages. From data
in Shilov (1973)

Age (days)	Blood sugar (mg/100 mg)	Liver glycogen (g/100 g)
3	30.0	0.60
4	40.5	1.04
5	—	1.17
6	52.5	1.04
7	54.0	0.57
8	59.0	0.58
9	—	0.40
10	65.0	—

lobin content promote the oxygen-carrying capacity of the blood and thus permit the nestling a greater capacity for heat production.

Energy substrates

The main energy base for oxidation is probably carbohydrate. Proteins are devoted largely to tissue growth, except in the event of food shortages. Carbohydrate reserves, often as glycogen in the liver, are easily mobilized and this is an advantage for short-term responses to cold stress. Table 8.5 shows for the Pied Flycatcher that, whilst sugars are initially carried in the blood stream in considerable quantities, they are gradually taken out of circulation and stored as liver glycogen. Liver levels were highest between days 4 and 6, a period of marked differential increase in liver size in other passerines studied (Ricklefs, 1975; O'Connor, 1977; see also Chapter 7). Several studies show the importance of such reserves for thermoregulation. Thus Blue Tit nestlings undergoing metabolic rate tests oxidized their flight muscles when their fat reserves were depleted (O'Connor, 1976). European Swifts adopted different tactics and entered nocturnal hypothermia on depletion of their reserves (Koskimies, 1950). Amongst precocial birds similar problems also occur (Koskimies and Lahti, 1964). Adequate energy reserves for oxidation are thus essential for sustained thermal control in young birds.

Hormonal control

Hormonal influences on chemical thermoregulation in young birds have been studied principally in the domestic fowl (Freeman, 1971) and in the Herring Gull (Hall, 1979). In mammals heat is generated by the breakdown of a certain type of fatty tissue (so-called "brown adipose tissue") into free fatty acids (FFA), under stimulation by the hormone noradrenaline. Although brown adipose tissue is absent in birds, newly-hatched chicks nevertheless also respond to cold exposure by increases in plasma FFA. In young up to 8 weeks old this can also be evoked by noradrenaline treatment, even though noradrenaline is a thermolytic agent in the fowl. A possible explanation is that the fatty acids serve not as a substrate for thermogenesis but as a metabolic uncoupling agent at the mitochondrial level. This would then cause the mitochondria to produce more heat than they would by the normal process of oxidative phosphorylation (Freeman, 1971).

Thyroid hormones have also been implicated in thermogenesis in the fowl and offer an alternative pathway for plasma FFA involvement. Thyroid activity in the Vesper Sparrow was at adult level or above throughout the nestling period. It peaked about the fifth day, a day or so in advance of full homeothermy, and declined slightly towards fledging (Dawson and Allen, 1960). Correlated changes in metabolic rates were observed.

The hypothalmus controls the rate of production of thyroid stimulating hormone and in the adult fowl surgical cutting of various areas of the hypothalmus leads to impaired response to temperature changes (Freeman, 1971). The hypothalmus also controls the temperature-dependent intake of water, of significance since evaporative water loss is essential in coping with heat stress (p. 131). In the hypothalmus surgery above, both thermogenesis and thermolysis were affected, depending on where the cuts are made. The development of nervous control of such thermoregulatory responses is therefore probably a key feature of the transition to adult patterns of response, though it is not known at what stage neural control is complete.

Heat production by shivering

Shivering is the principal short-term thermogenic response to cold. In domestic chicks, sustained shivering can be obtained immediately following hatching and before core temperatures can be affected. The implication is that shivering may be initiated as a peripheral nervous reflex. In House Wrens muscle tremor first appears after day 3 and is well established by day 6. In Black-capped Chickadees tested at a 21.1 °C ambient temperature, muscle tremor appeared on day 4 (Odum, 1942), well ahead of temperature stability and of plumage cover. The rate of tremor was higher in cool than in warm conditions. In House Wrens both the duration of the bursts of shivering and the rate of tremor within each burst increased with age to a maximum at 9–12 days, decreasing thereafter (Kendeigh, 1939). Since temperature stability was well established over this period, this pattern suggests a switch to other thermogenic processes less costly of energy than shivering (Shilov, 1973).

In precocial birds a major event for heat production is the switch to lung respiration on hatching, with an associated rise in metabolic rate. Shivering thermogenesis is probably more important in precocial hatchlings than in young altricials, since the former necessarily have better developed muscles at hatching. Aulie (1976) found that Willow Ptarmigan chicks responded to cold by shivering even from the day of hatching. At first they did so by activation of the leg muscles alone, but after day 3 activation of the pectoral muscles occurred as well. Between day 4 and day 13 the pectorals grew rapidly, oxygen consumption increased in parallel, and the chicks increased in ability to resist loss of body temperature. As in altricial species, the appearance of regulated heat production in precocial species is preceded by increases with age in erythrocyte number and haemoglobin content, as well as by changes in blood sugar and liver glycogen.

PHYSICAL FACTORS IN THERMOREGULATORY DEVELOPMENT

The second avenue of thermoregulatory development is that of control of heat losses. Of the four avenues of heat loss, convection and radiation are determined largely by the adults' choice of nest site and structure. However, conduction depends on three characteristics of the young bird – its plumage (including down) development, its surface–volume ratio, and its body temperature – and evaporative losses are determined by the ability of the nestling to pant (see below).

Plumage

In nestlings the development of the plumage is not a major factor in the onset of thermoregulatory control. In a comparative study of three altricial species metabolic intensity was not correlated at all with the stage of plumage development reached (O'Connor, 1975c). In many species homeothermic responses are already present before the feathers emerge from their quills. Shilov (1973) reports the results of experimentally shaving the plumage off Great Tits and Pied Flycatcher nestlings as they grew (Figure 8.5). He found that the nestlings still maintained homeothermy but at the price of increased metabolic rates. It seems, then, that the role of the plumage in development is simply to reduce the energetic cost of thermoregulation once established, rather than to allow its initiation.

Domestic chicks, ducklings, and other precocial young are characterized by their well-developed down. The thick layer of down serves to trap a layer of air around the small body of the chick, increasing both the insulation and the length of the conduction path to the environment. That it is the air trapped within the feathers, and not the feather material itself, that provides the insulation is shown by experiments on newly-hatched ducklings (Nye 1964). Wet birds lost heat more rapidly than did dry birds, particularly when the plumage was saturated through to the skin.

Amongst non-precocial species down is in general either present at birth or develops very rapidly only where there are special selective pressures for the young to be left unattended from an early age. Thus, in gulls both parents forage from an early stage in the nestling period, and the gull chicks hide in the vicinity of the nest during the parental absence. Their down ensures low

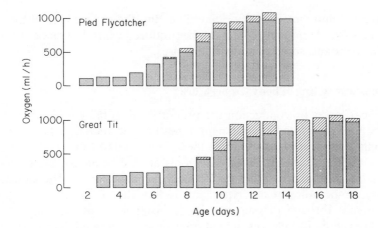

Figure 8.5 Oxygen consumption at 10 °C in relation to age in (top) Pied Flycatcher, *Ficedula hypoleuca*, and (bottom) Great Tit, *Parus major*, nestlings in normal condition (solid, foreground) and after shaving of their feathering (shaded, background). Redrawn from Shilov (1973).

heat losses in the absence of brooding. Similarly, many raptors and owls are left unattended by adults for longish periods from well before normal plumage development. These again show early down production. Many aerial insectivores, such as the European Swift (Lack, 1956) and House Martin (Bryant, 1975) also have a downy chick as protection against nest losses during spells of bad weather, when the adults may have to forage far afield even for their own survival. A similar adaptation is apparent in many seabirds encountering essentially the same ecological problem (Lack, 1968).

In growing birds the developing feathers are enclosed at least partially within their individual pins. At their base these open basally into a rich blood supply. The pins are organized in tracts in such a way that the basal sheath of one row is eventually covered by the expanding feathers of the row above, but until this happens both pins and blood supply are significant sources of heat loss.

Body size and surface–volume ratio

In growing birds, increase in absolute size decreases surface–volume ratio and this alone puts them into a more favourable energy balance. Purely as a result of the changes in surface–volume ratio, a growing bird can cut its relative heat losses. In some passerines this effect can reduce heat loss to some 40–50 per cent of their hatching-day values (O'Connor, 1975c). In domestic chicks the same principle accounts for the rise in body temperature in the first few days after hatching (Freeman, 1965). During this time the relatively inert yolk sac is converted into metabolically active tissue but the surface area of the chick stays the same. Hence more heat can be produced whilst the rate of heat loss remains constant, so body temperature rises.

Thermostatic setting of body temperature

Since heat losses by conduction are proportional to the temperature difference between body and environment, a nestling can reduce its heat losses by reducing its core temperature. Most species show a developmental increase in body temperature and this can be demonstrated to be an organized thermostatic strategy by the nestlings (O'Connor, 1975e). Thermostatic adjustment of body temperature of this type helps reduce the energy costs of nestling maintenance through the stages of poor insulation, but may also serve a second purpose. The physics of heat transfer are such that heat transfer from a brooding female will be greater, the larger the temperature difference between her and her young. Were adult and young to have the same temperature norms, a chilled nestling would return to that norm only asymptotically with time. As it warms up, the temperature difference between itself and parent gradually dwindles to zero, reducing the rate of transfer of further heat towards zero. With a difference in the temperature norms of adult and chick, heat transfer is maintained at a greater rate throughout brooding.

HEAT DISSIPATION

Heat stress is a serious hazard for those young birds who encounter it. Whilst the adults can insulate their nest and can use their brood patch to counter cold stress in their young by intensive brooding, there is little they can do to cool the chicks should ambient temperatures rise. They can shade the young from radiation but cannot protect them from convection nor from conduction. It therefore falls to the young themselves to counter heat loading.

Good insulation offers as effective a response to heat stress as it does to cold stress. Once the bird has cooled itself, e.g. by evaporative heat losses, any insulation retards the flow of heat back into the bird from its warmer surroundings, thus reducing the water costs of keeping cool. Young nestlings are therefore particularly at risk, since their insulation is as yet undeveloped. In addition, a small nestling has a low thermal inertia; that is, a smaller quantity of heat is needed to elevate its temperature by a given amount than is the case with larger nestlings. This too makes small nestlings more susceptible to heat stress.

The two main defences available to nestlings confined to the nest are hyperthermia and evaporative cooling. Hyperthermia, the elevation of body temperature above normal levels, is effective because it sets up a gradient from bird to environment, down which heat will flow by conduction and convection. Without this the nestling has to dissipate by evaporation of its water reserves not only its own metabolic heat production but also the heat it absorbs from its warmer surroundings.

For small nestlings hyperthermia is a particularly effective defence. Once they elevate their body temperature above ambient levels their small thermal inertia and poor insulation positively aid their further temperature control. In the House Wren, 3 day old nestlings can elevate their body temperature as high as 45.5 °C and 6 days later only to 42.1 °C (Baldwin and Kendeigh, 1932).

Evaporative cooling is the other main defence against heat loading. The latent heat of water evaporated is 580 cal/ml. This high value makes it energetically feasible for a bird actually to increase its heat production by working to move water across its respiratory surfaces and still show a net loss of body heat as the water is evaporated. The bird may then become limited by water reserves.

The rate of respiratory water loss varies with ambient temperature but in different ways at different ages. In the youngest nestlings water losses are due to simple respiration, and greater oxygen consumption at higher temperatures results in greater water loss. In older nestlings this has altered to an inverse relationship: at high temperatures heat production is reduced but heat dissipation by evaporation of water is increased. The exact timing of this change varies with the development rate of each species, e.g. at 7 days in Redwings (nestling period 12 days), at 10 days in European Robins (nest period 13 days), at 11 days in Pied Flycatchers (14 days), and at 14 days in Great Tits (18 days) (Shilov, 1973).

Water loss can be increased above normal respiration loss rates by altering the pattern of breathing either through panting or through gular flutter or both. In normal respiration counter-current heat exchange in the upper respiratory passages recovers a significant fraction of the water otherwise lost in respiration. If these passages are bypassed by expiring directly through the mouth (= panting) this water and its associated heat are not recovered. The result is an increase in heat loss. In several species this cooling is further enhanced by gular flutter, the hyoid apparatus and its associated musculature moving the moist gular region (Calder and King, 1974). For adults of several species it has been found that both panting rates and gular flutter frequency depend on the mechanics of the respiratory apparatus. When this apparatus is of a certain size, on its activation resonance occurs, i.e. panting and flutter are nearly self-sustaining and require only a small energetic input (= small rise in heat production) for their maintenance. Panting and flutter rates can be as high as 1000 movements per minute. Howell and Bartholomew (1962) showed experimentally that gular flutter was the main mechanism of heat loss in young Red-footed Boobies on Midway Island. When they temporarily taped together the two mandibles to block gular flutter (Figure 8.6), body and foot temperatures rose at the same rate until body temperatures reached about 44 °C, when the chicks showed distress symptoms. On removal of the tape, gular flutter began immediately and temperatures fell sharply.

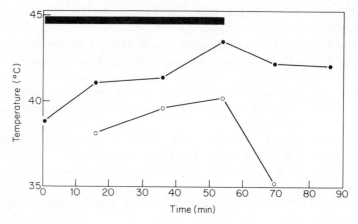

Figure 8.6 Effects of preventing gular flutter in a young Red-footed Booby, *Sula sula*. The bill was taped shut for the period indicated by the horizontal bar and cloacal (●) and foot (○) temperatures were measured as shown. Substratum temperature was about 35 °C. After Howell and Bartholomew (1962).

Even newly hatched young can show a panting response to thermal stress. Morton and Carey (1971) have described how very small chicks of the White-crowned Sparrow pant when exposed to excessive insolation. Passerines have only the ability to pant, but in several seabird groups regularly nesting in hot climates (e.g. cormorant and pelican families) gular flutter is also used from hatching.

RESISTANCE TO CHILLING

The absence of effective temperature regulation in very young altricial nestlings can result in their being chilled to unusually low body temperatures should the parents be prevented from brooding them for some reason, such as proximity of a predator. Yet young birds show an extraordinary resilience in the face of such chilling. A 2 day old Vesper Sparrow experimentally left at a temperature of 18–20 °C for 8 hours recovered fully (Dawson and Evans, 1960). Similar recoveries have been documented in Field and Chipping Sparrows, Horned Larks, House Sparrows, and House Wrens. Such tolerance is probably rarely required in nature since parental brooding will normally keep the young in near-homeothermic conditions.

One possible exception to this occurs with aerial insectivores, in which the young may be temporarily abandoned by their parents when the latter are forced by bad weather to feed far from the nest to ensure their own survival, let alone that of their young. Such young may display special adaptations. After several days without food young European Swifts show nocturnal hypothermia. Their body temperature drops from the normal 38–41 °C almost to nest temperature (21 °C) overnight and recovers again the following morning. The cycle was repeated over successive days without food (Koskimies, 1950). Such an adaptation saves overnight energy expenditure but allows the young to remain sufficiently active to take food from their parents should they return during the day. Its effect therefore is to maximize the nestling's chances of being alive on the next good feeding day. Similar hypothermia occurs in Leach's Storm Petrel on the verge of death from starvation. Nestlings of most species, however, are unlikely to be able to enter partly-regulated torpidity of this sort.

SOCIAL THERMOREGULATION

The timing of endothermy in natural nests differs from those determined experimentally in the laboratory, for two reasons. First, the presence of the nest and its insulation around the young birds provides a within-nest microclimate more favourable than that outside the nest. Once the interior of the nest has warmed up (through brooding or as a result of nestling heat production) the nestlings need to produce less heat to remain in thermal equilibrium than if they were exposed to ambient conditions (O'Connor, 1975d). Second, the presence of brood mates in a nest allows the young to huddle together, reducing the proportion of the brood surface exposed to the air and thus reducing the individual heat production needed for stability in body temperature. Consequently, young birds in large broods may be able to produce enough heat for endothermy earlier than in broods of fewer young losing proportionately more heat (Figure 8.7).

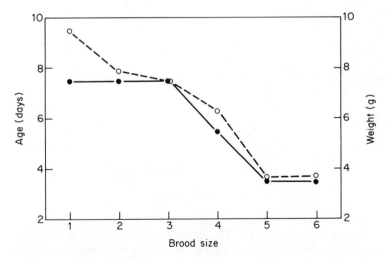

Brood size

Figure 8.7 Age (●) and weight (○) at which effective temperature regulation was achieved by nestling House Wrens, *Troglodytes aedon*, in broods of different size. Data from Dunn (1976).

Such a relationship was first suggested by Royama (1966a), who studied Great Tits. In broods of three chicks, nestlings aged 7–13 days individually consumed 1.6–1.9 g food each day, but those in broods of 13 consumed only 0.6–0.8 g each. Nestlings in the large broods were lighter than those in small broods but to a much smaller extent than suggested by the more than twofold difference in food consumption. That is, proportionately more of the food eaten by the chicks in the large broods was expended on growth, and proportionately less was expended on keeping warm, than in the small broods. Indeed, the adult female with a brood of three brooded much more than did the females with the larger broods.

Experimental measurement of oxygen consumption by broods of Great Tits has confirmed that chicks in large broods spend less energy on heat production (Mertens, 1969, 1977a). Metabolic rates were strongly dependent on body weight according to an allometric equation with exponent 0.672 (Figure 8.8). Thus when brood weight doubles, whether by doubling the number of nestlings or by doubling each nestling's weight, metabolic costs increase by only 59 per cent ($100 \times 2^{0.672}$) and not by 100 per cent. This saving significantly alters the relative costs of rearing broods of different sizes,

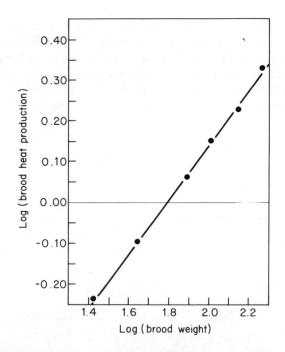

Figure 8.8 Heat production (in kilocalories per hour) by broods of Great Tit, *Parus major*, nestlings at 12 °C in relation to brood weight. Redrawn by permission of The British Ornithologists' Union from Mertens (1969).

since halving brood size does not halve the maintenance costs of the total brood (e.g. Hails and Bryant, 1979; Perrins, 1979). In a similar study of 11–12 day old Blue Tits, oxygen consumption was more than halved when brood size was increased from one to eight in experiments conducted at 15 °C (O'Connor, 1975d). However, it is not known whether these nestlings were in their thermoneutral zone (i.e. metabolizing at a basal rate) during these experiments.

In very large broods of titmice metabolic intensity is higher than in moderately-sized broods (rather than being lower still) (O'Connor, 1975d). This effect is due to the onset of hyperthermia (overheating) when the individual nestlings in the brood have minimized their heat production (i.e. down to basal metabolic rate), yet still have too great a production to dissipate by physical means. At this point evaporative water loss must be invoked, with respiration rising in order to carry away water vapour. It is therefore conceivable that very large broods might occasionally fail through inability to dissipate enough of their heat burden. In one experimental investigation on this point, mortality rose steeply in unusually large broods held at high ambient temperatures (Figure 8.9). At lower temperatures mortality was less severe. Most of the deaths occurred in the first 4 hours and larger broods suffered at lower temperatures than did smaller broods (van Balen and Cavé, 1970). In natural broods such mortality may be avoided if the nestlings can move away from each other, increasing the average surface–volume ratio of the brood. When Mertens (1977a) observed a brood of nine Great Tits reared in a nest-box with hollow metal walls, where the

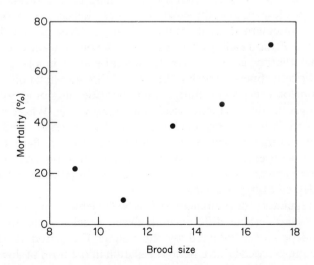

Figure 8.9 Mortality in 10–12 day old Great Tit, *Parus major*, broods after 12 h at 35 °C. Data from van Balen and Cavé (1970).

temperature could be varied experimentally by circulating hot or cold water through the walls, he found that the young began to pant at temperatures between 25 and 30 °C. As they did so they started more preening and stretching, behaviours increasing their average separation. In the wild, too, the nest consists initially of a small cup in a bed of moss and other material, thereby keeping tiny young together; but as the young grow their weight gradually flattens the nest material until older chicks are free to spread over the full cavity floor. Clutch size in several species (including Great Tits) is correlated with nest-box area, which may be an adaptive trait for birds at risk from hyperthermia. In addition, the construction of Great Tit nests from moss may itself be significant: moss readily absorbs the water released in panting, thereby becoming more conductive to heat. This tends to reduce the heat stress on the nestling. In the smaller titmice species there is less risk of hyperthermia (because each chick has a large surface–volume ratio) and less moss is used in the nests than is the case for Great Tits (Mertens, 1977b).

SUMMARY

Altricial and precocial species differ in their thermoregulatory development. Altricial species acquire their initial temperature control some days after hatching when improvements in thermogenesis first allow them produce heat faster than they lose it. These improvements are founded on developmental changes in respiration and oxygen transport, in the availability of energy substrates, and in the integration of neural and hormonal control of heat production, and are well correlated with nestling growth rates. Thermoregulation by nestlings is initially costly of energy because of poor insulation afforded by vascularized quills, but as plumage develops and the surface–volume ratio of the nestlings becomes more favourable these costs reduce. Early homeothermy is to a lower temperature than in older nestlings, a thermostatic adjustment reducing the cost of thermoregulation. Broods can achieve homeothermy some time earlier than the physiology of individual nestlings suggests, due to mutual insulation by the young. Where nestlings are particularly likely to go unbrooded, the young hatch with or rapidly acquire a thick down and may be able to enter temporary torpor. Parental brooding cannot cope with excessively high ambient temperatures and young of hot regions show capacities for evaporative water loss by panting and gular flutter and by a degree of hyperthermia.

Thermoregulatory development is less well defined in precocial species because of considerable differences between species as to the degree of thermal independence attained by the young. Compared to the altricial species, young precocials have better insulation in the form of down and have greater capacity for heat generation because of their greater endowment with functional muscles. They also hatch with the appropriate physiological controls developed embryonically.

CHAPTER 9

Feeding

Some young are not fed by their parents at all. The most striking examples are the megapodes, in which the young forage independently from the moment they break free of the mound of fermenting material in which the eggs are incubated (Chapter 2). Another large category of self-feeding young comprises most ducks and shorebirds. These follow the parents but depend on them only for brooding and/or defence. Species such as the Quail lead their young not just to feeding areas but to individual prey items, and others may make food available to the young by scratching the ground, as do chickens, or by cutting vegetation, as do some swans. Yet another group of precocial young are actively fed by the parents, as in the grebes. Finally, in Cuckoos and other brood parasites the parents do not feed their young, this duty falling on the foster parents.

The remaining species depend on the adults to bring food to the nest and to feed the young there. With altricial young the parent almost invariably places the food into the nestling's mouth but in other groups such as gulls and herons the young can, once a few days old, pick up food regurgitated on the ground in front of them. Birds of prey initially depend on direct feeding, but at a later stage the young can tear a food item to pieces for themselves and the adults merely deliver prey to the nest.

BEGGING BEHAVIOUR

Most young have characteristic begging behaviours, normally elicited by stimuli associated with the arrival of the parents. Jarring of the nest in tree-nesting species, darkening of the nest entrance for domed or hole-nesters, even air currents (as in the Chimney Swift, where the adults may have to flap their way down a vertical shaft to reach the nest), suffice to evoke begging from altricial young. But in the course of their development the young become more selective as to the stimuli to which they will respond, largely through conditioning to parental characteristics and through habituation to non-significant stimuli. In many species this process is reinforced by the adult giving a distinctive feeding call as it arrives.

Begging in altricial hatchlings consists of simple gaping, perhaps accompanied by head waving and wing movements. The mouth is often brightly coloured and sometimes marked with special 'feeding targets' to aid the adult to deliver its food load accurately. Gaping is usually oriented vertically but in some species where the nest entrance is lateral, as in the Cactus Wren, begging is oriented laterally as well (Ricklefs, 1966). In the estrildine finches, the entrance is below the nest platform level and the nestlings beg with their heads downwards (Welty, 1975).

Older birds have less stereotyped begging behaviour, often incorporating vocalizations previously absent. Such calling serves to signal their food needs to the parents. In Finland, von Haartman (1949) conducted experiments to demonstrate this point. A drawer in the base of a nest-box allowed easy replacement of young Pied Flycatchers, without the parents noticing, so that the parents could be given a succession of young at a constant hunger level. Under these conditions the adult feeding rates rose over 3 hours to around 30 visits per hour instead of the 12 visits paid per hour to young in unmanipulated nests. A second experiment used a two-compartment nest-box. The adults could enter one compartment which contained a single young, but could only hear (and not see into nor visit) the other compartment, which held six hungry young. Under these conditions the adults more than doubled their feeding rates to the single young they could feed, attempting to feed it even when it was satiated. The first experiment shows that it was the behaviour of hungry nestlings which stimulated parental feeding, the second that it was their calling which was the crucial behaviour. Similar conclusions have been reached with Ringed Turtle Doves: deafened adults fed their young less efficiently than normal birds, their chicks at 2 weeks weighing only 56 per cent of the normal weight (Nottebohm and Nottebohm, 1971).

Precocial young of all ages take a more active part in feeding than do young altricials. Precocial young often show marked sensitivity to the colour of the parent's bill. This colour is usually (though not always) more effective in evoking a response than is any other colour (p. 219). In gulls, the chick response is to peck at the parent's bill until food is regurgitated into the bill where the chick can reach it. In various species of heron the chick is allowed to catch the parent's bill in its own once the adult is ready to regurgitate, and the food mass then passes across to the young. Young pelicans stick their heads deep into the parent's gullet to feed on the shrimp and fish mass there. In hummingbirds and in some swifts it is the adult who inserts its bill and head into the mouth of the young before transferring the feed.

If the nestlings fail to beg when an adult arrives at the nest, the parent may call softly and preen or otherwise touch the young. If this is not sufficient to induce the young to beg the adult may swallow the food itself or fly off with it still in its beak. Such events are usually followed by a longer interval than usual before the next visit.

The order in which the young are fed is very much on a first-come, first-served basis. In asynchronously hatched broods the largest chick begging is fed preferentially until it is satisfied. Presumably the largest chick is bigger and therefore more conspicuous to the adult, but it may also be more alert and faster to respond to the arrival of a parent simply because it is more advanced in development. For young Pied Flycatchers it has been suggested that the youngest chicks of asynchronously hatched broods catch up in their development because the constant stimulation provided by the older siblings within the confines of the nest cavity accelerates their sensory and motor development by keeping them more alert and motivated to beg (Khayutin and Dmitrieva, 1978). But the sharing out of feeds within a brood is more generally uneven: in one Wall Creeper nest, for example, the 81 feeds seen going to four colour-ringed young over the last few days in the nest were divided 24, 20, 20, and 17 respectively (Löhrl, 1975). Young Black-crowned Night Herons at the 'brancher' stage fight amongst themselves for the highest position on the branches around the nest, young in this position being fed first (Maxwell and Putnam, 1968). Rivalries of these types may result in the death of one or more of the smallest nestlings in the event of food shortages, leaving the favoured young to survive. This is an evolutionarily more acceptable result than would be the total loss of the brood, had food been shared out evenly but inadequately (p. 12).

Because begging is usually noisy or conspicuous (or both) it involves an element of danger for the brood. Broods of hungry young Great Tits may be heavily predated by weasels in years when food is scarce, probably because the predators are attracted to the nest-boxes by the begging calls (Dunn, 1977). This type of evidence has induced speculation that begging calls evolved in a sort of Russian roulette between parents and offspring. Because the two generations differ as to how they view the relative value of parental care (Chapter 11) a hungry young at risk of death might in effect seek to extort greater feeding effort than the parent is willing to provide, thus endangering the whole brood by making loud begging calls. In this way the cost to the parent of not foraging harder is raised above that it would experience by losing just the single hungry chick. Dawkins (1976) provides a critical review of this argument.

There is, however, another way in which unsuccessful begging leads to increased mortality (Hunt and McLoon, 1975). Figure 9.1 shows the activity patterns of Herring Gull chicks before and after a bout of begging. The arrival of an adult at the nest usually stimulates a burst of begging. If the adult can feed the chick it quietens again and continues to stay close to a parent. Unfed chicks, however, continue to be active and wander about the territory several feet from the nearest adult (Figure 9.1). Chicks at or beyond the territory boundary may be attacked by neighbouring adults and the majority of such attacks were experienced by unfed wandering chicks. This may be the explanation for the higher mortality of slow-growing chicks in gull broods when food is inadequate for all three young. It would also explain why large colonies of gulls with their smaller nest territories suffer greater mortality than do small ones.

ROLE OF ADULTS AND HELPERS

In a number of species nestlings are fed by birds other than their parents. In some cases, e.g. House Martins, these helpers are the offspring from an

140

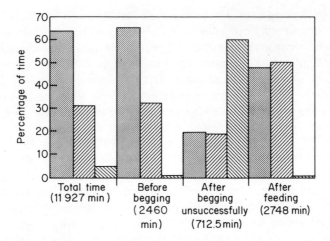

Figure 9.1   Activity patterns of Herring Gull, *Larus argentatus*, chicks during 20 min periods prior to begging and after feeding or begging unsuccessfully. *Key*: ▓, resting; ▨, activity category I; ▩, activity category II. Activity categories I and II denote movements within 1.2 m (4 feet) of and more than 1.2 m (4 feet) from the nearest parent. Redrawn with permission from Hunt and McLoon (1975).

earlier brood of the season, but in other highly social species the helpers form part of a territorial flock, members of which may even lay communally. In some such species, such as the Pukeko, studied by Craig (1975), this propensity is so prevalent that older chicks may even feed younger chicks of the same brood! Such helpers may contribute significant amounts of food to the nestlings. In the Mexican Jay, for example, eight out of 53 helper–brood combinations observed over 2 years involved helpers providing 15 per cent or more of the food received by the brood, and 26 of 53 involved at least 5 per cent contributions by helpers; 46–48 per cent of all nestling food was provided in this way (Brown, 1972). In the congeneric Florida Scrub Jay, helpers provided about 30 per cent of the food (Stallcup and Woolfenden, 1978). In these species the food thus brought does not seem to accelerate growth or otherwise improve nestling condition, though starvation is rare. Rather it allows the parents to reduce their own feeding efforts, with consequent improvements in survival: mortality among breeding birds which had help the previous year averaged 13 per cent whilst for those without helpers it averaged 20 per cent.

KINDS OF FOOD

Although adult birds feed on a wide variety of food types, young birds are almost invariably reared on protein-rich diets, even if this means they have to be given foods very different from the adult diet. The granivorous House Sparrows studied by Kalmbach (in Welty, 1975) averaged 96.6 per cent vegetable matter and 3.4 per cent animal matter in adults but 31.2 per cent vegetable and 68.1 per cent animal matter in nestlings. Seel (1965) showed that nestlings fed on bread rather than on protein-rich foods were under-weight for their ages.

Just how seriously a low protein content affects growth is illustrated in Figure 9.2. A barley meal diet with 13.5 per cent protein allowed Mallard ducklings to increase their body weight over 14 days only by 60 per cent, whilst an invertebrate diet (51.8 per cent protein) allowed a 930 per cent increase (Street and MacDonald, 1977). Captive waterfowl generally have a minimum protein requirement of 18–22 per cent in their diet (Scott, 1973) and in the wild this is reflected in a high consumption of invertebrates during the period of peak growth. Downy Black Ducks averaged 88 per cent inverte-brates in their diet; older, part-feathered juveniles took 91 per cent but fully-feathered juveniles reduced the percentage to only 43 per cent (Reinecke, 1979). Invertebrates thus provide the major source of protein during times of peak growth: crude protein constituted 49, 37, and 26 per cent

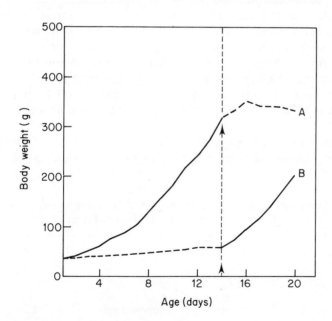

Figure 9.2 The effect of dietary protein on the growth of Mallard, *Anas platyrhynchos*, ducklings. Groups A and B were initially fed on high protein (——, invertebrates; 51.8% protein) and low protein (----, barley meal; 13.5% protein) diets respectively and their diets were interchanged on day 14. After Street and MacDonald (1976).

of the dry weight consumed by the three age classes. In the Black Duck this dependence on invertebrates lasts about 6 weeks but other species differ in this respect. Juvenile Gadwall and American Wigeon do so for 2 weeks; species such as Pintail, Mallard, and Redhead are similar to the Black Duck; others such as Lesser Scaup and Goldeneye are virtually exclusively invertebrate feeders. In the Canvasback, invertebrates decreased as a percentage of the diet from 99 per cent in ducklings aged 0–5 weeks through 86 per cent in young aged 5–9 weeks to only 38 per cent at fledging (Bartonek and Hickey, 1969).

Some species do indeed rear their young on low protein foods but in such cases special adaptations have evolved. Thus the manakin *Manacus manacus* is reared on fruit but is hatched from a disproportionately large egg (2.6 g instead of an expected 1.9 g). The incubation period is unusually long (18–20 days instead of an expected 15 days for a 2.8 g egg; Rahn and Ar, 1974), which Lack (1968) suggests is an adaptation allowing the egg to produce a nestling at an advanced state of development and capable of processing fruit. In adult birds, the gut size of species feeding on plant material is unusually large (Prys-Jones, 1977), a point supporting Lack's view.

The extreme of specialized food delivery to the young has been achieved by columbids, some seabirds, and one or two other groups (Table 9.1). These birds secrete a special milk (in the pigeons) or oil (in the seabirds) rich in proteins and fats. The liquid is regurgitated to the young during feeding. By producing such secretions a pigeon may bypass the normal procedure of feeding young on an animal diet, the adult physiology in effect taking on the food processing which the young are poorly equipped to tackle. Ashmole (1971) has suggested that the secretion of stomach oil in petrels is an adaptation to reduce the weight of food carried on the long journey back to the nest.

Few birds rear their young exclusively (or even wholly) on fruit (Morton, 1973). Few plants produce their fruit during the avian breeding season in the

Table 9.1 Composition and origin of fluids fed to their young by various bird species. From Welty (1975), Murton and Westwood (1977), and Prince (1980)

Species	Fluid type	Origin	Composition (%)	
			Fat	Protein
Pigeons and doves	'Pigeon's milk'	Crop wall cells	7–13	13–19
Flamingos	Fluid	Oesophagus	18[a]	58
Emperor Penguin	Fluid	Crop	29[a]	59[a]
Petrels	Stomach oil	Proventriculus	—	—
Albatrosses	Stomach oil	Dietary	1–27	5–9

[a] Percentages by dry weight.

Table 9.2   Average composition of tropical plant and insect foods. Calculated from data in Morton (1973)

	Fruits[a]	Insects[b]
Water[c]	80.9	67.2
Fat[c]	2.7	3.4
Carbohydrate[c]	12.6	2.3
Protein[c]	1.3	17.7
Energy (kcal/100 g)	74.3	141.3
Protein–energy ratio (g/kcal)	0.017	0.125

[a] Average of nine species, not including oil palm.
[b] Average of data for ants, beetles, caterpillars, crickets, and grasshoppers.
[c] Weights per 100 g sample. The totals do not add up to 100 per cent because the data presented are themselves averages from skew data distributions.

Temperate Zone, so frugivory is viable only in the tropics. Morton suggests that the limited use made of fruit as nestling food in the tropics is a consequence of the interaction between the dietary quality of fruit and predation intensity. Tropical fruits provide only slightly fewer calories than does the same weight of insects, but their protein content is extremely low (Table 9.2). Nestlings fed on fruit are thus limited by protein supply and their nestling periods are correspondingly prolonged (Table 9.3). In the high-predation regime of the tropics any prolongation of the nestling period reduces overall breeding success. Morton calculated that 39 per cent of nests otherwise surviving would be predated if the dependency period of nestlings were lengthened by 9 days to allow frugivory. Consequently frugivory is viable only with very safe nest sites, but if these are available the abundance of fruit can then permit broods of three or even four young, as in certain

Table 9.3   Nestling periods in various neotropical passerines in relation to diet. From Morton (1973)

Diet class	Oscine passerines (N)	Non-oscine passerines (N)	
Insectivores and granivores	12.0 (13)	Open-nesting insectivores 11.8	(9)
		Hole-nesting insectivores 19.5	(10)
Omnivores	14.2 (15)	20.3	(4)
Frugivores	22.3 (4)[a]	33.0	(1)[b]

[a] All tanagers of genera *Euphonia* and *Chlorophonia*.
[b] Bearded Bellbird, *Procnias averano* (Snow, 1970).

tanagers (Morton, 1973) and in the cave-dwelling South American oilbird (Snow, 1962). Amongst those species feeding young with both insects and fruit, the fruit is not brought to the young until they are a few days old (and presumably past their peak protein requirement).

Frugivorous birds provide an extreme example of a general point, that the supply of calories alone is not all that is required for the successful growth of young. Other dietary constraints can affect the young, even those of insectivorous species. In Holland, for example, one leatherjacket species accounts for 38–79 per cent of the foods brought to Starling nestlings, reaching higher proportions in the largest broods. The leatherjackets are most readily obtained in pasture in the middle of the day but the adults instead spend this period searching a remote saltmarsh area for a caterpillar which takes much longer to collect but apparently supplies some dietary component missing from the monospecific leatherjacket diet (p. 145). Feeding experiments in which supplies of leatherjackets and caterpillars were offered to the parent birds in various ratios showed they preferred the caterpillars but used the leatherjackets as a source of easily acquired energy, for example, when feeding large broods or particularly hungry young. Moreover, when under stress with a large brood the females reduced their own consumption of caterpillars in favour of bringing them to the brood, they themselves taking a pure leatherjacket diet (Tinbergen, 1981). Such effects are less likely to be found in species normally taking a varied diet. In the aerial-feeding House Martin, for example, a wide variety of aeroplankton are taken (Bryant, 1973) and the nestling diet selected at any moment is largely determined by the mean energy value of the individual insects available.

VARIATIONS IN THE DIET OF YOUNG

Nestling diets are largely a characteristic of each species but regional and seasonal differences do occur. Orians (1973), comparing the prey spectra of tropical and temperate colonies of Red-winged Blackbirds, noted that the former were dominated by herbivorous insects which were scarce in the latter, these being biased instead towards emergent aquatic insects such as damselflies. Since the herbivorous insects had to be captured one at a time, the feeding rate of the nestlings was reduced. In other species seasonal variations in the diet have been reported, again with reductions in growth rate when less suitable foods are used.

Nestling diets are often more varied in the early part of the nestling period than later, reflecting partly an early need for particular elements or nutrients when tissue synthesis is high and partly the later need for energy for maintenance costs on achieving thermal independence. Thus the Carrion Crow brings its young quantities of bees, flies, and spiders in the early half of the nestling period but does not do so in the latter half, and the proportion of tipulid and lepidopteran larvae in the diet also increases in this way (Yom-Tov, 1975).

It is intriguing to note just how widespread is the habit of bringing spiders to tiny nestlings. Titmice regularly bring spiders to their young, usually doing so in the middle of a sequence of caterpillars (Royama, 1966a). In the light of the abundance of caterpillars in the area, this habit suggests that the spider contains some trace nutrient which the lepidopterans lack. Trace deficiency illnesses are not unknown among young birds. Young Willow Grouse reared on artificial Ptarmigan food become ill from an ascorbic acid deficiency. The chicks synthesize some ascorbic acid, but their high requirements for it necessitate external supplements (I. Hanssen in Spidso, 1980). Amongst the plants preferred by the chicks when feeding in the wild is the blueberry, a species rich in ascorbic acid. Until about 10 weeks old their bills are too weak to allow them to break off willow twigs and other sources of the adult diet (Stokkan and Steen, 1980).

Wall Creepers make special efforts to bring a diversity of food to their young, presumably to maintain a nutrient balance. Feeding intensity fell, in the case of one pair breeding in an aviary, if only one food type was available, but increased again immediately a new prey type was provided (Löhrl, 1975). Even so, certain species of moth were always rejected and others were eaten by the adults but never taken to the young.

Although the adaptive value of a diverse diet is not understood, research on Starlings has shown that an excessive proportion of leatherjackets in the diet has adverse effects (Kluyver, 1933; Tinbergen, 1981). An excess of leatherjackets caused the nestling faeces to become wet and loose instead of forming a compact mucous sac. The adults were then unable to remove the faeces without also removing part of the nest material, and the young ended sitting in their own faeces in the bottom of the nest-box! Being permanently wet, such nestlings had higher metabolic rates and suffered greater mortality than is usual. This effect does not occur with all monotonous diets, however. Starling young fed largely (96 per cent by weight) on caterpillars (94.4 per cent of them a single species) grew well without mortality (Tinbergen, 1981).

Boecher (1967) suggests that young Arctic Terns and other seabirds may have a special need for foods of high caloric density until their thermoregulatory abilities are fully developed. He found that young less than a week old were fed almost entirely on fish, with less energy-rich crustaceans being brought only to chicks older than this. On this basis young chicks might be profligate of energy whilst young, yet grow successfully. This would be especially useful in northern areas with colder climates.

Precocial young feeding themselves are free to make their own choice of prey items from those actually available to them, though this choice is made in the feeding grounds to which they have been brought by their parents. In Iceland ducklings of different species each took slightly different foods (Bengston, 1971a), thus showing a degree of ecological separation of these species even at the duckling stage. A few species show an increase in the proportion of insects in their diet with age simply because they become more adept at catching insects: Lavery (1970) found that young Australian Wandering Whistling Ducks increased their relative intake of insects nearly threefold in this way.

Differences in diet are associated also with differences in bill structure (Collias and Collias, 1963; Partridge, 1976a). In the Ruddy Duck the young have broad flat bills with many well-developed ridges adapted for straining food, whilst young of the Blue-winged Teal and the Wood Duck have relatively narrow bills with poorly developed internal ridges. Such bills are better suited to use in pecking than in straining. Mallard and other generalist feeders are intermediate in their bill shape and structure. Collias and Collias found that the distribution of ducklings amongst the various habitats in a large Manitoba marsh was linked with the susceptibility of individual prey species to these different feeding methods.

CONSUMPTION OF GRIT AND MINERALS

For many (and perhaps for all) birds, particles of grit or other hard materials are needed in the stomach to assist the grinding of food items into smaller, more digestible, pieces. Such particles are as necessary to young birds as to adults and have to be brought to them by their parents. The behaviour has been reported both of insectivorous species such as titmice (Royama, 1966a) and of seed eaters such as the Bullfinch (Newton, 1967). Perhaps related to this is the habit of the African Stanley Crane of feeding the egg shell to the newly-hatched chicks, though a build-up of calcium reserves for skeletal development could also explain this habit (Walkinshaw, 1963).

For Lapland Longspur nestlings in Alaska the major food items are calcium-deficient craneflies (0.08 per cent dry weight in calcium) and various sawflies (0.12 per cent dry weight in calcium) (Seastedt and MacLean, 1977). Pieces of lemming bones and teeth and egg shell are therefore fed to the nestlings as a calcium supplement. Such dietary supplements were much more frequent (present in 60 per cent of all nestlings older than 3 days) in this altricial species than in the precocial sandpipers (8.2 per cent with lemming remains) examined by S. F. MacLean (1974), probably because the female sandpipers can provision their eggs with calcium to ensure that their young hatch in an advanced state. Other evidence that calcium is particularly needed by altricial young comes from Douthwaite's (1976) study of the Pied Kingfisher on Lake Victoria. Adult kingfishers regurgitate three of four pellets of fish bones each day but much of the bone consumed by chicks is digested to meet their calcium needs.

Fledglings also have to learn to recognize grit for themselves. Adult Chipping Sparrows have been noticed picking up grit and feeding it as though it were food to their newly-fledged·offspring, who then pecked at gravel for themselves (Crook, 1975).

WATER REQUIREMENTS

Nestlings are of course entirely dependent on their parents for whatever water they need. In fact many nestlings seem to rely completely for their water needs upon the water content of the food, though Bullfinches and other species regurgitating mucus-bound food balls to their young probably thereby transfer some additional water in the mucus. For most nestlings the proportion of body water to lean dry weight actually decreases during development, partly as the result of water elimination from muscle tissue during fibril development and partly because the blood supply can be proportionately smaller in more mature birds (O'Connor, 1977; Ricklefs, 1979a).

Two groups of birds have special problems with respect to water. Nestlings or chicks reared in hot areas need unusually large quantities of water for evaporative cooling and may, like young Darters, have a special begging behaviour when seeking water rather than food. Most precocial species can travel to water, but the Namaqua Sandgrouse rears its young in the Kalahari Desert perhaps 50 miles (80 km) from the nearest waterhole and the ventral feathers of the adult male have been specially modified to carry water. On immersion in pools of water the feathers soak up 25–40 ml of water and retain sufficient during the male's flight back to the chicks to allow them a water intake of about 40 per cent of these quantities. Other species are more limited in such desert areas: the Mountain Quail, for example, must remain no more than a mile (~ 1.6 km) from water since the chicks require water soon after hatching (Johnsgard, 1973).

The second group of young facing water problems are those chicks and nestlings reared on fish diets. For these the water in their diet is largely saline and poses difficulties over salt balances. In the tubenoses special glands allow the elimination of the excess salt, but for other species fresh water is needed to restore the balance. One possibility here is that the large yolk and lipid reserves of newly-hatched gulls provide the chicks not with energy but with water: each gram of fat may yield as much as 0.96 g of water in the course of its metabolic oxidation (Ricklefs et al., 1978).

FEEDING RATES

Feeding rates increase gradually through the nestling period for most species (Figure 9.3), partly because the parents feed their young more regularly as the adults make the transition from incubation behaviour to nestling care but mainly because the energy requirements of the young increase with age. One reason for the delay in the onset of a full feeding rhythm by the adults is that the non-incubating parent does not immediately learn of the hatch, though once discovered the nestlings may then be fed within minutes (Skutch, 1976). Skutch's data for tropical species show median times to first feeding of 27 minutes for feeds by females and of 169 minutes for feedings by males (usually the non-incubating bird) (Skutch, 1976). One consequence of such delays in an extended hatch is that young hatching late are fed more intensively than were their early siblings when the same age; their extra growth then tends to synchronize the development of the brood as a whole and allows the brood to fledge as a unit despite the hatching spread (O'Connor, 1975a).

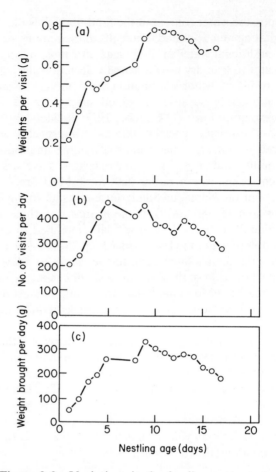

Figure 9.3 Variations in the feeding pattern at a Starling, *Sturnus vulgaris*, nest, showing changes with nestling age in (a) the size of prey item brought to the young, (b) the visiting rate to the nest each day, and (c) the product of these two = daily food delivery to the brood. After Tinbergen (1981), by permission of The Netherlands Ornithological Union.

Feeding rates vary diurnally in some species (Figure 9.4). The usual pattern of a mid-day lull is probably due to the warmer temperatures during these hours, the nestlings consequently requiring less food. The evening peak may well be a stocking-up of overnight reserves, as in adult birds. In broods of small young, feeding is often more intensive in the warmer afternoon than in the morning, the higher temperatures allowing the female to hunt food for the young instead of brooding them (Gibb, 1955; Best, 1977). The pattern may also change over the course of the nestling period. Gibb (1955) found for

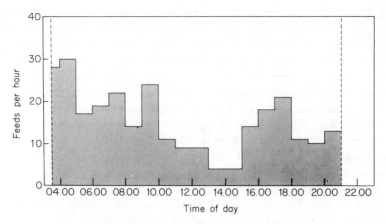

Figure 9.4 Diurnal variation in feeding frequency by the Robin, *Erithacus rubecula*, bringing food to a brood of six 12 day old young. After Boetius (1949).

Great Tits that parents of large broods, but not those of small broods, gradually fed relatively less each afternoon as the nestling period progressed, apparently due to sheer exhaustion from feeding effort. Boetius (1949) similarly suggested that adults of the various passerines studied by him tired in the afternoon. In other species, however, the daily feeding pattern is at least partly determined by a rhythm of food intake associated with the nestling or chick (Chura, 1963; Kushlan, 1976).

In arctic species that breed in continuous daylight, feeding rates fall off during the periods of lowest light intensity (Karplus, 1952). However, these periods are also closely correlated with the periods of lowest temperature when brooding is most intensive, so both light intensity and ambient temperatures are involved. Light intensity may here affect the detection of prey by adult birds.

The young of some of the larger seabirds are fed only a single large meal each day. This is usually associated with remote feeding grounds from which it is uneconomic for the adult to return with smaller loads. For the same reason the meal is frequently delivered in late afternoon or evening (Diamond, 1973; Brown, 1975) and is often associated with a changeover between parents attending the nest.

Weather and feeding rates

Weather conditions significantly affect feeding rates in some species. Roseate Terns have difficulty catching fish for their young as the wind freshens and prevents them spotting fish in the sea (E. K. Dunn, 1975): their young consequently put on less weight and may even lose weight. Similarly

Guillemots reduce their feeding rates by 28 per cent as sea conditions change from calm to rough (Birkhead, 1976) and in bad weather Brunnich's Guillemots do not even leave the colony to look for food (Tuck and Squires, 1955). The daily growth of European Swift nestlings is favoured by warm or sunny weather and depressed by wet and by windy weather, largely because of the effects these conditions have on aeroplankton abundance and distribution (Lack and Lack, 1951). Even amongst fruit-eating birds in Panama rainfall interferes with feeding: in light rain feeding continues but at a much reduced rate, in moderate rainfall it reduces still further, and in heavy rain ceases altogether (Leck, 1972). As most tropical land-birds breed in the wet season (so that young hatch at the period of peak food abundance) the adults must lose 1–2 hours (and often longer) of feeding time due to daily outbursts of torrential rain. This restriction may be one of the factors responsible for small clutch size amongst tropical birds (Foster, 1974b).

Low temperatures and rain are also important for their effects on brooding. Where both parents feed nestlings, one may remain on tiny young to brood them whilst its mate forages. But with older young not yet able to thermoregulate, the demands of brooding may conflict with the nestlings' food needs. Female Field Sparrows in Illinois increased their brooding time as ambient temperatures fell below 10 °C, from about 15 minutes brooding per hour above this to more than 30 minutes per hour below (Best, 1977). Since the female's contribution to feeding decreased the more she brooded, the nestlings received less food in cold weather. Similar behaviour has been recorded of the House Wren (Kendeigh, 1952) and the House Sparrow (Seel, 1960). Older nestlings who have developed the ability to thermoregulate may not be as severely affected by the cold weather, since both adults are free to forage.

Altricial species are buffered against the effects of cold and rain by the shelter provided by the nest, but precocial chicks are more vulnerable in such weather. First, they require increased brooding and, second, they are also prevented from foraging. The timing of bad weather is fairly critical: Marcstrom (1960) found that Capercaillie chicks were least sensitive to poor conditions in their first few days after hatching, when yolk and fat reserves were high, but they became more sensitive – and suffered heavy mortality – later in their first week when their yolk sacs were consumed. In Blue Grouse, on the other hand, chick survival was unaffected by the reduction in feeding time brought about by the need for additional brooding in cold or wet weather (Zwickel, 1967).

SIZE OF FOOD ITEMS

The size of food items fed to young can vary substantially even within a single day, but this is usually offset by a corresponding change in feeding frequency to keep the overall food intake at the required level (Figure 9.5). Birds feeding close to the nest tend to bring individual food items to the nest at high

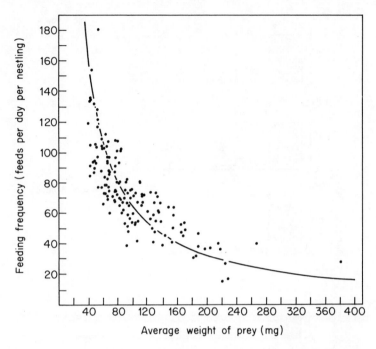

Figure 9.5 Relationship between feeding frequency and prey weight for Great Tit, *Parus major*, broods 8–17 days old. The line describes the hyperbola for daily food delivery of 6487 mg to each young. Redrawn by permission of The Netherlands Ornithological Union from van Balen (1973).

frequency, whilst those feeding far away bring large food loads at long intervals. Also, birds feeding on foods small relative to their body size tend to bring back several items per visit, whilst those with larger foods bring back one or a very few items. As the young grow, however, the mean size of the foods they receive increases in many species, probably because the larger young can handle bigger items. Thus very young Bullfinches are fed fragments of seeds already crushed by the adults, but older nestlings receive the seeds intact (Newton, 1967). Similarly, young Blackbirds are given only segments of earthworms where older birds are given entire worms. In other species older nestlings continue to receive small items (which may be particularly abundant) and it is the size range of prey – particularly as to the inclusion of large items – which increases with nestling age (Figure 9.6). Such changes reduce the amount of time the adult has to spend in seeking small food items or in breaking large items into smaller pieces, and thus allow a higher feeding rate. For this reason many altricial young show marked differential growth of food-processing organs (such as bill, gizzard, and liver)

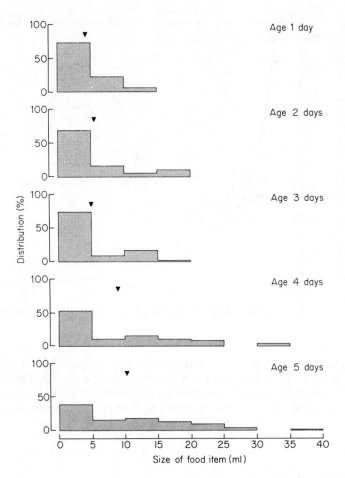

Figure 9.6 Size distribution of food items brought to Field Sparrow, *Spizella pusilla*, nestlings at different ages. Triangles indicate mean values. Based on Best (1977).

during their first days as nestlings, thereby increasing their capacity to ingest food (Chapter 7).

Adults actively select only the larger of the items they encounter for use as nestling food, consuming smaller ones encountered themselves. This is optimal because, when foraging at a distance from the nest site, the adults incur a fixed travel cost in time and in energy with each trip back to the nest to feed the young (Hartwick, 1976).

ENERGETICS OF PARENTAL FEEDING

The energy costs of feeding young in House Martins have been measured by Hails and Bryant (1979) and their work has produced some striking revelations. Both abundance and quality of the food supply decrease seasonally. Hails and Bryant measured the rates at which males and females contributed to the feeding of their first brood and found that, as total brood mass increased, both sexes increased their delivery of food to the nest. The male, however, responded at a more rapid rate. A point of interest is that feeding rates were proportional not to brood weight but to that weight raised to a two-thirds power. Use of this function allowed for the mutual insulation within the brood (Mertens, 1969; see also Chapter 8) and its linearity with feeding rate suggests a useful generalization for future studies. By injecting the adults with isotopically labelled water (D_2O^{18}), Hails and Bryant were able to measure the expenditure of energy by the adults. They found that female energy expenditure was independent of brood size, despite her increased feeding of large broods, whilst male energy expenditure increased with brood weight (Figure 9.7).

These experiments were repeated with second broods, with interesting results (Figure 9.7). Female metabolic costs remained independent of brood weight but now at a higher level and male costs continued proportional to metabolic brood weight but also at a higher level. At constant brood size the

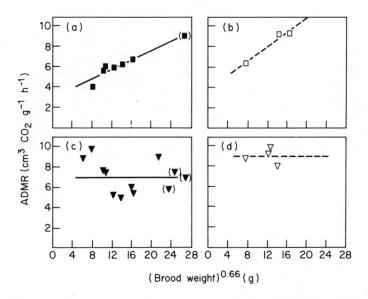

Figure 9.7 Average daily metabolic rate (ADMR) of adult House Martins, *Delichon urbica*, rearing broods of different metabolic needs. (a) Males rearing first brood; (b) males rearing second brood; (c) females rearing first brood; (d) females rearing second brood. Points in brackets indicate experimentally enlarged broods. Redrawn with permission from Hails and Bryant (1979).

combined (male plus female) costs of the second brood were 35 per cent greater than when reared as a first brood. A seasonal decline in average clutch size (from 3.5 to 2.9) restricted the average increase actually experienced to 30 per cent. Thus, although second broods were smaller, the adults still had to work harder to rear the second brood.

Direct measurements of this type are rare but revealing. Hails and Bryant estimated the average daily metabolic rate of House Martins feeding young to be 3.9 times their standard metabolic rate, with wide variation (2.2–5.3 × SMR). Similar data for Purple Martins and for Mockingbirds feeding young yielded figures of 2.9–3.4 × SMR (Utter and Le Febvre, 1973). The available data are reviewed by Drent and Daan (1980) who suggest that parent birds can sustain a maximum energy output of about 4 × BMR, i.e. about a 50 per cent increase over their non-reproductive existence energy of around 2.6 × BMR.

SUMMARY

Altricial and semi-altricial young depend on their parents for food and, within limits, the rate at which their parents feed them is set by intensity of begging. Feeds are generally given to the most actively begging nestling and, over time, this may establish a brood hierarchy. Young are fed on protein-rich diets, except where special adaptations to diets of poor quality have evolved, and the available evidence suggests the adults deliberately diversify the nestling diet to some extent. Feeding rates and size of food item vary substantially even within a species. The adults are limited in the amount of energy they can mobilize whilst feeding young; within this limit they optimize their foraging efforts in respect of size of prey item taken back to their nest. In communally breeding species the presence of helpers allows the young an adequate feeding rate without working the parents to this foraging limit; thus the adults survive better. Precocial young take food items according to their abundance and suitability to the birds' bill morphology.

CHAPTER 10

Imprinting

Very young birds display social responses to a wide variety of objects remote in appearance to their own species but after responding to one of these objects the bird may develop a preference for it and ignore or reject the other stimuli. This process of restriction of social preferences to a specific class of objects is known as 'imprinting' (Bateson, 1966).

THE 'FOLLOWING RESPONSE'

Imprinting has been studied most extensively in domestic chicks and ducklings and in a few other precocial species. Young of these birds are able to walk shortly after hatching and can therefore approach their parent if separated from her or follow her if she moves away. Hence the 'following response' can be used to study imprinting in experiments in which the young bird is initially exposed to a single stimulus and its subsequent preference for that object is later assessed as strength of following response. Bateson and Wainwright (1972) have developed for the experimental study of the filial response a most ingenious device (Figure 10.1) which illustrates the nature of response and its measurement. Experimental subjects (usually domestic chicks or ducklings) are placed individually in an activity wheel positioned mid-way between two test stimuli, typically flashing lights patterned with different colours or shapes. The activity wheel is free to move along its carrying track in response to the chicks' activity, the idea being to measure the strength of the approach response of the bird to each of the stimuli. Were the chick allowed to approach the initially favoured stimulus it would experience a larger angle of viewing the nearer it got to the selected object, thus increasing any preference it has for approaching a conspicuous object. In the experimental apparatus, therefore, the activity wheel is reverse-geared to its bearing wheels on the carriage track. Then any locomotion towards a preference stimulus bears the chick steadily away from the object of its approach and towards the less preferred one. At some point its resulting closeness to the other light may evoke approach but when the chick turns round in the cage and attempts this it is now borne away in the opposite direction. The balance point is thus a 'behavioural titration' as to where the bird's readiness to approach one stimulus is matched by its readiness to

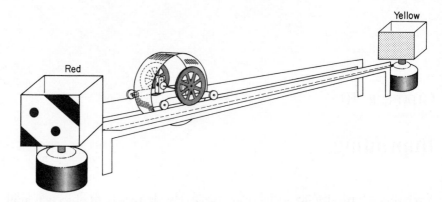

Figure 10.1 Test apparatus used by Bateson to determine imprinting preferences in young birds. The bird is placed in the running wheel placed mid-way along a track between the two test stimuli. The wheel is reverse-geared to its bearings so that attempts by the chick to approach either stimulus result in a weakening of that stimulus by the chick being carried further away. Based on Bateson and Wainwright (1972).

approach the other (Bateson and Wainwright, 1972). Chicks could therefore be exposed to potential imprinting stimuli under the desired test conditions (age of bird, environmental conditions, etc.) and its approach response in the choice situation subsequently measured.

Early studies of the imprinting phenomenon considered it to be disting-uished from other forms of learning and conditioning by four characteristics (Lorenz, 1935): (1) it could take place only within a restricted time period during the development of the bird, a period Lorenz christened the 'critical period' by analogy with phenomena then being established in embryological research; (2) preferences formed by imprinting were irreversible and never subsequently forgotten by the bird; (3) it involved learning of species-specific characteristics rather than those of particular individuals; and (4) the process was complete and the imprint formed before the young bird was called upon to use the information thus acquired. Historically these four ideas have been important for the development of research in imprinting, but only the first two have survived – and then only in much modified form – in the modern understanding of the phenomenon.

SENSITIVE PERIODS

Susceptibility to imprinting is, almost by definition, not age-independent: the proportion of ducklings or chicks which develop a filial response to a moving object to which they are exposed for a short period at first increases and then decreases, all within a few days (or even hours) of hatching. Ramsay and Hess (1954) allowed incubator-hatched Mallard ducklings to spend either 10 minutes or 30 minutes with *papier-maché* models of adult males; the birds were tested 5–70 hours later for preference for this model over that of females. Imprinting most frequently occurred when the ducklings were 13–16 hours old, with no imprinting occurring with very young ducklings nor with those older than 28 hours. Bateson (1966) reviews other evidence for the existence of such finite 'sensitive periods' during which imprinting can occur. Sensitive periods restrict the development of other behaviours in young birds – song development (Chapter 14), the learning of feeding techniques (Davies and Green, 1976), and the development of impaling behaviour in shrikes (Smith, 1972) – and are therefore not peculiar to social imprinting.

Onset of sensitivity

The onset of the sensitive period is associated with the development of the birds to a stage at which they can learn the characteristics of the training object. Gottlieb and his colleagues found that the sensitive period was more satisfactorily measured in terms of age from the beginning of embryonic development than in terms of age since hatching (Gottlieb and Klopfer, 1962). Spontaneously hatching eggs show substantial variation in the developmental age of the young they yield, so for these experiments the eggs were held at low temperatures for several days before incubation. This treatment kills advanced blastoderms but is survived by those in which extensive cell division has not occurred. Birds hatched from these eggs are thus alike in developmental age.

At least three factors could account for this tight coupling of sensitive period and developmental state. First, chicks improve in walking ability as they get older, so they may be able to respond to stimulation more effectively. Second, retinal organization continues to mature post-hatching (Chapter 5), thus improving the ability of the young to respond to visual stimulation. Third, young chicks increase in wakefulness over the 12 hours post-hatching and so may be more ready to respond (Tolman, 1963).

If the state of the arousal of the young were important in promoting its response to imprinting stimuli, one might expect the presence of auditory cues to enhance the response. The auditory system matures in the embryo several days earlier than does the visual system (Gottlieb, 1968) and in fact auditory stimuli reach peak imprinting effectiveness earlier than do visual stimuli (Figure 10.2). Other studies show that ducklings and domestic chicks are more responsive to auditory than to visual ones and even more responsive to the combination of the two (Gottlieb, 1971; Storey and Shapiro, 1979). Alone, these experiments might show merely that auditory stimuli serve as approach evokers. However, Storey (in Storey and Shapiro, 1979) has demonstrated the arousal properties of the auditory cues for ducklings. Young aged between 1 and 3 days approached a visual model more closely in the presence of a non-localizable maternal call than in the absence of such stimulus. Older ducklings (aged 4–6 days) preferred a particular visual model, even in the absence of maternal calling.

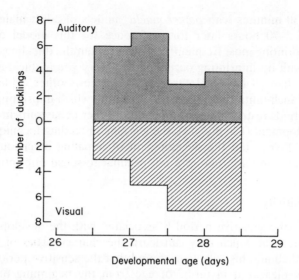

Figure 10.2 Number of Peking ducklings, *Anas platyrhynchos*, responding to auditory and visual stimuli in imprinting tests, as a function of developmental age. Redrawn with the permission of Springer Verlag from Gottlieb and Klopfer (1962).

Termination of sensitivity

Young birds eventually learn the characteristics of any environment in which they are reared and thereafter avoid dissimilar objects. Figure 10.3 shows that the older a bird is when first exposed to a moving object the more likely it is to avoid the object and the less likely it is to respond to it socially. Early workers considered that such changes were associated with fearfulness on the part of the young, fear developing independently of experience. In reality they are due to the young becoming familiar with their environment and avoiding unfamiliar objects. Bateson (1973) reared domestic chicks in isolation in two types of pen: one was painted with black and white vertical stripes, the other with yellow and red horizontal stripes. The chicks were then tested as to their avoidance of moving boxes with these patterns on them. They avoided the unfamiliar boxes for much longer than they did the boxes of familiar colour and pattern. That is, the young birds had learned the characteristics of their home pen and were avoiding objects which they could subsequently identify as different.

RICHNESS OF EXPERIENCE

The degree to which young birds may be imprinted on an object depends in part on the duration of their exposure to that object and in part on the variety

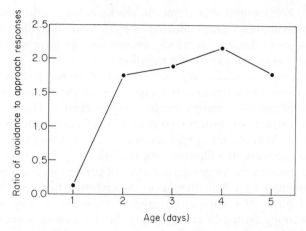

Figure 10.3 Increase in avoidance responses by domestic chicks, *Gallus domesticus*, presented with an unfamiliar moving box on successive days from hatching. From data in Bateson (1973).

of stimuli to which they are exposed during training. Even the use of the visual pathways of the nervous system enables chicks to learn more. In one experiment (Bateson and Wainwright, 1972), domestic chicks were placed in isolation either entirely in darkness or in darkness except for a 30 minute exposure to a constant white light. Half of each group were subsequently trained with a red flashing light and half with a yellow flashing light, in each case for a 45 minute period. After a further hour in darkness they were tested for preference between the red and yellow lights. Chicks exposed to the constant light source were found to approach the flashing light to which they had been exposed more rapidly than did chicks kept in darkness prior to training. Visual stimulation is thus shown to have promoted learning.

Environmental stimulation can be effective in this way even if administered prior to hatching. Chicks hatched from eggs exposed to light during incubation took significantly longer to approach a moving object than did dark-incubated chicks (Dimond, 1968). These tests were conducted on the second day of post-hatching life, so the chicks had time to learn something of their environment. Their pre-hatching stimulation would thus promote greater learning of the familiar on the part of the light-incubated chicks, yielding the greater frequency of avoidance noted.

The importance of stimulation in promoting learning is also indicated by data obtained by Klopfer (1967a). He examined duckling preferences between a strikingly patterned model of an adult duck and a plain model. When comparing the preferences of ducklings with and without exposure the previous day, he found that the preference for the patterned model was greater in birds exposed to either model than in naïve birds. Similarly, Graves and Siegel (1968) found that domestic chicks given gentle stroking subsequently discriminated against moving objects more than did unstimulated chicks. It is also relevant that chicks reared socially have shorter sensitive periods than have chicks reared in isolation.

The difficulties of quantifying the amount of stimulation received by a young bird under these treatments mean that such experiments provide only qualitative evidence as to the greater learning postulated. However, domestic chicks given variable exposures to constant light in experiments such as that of Bateson and Wainwright (1972) above do show a subsequent gradation of response on exposure to a flashing light (Figure 10.4). Exposure of chicks to the training stimulus for longer periods does of course increase the preference shown subsequently for that stimulus in choice tests (Bateson, 1974), but the results in Figure 10.4 imply that greater exposure to *any* environment enhances learning during later imprinting. Stimulation of a variety of types promotes the formation of neuronal connections within the central nervous system of the developing bird (Horn *et al.*, 1973a).

CHARACTERISTICS OF IMPRINTING OBJECTS

Young birds do not imprint equally well on all test objects but the surprising finding is that models of natural parents are no more effective than other stimuli (Hess, 1959). Moving or intermittently appearing objects are more

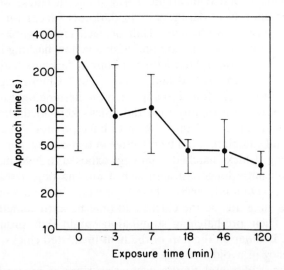

Figure 10.4 The effect of exposure time with a constant light on the subsequent readiness of domestic chicks, *Gallus domesticus*, to approach a flashing rotating light. Median and interquartile ranges with eight birds in each group are shown. Redrawn from Bateson and Seaburne-May (1973).

effective, however, than are static ones; conspicuous objects are more effective than are inconspicuous ones; and objects accompanied by auditory cues are more effective than those offering solely a visual stimulus. Such features are of course consistent with the natural context of imprinting. A young duckling will first encounter its mother as a large conspicuous object providing maternal calls. The young bird is therefore responsive to suitably adaptive cues to which to narrow its social attachments. However, this by itself is inadequate, for the early views a hatchling receives of its mother will not be those it obtains later, when feeding away from its mother. That is, the chick must develop perceptual constancies with respect to distance, angle of viewing, and so on, even within the constraints of the imprinting process. Figure 10.5 shows what seems to happen in meeting these requirements (Bateson, 1973). With increased exposure to the imprinting object the acceptability of totally unfamiliar objects decreases sharply but the acceptability of slightly novel objects rises over even that of the imprinted object. Young birds might therefore actively work to expose themselves to novel views of the imprinting object. Bateson (1973) found that young chicks and ducklings reared in plain grey boxes were much more active than were birds reared in conspicuously patterned boxes, until presented with a conspicuous stimulus.

These findings point up one of the major differences between our present understanding of the imprinting process and the early ideas behind the

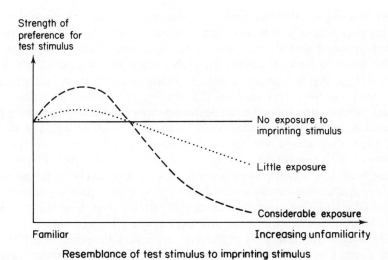

Figure 10.5 A series of hypothetical curves showing how a young bird's preference for a slightly novel test stimulus and avoidance of quite unfamiliar stimuli alter with imprinting exposure. After Bateson (1973).

concept. The young bird does not imprint with every detail of the stimulus in one fell swoop, thereafter remaining immune to further impression. Rather, the young bird has a predisposition to learn certain information and to work to learn more in detail. In Bateson's (1973) analogy, the process is not so much one of imprinting as of sketching: an outline is quickly prepared and finer details within the sketch added later.

SEXUAL IMPRINTING AND MATE RECOGNITION

Since the following response of young precocials is functionally short-lived, it is impossible to measure effectively the duration of any preferences formed during imprinting. With sexual imprinting, on the other hand, functional expression of any preferences formed is necessarily delayed until adulthood. Consequently, sexual imprinting has been the main area of investigation of the longevity of imprinted preferences.

Imprinting has been shown to be an important determinant of species recognition in birds. Lorenz (1935) first suggested that the learning of parental plumage characteristics served to enable the young birds subsequently to recognize members of their own species. For this purpose the 'classical' features of imprinting – rapidity, permanency, species specificity – are desirable qualities of the recognition process, since the young must learn the relevant features of the adults whilst dependent on their parents and retain them for months or even years until they themselves become sexually mature.

Cross-fostering experiments with estrildids

Cross-fostering of estrildine finches has shown that these birds do in fact form mating preferences in this way (Immelmann, 1970). The main experiments were performed with Zebra Finches and Bengalese Finches. Males of each species fostered under the other invariably subsequently courted females of the foster species; conspecific females were solicited only if the foster females were very inactive. This suggests that mating preferences may be formed during juvenile development. These preferences proved highly stable: birds isolated from the foster species could be induced to breed with conspecifics but, even after almost 4 years of such breeding experience and the joint rearing of up to nine broods, cross-fostered male Zebra Finches still preferred Bengalese Finch females when given a choice. Brosset (1971) found rather similar cases amongst cross-fostered columbids, with imprinting determining the initial orientation of sexual activity and with experience determining later orientation but never completely overcoming the effects of imprinting.

To establish the timing of preference formation Immelmann (1970) fostered Zebra Finches under Bengalese Finches but separated them from the foster parents at different ages between independence and sexual maturity. Those separated when 40 days or older remained firmly imprinted on the foster species, but those separated earlier and placed with Zebra Finches were variable in behaviour, courting one or other species or both species on an individual basis. Thus the sensitive period for sexual imprinting was over by day 40. Figure 10.6 summarizes the experimental results as to the onset of the sensitive period. For these experiments young Zebra Finches were initially reared by their own parents but were transferred to Bengalese Finches for completion of the rearing period. If this transfer took place before day 15 the young males imprinted completely on the Bengalese Finches, but if transferred after day 20 they had already imprinted on Zebra Finches; between these ages results varied from bird to bird. Note that the cut-off at 20

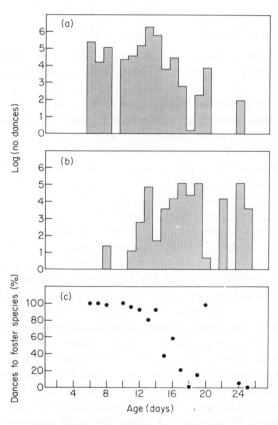

Figure 10.6 Frequency of courtship dances by fostered male estrildid finches that were addressed to (a) females of the foster species and (b) conspecific females in relation to the age at which the natural parents were replaced by the foster species. In (c) is shown the percentage of all dances addressed to the foster species. From data in Immelmann (1970).

days in these experiments differed from the 40 days of the previous set. That is, imprinting of Zebra Finch males on their natural parents cannot be reversed after day 20 but imprinting of Zebra Finch males on Bengalese Finch foster parents can be reversed in favour of conspecifics until day 40. There is thus some genetic factor predisposing the young Zebra Finch to imprint on its own species.

How strong is this predisposition? Immelmann conducted various experiments in which the young bird had the opportunity to imprint on conspecifics whilst being reared by foster parents. He found that the strong social attachment between (foster) parent and offspring is dominant. Even the social relationships which develop between conspecific siblings had little effect on the preference formed for the species of the foster parent.

Sexual imprinting in the Lesser Snow Goose

In other species mating preferences develop in more 'open' fashion, particularly in species with stronger family bonds than hold amongst estrildids. Studies of Arctic geese have shown that sibling plumage can modify the effects of parental pattern in the selection of mates in later life.

The Lesser Snow Goose has two colour morphs – one 'blue', the other 'white'. Mixed matings between the two colour phases do occur but they are less frequent than would be expected by chance. In colonies on Hudson Bay in the Canadian Arctic both phases are common but only 15–18 per cent of the pairs are mixed, where 35–41 per cent would be expected with random mating (Cooke, 1978). Birds with white parents tend to choose white mates, those with blue parents blue mates, and those with one white and one blue parent choose mates of either colour in rough proportion to the abundance of each colour phase in the pair-formation areas. This pattern of mate selection is the outcome of learned preferences for mates of the young goose's familial colour. Naïve young were reared in families of white or blue or 'pink' goslings (the last were experimentally dyed white birds) accompanied by a foster parent of the same colour. When tested in a three-way choice arena after 2 and after 10 weeks, they selectively approached birds of their own colour. If the colour of the foster parent was changed after 4 weeks, the gosling's preference also changed (Cooke et al., 1972). When further experimental flocks were set up but with white goslings fostered by blue parents and vice versa, the choice in favour of parental colour weakened (in tests at 9 and 11 weeks after flock formation). In broods with one parent of each type the choice was random (Cooke and McNally, 1975). Moreover, in large flocks goslings chose to associate with their 'siblings' where possible, otherwise tending to associate with birds of their own colour, and this provided a second influence. This was shown by examining the mate preferences of birds from pure (gosling and parent colour the same) and mixed (gosling and parent colours different) families. In only two of 13 cases did a bird from a pure family mate with a bird of non-familial colour, but eight of 14 mixed family birds paired with a mate of non-parental colour (Cooke et al., 1976). These experiments thus show that assortative mating in the wild could be explained by young birds choosing mates on the basis of familial colour. In fact, field data confirm that the young tend to mate in just this way (Cooke *et al.*, 1976).

If mate choice were always based on traditional, learned preferences, the two morphs would eventually speciate. In practice, however, other factors intervene: (1) a degree of intraspecific nest parasitism takes place; (2) some females are raped by males not their mate; (3) some goslings wander and attach themselves to foster broods; and (4) some birds cannot find mates of the preferred colour. Each of these factors can operate to prevent speciation.

Imprinting and avoidance of inbreeding

Imprinting may have evolved to allow birds to avoid inbreeding. Species-specific selection pressures will make for a high degree of adaptation and thus for conservation of species characteristics. Such requirements can be genetically programmed. On the other hand, genetic variability maintained by outbreeding is a valuable reserve of genetic diversity when adaptability to a changing environment is needed, but is also a process which results in some individuals being of sub-optimal phenotypes. Imprinting provides a process by which these conflicting demands might be optimized (Bateson, 1978). Imprinting might, by allowing the recognition of close kin, adjust the preference of a bird to its own species to select for mates of slightly different characteristics from those to which it was exposed in its own early life.

Bateson found that the mating preferences of male Japanese Quail were consistent with this idea. Given a choice between a familiar female and an unfamiliar female of the same brown plumage type, the males spent 75 per cent of their time alongside the novel female and directed 62 per cent of their copulations to it. In contrast, when given a choice between unfamiliar females of brown and of white plumage types respectively, the brown males spent only 16 per cent of their time with the white-plumaged bird. Moreover, given a choice between a familiar brown-plumaged female and an unfamiliar white female, only 6 per cent of the copulations were directed at the latter. Thus the males preferred the unfamiliar female of the correct plumage type in preference to one of foreign plumage type. That this preference was influenced by experience with siblings, and not as a response to plumage colour, is shown by equivalent preferences for white over brown plumages shown by males reared with white-plumaged females (Gallagher, 1976).

Such selection for imprinting as an inbreeding avoidance process requires that sexual imprinting should take place at a time when females have developed sufficiently adult appearance for the males to recognize them again in later life. Figure 10.7 shows how in a number of different species the sensitive period does in fact coincide with this stage of plumage development on the part of the females. Another consequence of the theory is that there must exist different mechanisms of species recognition in males and females of those sexually dimorphic species where only one parent attends the brood. In Mallard females preferences are virtually independent of early experience,

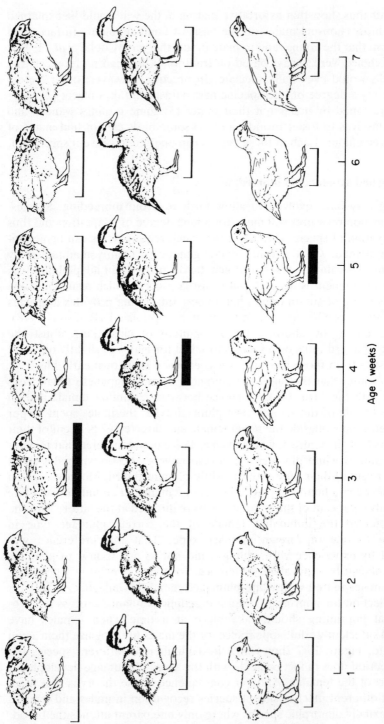

Age (weeks)

Figure 10.7 Changes in external appearance with age in individual Japanese Quail, *Coturnix coturnix*, Mallard, *Anas platyrhynchos*, and domestic fowl, *Gallus domesticus*, young at 1 week intervals from hatching. Note that the 10 cm scale is reduced with older birds, to keep scale constancy. Heavy black bars indicate the approximate start of sexual imprinting (see text). After Bateson (1979).

whilst males imprint on their mother (Schüz, 1963). In monomorphic species, on the other hand, male and female plumages are identical and sexual imprinting by females is consistent with heterosexual mating. In such species, females do indeed display mating preferences contingent on early experience (Schüz, 1963).

A third prediction one might make from Bateson's (1978) avoidance-of-inbreeding theory is that imprinting should be most prevalent in those families undergoing rapid evolutionary development. If female plumage characteristics vary in response to changing selective pressures, the males can track the changes across generations by learning their maternal (by definition, successful) plumage variants afresh each time. Immelmann (1975a) notes that sexual imprinting is particularly prevalent within certain groups, examples being ducks and geese, gallinaceous birds, pigeons and doves, and the estrildid finches. These are the groups independently regarded as displaying extensive adaptive radiation, with several closely related species of similar appearance occurring sympatrically. He suggests that the open ontogeny of species recognition provided by imprinting may be one of the preconditions for rapid evolutionary change.

IMPRINTING AND RETENTION OF INFORMATION

Immelmann (1975a) summarizes the available studies as indicating the crucial feature of sexual imprinting to be the retention rather than the acquisition of information. Zebra Finches imprinted on fostering Bengalese Finches were able to mate successfully with conspecifics when exposed to them in later life, but even such matings did not reverse the early-formed preference for courtship of Bengalese Finches; this reappeared when the imprinted Zebra Finches were allowed a choice. Such a stable memory is in contrast to preferences formed by adult birds, preferences which can be lost within a couple of weeks in the absence of reinforcing exposure to the mate. One possible basis for this lies in anatomical changes in the brain during the period of imprinting, for in birds, as in other animals, the numbers of neurons and synapses continue to increase for some time after birth.

Biochemical correlates of imprinting

Biochemical evidence for the reality of such changes has been obtained in a series of elegant experiments by Bateson, Horn, and Rose (Bateson et al., 1969, 1975; Horn et al., 1973a). Rates of protein and RNA synthesis in the central nervous system of young animals can be determined by measuring the uptake of amino acids previously labelled with radioisotopes. Domestic chicks were trained with a flashing light for 60 minutes, injected with tritiated (i.e. radioactively labelled) lysine, exposed to the flashing light for a further 45 minutes, and then killed after a period in a darkened incubator and a test of their approach response to the stimulus. The imprinted birds showed significantly high incorporation of the labelled molecules into the forebrain roof (Bateson *et al.*, 1969). Protein synthesis there was therefore greater during the experimental exposures than in control birds. To test whether this was due to a general non-specific side effect of visual experience, Horn *et al.* (1971, 1973b) conducted a further experiment. By cutting the forebrain commissure they isolated the two halves of the brain, so that information acquired by each eye was stored independently in different parts of the brain instead of being shared as in intact chicks. Covering one eye of each chick with a patch allowed the birds be imprinted through only one eye. By injecting [³H]uracil prior to training they detected a 15 per cent increase in protein and in RNA synthesis in the forebrain roof on the side controlling the imprinted eye, thus providing a direct biochemical link with the imprinting process as such. In a further set of experiments they established biochemical correlates of increases in exposure times during imprinting (Bateson *et al.*, 1973). Birds were trained for periods of 20, 60, 120, or 240 minutes on the first day after hatching and the uptake of radioisotope-tagged uracil was measured

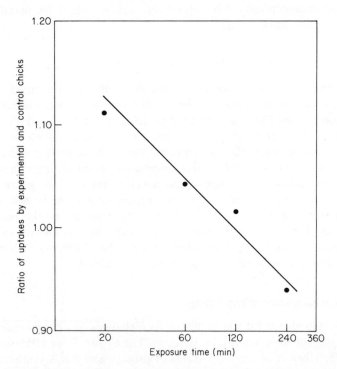

Figure 10.8 Uptake of radioactively labelled uracil into the forebrain roof of domestic chicks, *Gallus domesticus*, during a 1 hour imprinting exposure on day 2, in relation to duration of exposure to the same stimulus on day 1. Uptake is measured here as a ratio with the uptake of control chicks exposed appropriately on day 1 but kept in darkness for 1 hour on day 2. Computed from data in Bateson *et al.* (1973).

during a further 60 minute training session the following day. Since learning during imprinting was thought to be cumulative (see above), birds given longer training the first day should have less to learn the second day and should therefore incorporate less uracil, if this was indeed linked to learning. Figure 10.8 shows that this inverse relationship was established as expected. More recently, the same team have been able to demonstrate a direct correlation between the imprinted preference for the familiar object and the incorporation of ^{14}C-labelled uracil into the forebrain roof. The link between protein synthesis in the central nervous system and the intensity of the imprinted preference seems finally to have been closed.

SUMMARY

Young precocial birds hatch with a predisposition to learn the characteristics of certain social objects – normally the mother and siblings – and to develop social preferences for those objects to the exclusion of others. The process is aided by environmental stimulation and by certain characteristics, notably movement, of the imprinting object. As imprinting proceeds the young reject totally unfamiliar objects but respond to slightly novel ones, thereby developing perceptual constancies with respect to distance and viewpoint, as appropriate for a filial response to the mother. Cross-fostering experiments show that similar preferences are formed by altricial species and subsequently used for mate selection: the rejection of unfamiliar plumaged birds avoids hybrid breeding and the preference for slight novelty avoids breeding with immediate kin. Such preferences are formed in the nest or very soon after departure and this timing coincides with the acquisition of adult-type plumages by siblings. These preferences are very stable and are associated with chemical changes in the forebrain. Such a process of mate recognition is appropriate to species udergoing rapid evolutionary changes and is in fact most prevalent amongst such groups.

CHAPTER 11

Parental care and family life

All young birds except the superprecocial megapodes spend some time in a family unit, interacting with at least one parent or foster parent and possibly also interacting with one or more siblings (in some cases from a previous brood). The amount of parental care which a young bird receives varies widely between species and generally involves components relating to food (Chapter 9) and temperature regulation (Chapter 8), body care and nest maintenance, defence against predators and the acquisition of the skills needed for independent existence.

PARENTAL ASSISTANCE WITH HATCHING

Eggs often hatch unaided, but assistance from the parents is not uncommon. Eastern Bluebird females, for example, speed the process by pecking at the shells and forcibly removing the young (Hartshorne, 1962). Sanderlings, Stone Curlews, and other waders similarly assist their chicks in hatching (Nethersole-Thompson, 1951; Parmalee, 1970).

Disposal of hatched egg shells from the immediate vicinity of the nest is a well-developed component of parental care in a variety of species and is most pronounced about the time the eggs would normally hatch and yield young (Figure 11.1). In some species the shells are built into the nest material; in others they are eaten, in whole or part, and in most species the shells are picked up and dropped some distance from the nest. There are no clear taxonomic patterns to the behaviours used, though eating of the shells may be commoner in dry seasons, when shelled invertebrates such as snails may be a scarce source of calcium (Nethersole-Thompson and Nethersole-Thompson, 1942).

Shell removal tendencies are not constant over the nest cycle and the patterns differ between altricial and precocial species. In the former the removal tendency increases throughout the nesting cycle, while in the latter the tendency has been found to increase at laying and to remain high and constant throughout laying and incubation periods (Montevecchi, 1974). The difference is related to the difference in risks experienced by the two groups. Altricial species may be open to reduction in the success of still unhatched eggs in the nest if the hatched shells slip over and 'cap' these intact eggs:

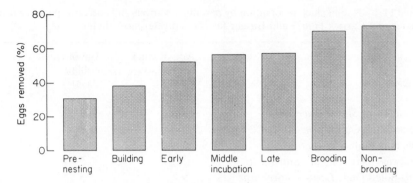

Figure 11.1 Rates of removal of egg shells placed on Ringed Turtle Dove, *Streptopelia risoria*, nests at different stages of the nesting cycle (Montevecchi, 1974).

Montevecchi found that five of 24 shells not removed after 1 hour had interfered with other eggs in this way. Precocial species, on the other hand, are generally ground-nesters and may rely on egg colour for camouflage during predator intrusion into the vicinity of the nest. A conspicuously white shell membrane within a hatched shell effectively draws the attention of the predator to the nest (Tinbergen, 1963, 1972).

Chick recognition by parents

Birds with altricial young have little need to recognize their offspring whilst the young are small, since the young cannot move from the nest. But once the young have fledged, the parents need to recognize their own chicks as individuals; they are otherwise liable to wind up feeding another pair's offspring. Individual recognition of the young can therefore develop over the nestling period by some form of learning process. Davies and Carrick (1962) found that recognition develops faster the shorter the period the young spend at the nest (Table 11.1). Peek *et al.* (1972) found that female Red-winged Blackbirds responded to the nest site rather than to the nest or its contents until the young were 7 days old. When older nestlings were experimentally transferred to adjacent nests and replaced with other young of the same age the female followed her own young to the new nest. Since the nests were well hidden in dense marsh vegetation, the females must have used the calls of the offspring as individual identifiers. Hungry young called loudly about once an hour and sonograms of the calls showed that individual young had relatively invariable calls quite different in structure from those of other young.

The dominance of ecological factors in determining the speed with which adults come to recognize their young individually is emphasized by a study of Herring Gulls by von Rautenfeld (1978). Ground-nesting Herring Gulls will

Table 11.1   Age of chick recognition by parents of various gull and tern species. From Davies and Carrick (1962) and Burger (1974) and references therein

Species	Age of chick recognition (days)	Age of chick mobility (days)
Arctic Tern, *Sterna macrura*	2–3	1–2
Black-billed Gull, *Larus bulleri*	2–3	2–3
Laughing Gull, *Larus atricilla*	6	2–3
Sandwich Tern, *Sterna sandvicensis*	2–3	2–5
Ring-billed Gull, *Larus delawarensis*	5	3–5
Crested Tern, *Sterna bergii*	2	A few days
Sooty Tern, *Sterna fuscata*	c. 4	A few days
Black-headed Gull, *Larus ridibundus*	10	10
Franklin's Gull, *Larus pipixcan*	7–14?[a]	12–20+[a]
Brown Noddy, *Anous stolidus*	c. 14	c. 14
Kittiwake, *Rissa tridactyla*	Never	35–55 (fledges)[b]

[a] Earliest ages; chicks at risk of getting lost in marsh or of drowning if they leave the nest early.
[b] Chicks remain on cliff nest until fledging; otherwise lost or killed in fall from cliff.

not accept foreign young more than 4 days after hatching, but von Rautenfeld found that pairs using cliff nests would accept chicks a week or more old. Such intraspecific cliff-nesting populations are under no selection for early chick recognition, a point made by Cullen (1957) in explaining the anomalously late (for gulls) recognition of their own chicks by cliff-nesting Kittiwakes. Franklin's Gulls are also late in recognizing their chicks, a delay associated with prolonged use of a well-built marsh nest (Burger, 1974).

Parental recognition of young is well developed in species in which young gather in creches but continue to be fed by adults. Penguins call their chicks out of the creche to be fed, with each chick responding only to its own parent (Penney, 1962), and this also happens in Royal Terns and in Sandwich Terns (Buckley and Buckley, 1972). In Royal Terns, experiments suggested that most chicks were recognized principally by voice, although chick plumage varies substantially amongst individuals of this species.

PARENT RECOGNITION BY YOUNG

The ontogeny of parental recognition by chicks has been studied in some gulls (Beer, 1970; Evans, 1970a,b). Laughing Gull chicks show evidence of being able to discriminate their parents' voices as early as 24 hours after hatching, but their discrimination increases sharply between the fourth and sixth day post-hatching.

Similarly, in experimental choice tests Ring-billed Gull chicks showed increased discrimination of the parental call with age, with 58 per cent of 2-day-olds approaching the parental call, 67 per cent of 3-day-olds doing so, and 71 and 79 per cent doing so on days 4 and 5 respectively. This improved recognition is very closely correlated with the mobility of the brood at these ages, the proportion of young outside the nests being 0.10, 0.38, 0.66, and 0.87 on each of days 2 through 5, respectively (Evans, 1970a). Parental recognition develops even more rapidly with age in the Black-billed Gull, whose chicks may leave the breeding territory when only 1–2 days old (Evans, 1970b), and is delayed (to about 16 days) in the marsh-nesting Franklin's Gull, in which broods are confined to the nest platform until 25–30 days old (Burger, 1974).

Parental recognition develops even earlier in the Common Guillemot, whose chicks hatch out in crowded breeding ledges (Tschanz, 1968). Here the family is held together by the chick's responsiveness to parental calls learned whilst still in the egg. This learning develops shortly before the egg pips and continues at least until the chick leaves the ledge.

PARENTAL BROODING

Brooding of altricial young is provided by the parent in proportion to the needs of the young, decreasing as the young grow and become better able to maintain their own temperature and increasing or decreasing as the ambient temperature falls or rises. Brooding also persists longer in small than in large broods (cf. Dunn, 1976). In several species the frequency of nocturnal brooding is also linked to the needs of the young. In Pied Flycatchers, however, the brooding phase is not endogenously fixed but depends on the stimuli available. Nestlings normally acquire thermoregulatory capacities about days 5–6 and nocturnal brooding by the females persists to days 6–7, but repeated replacement of the young as they reach 5 days of age led to extended brooding, to a maximum of 23 days or about four times the normal duration (Winkel and Berndt, 1972)! Feathered young substituted for younger unfeathered young similarly received extra nocturnal brooding (and more so the younger the nestlings they had replaced), but the total brooding time of the female was nevertheless reduced below normal.

Precocial young are not immediately independent of parental brooding, though their need for it is essentially that for long-term temperature regulation. The period of such brooding varies substantially between species, even within a single family. Amongst ducks it may last for 2–21 days until full thermoregulatory ability is established (Koskimies and Lahti, 1964; Kear, 1970a). The amount of brooding needed probably varies with the natural cold-hardiness of the species concerned, for diving ducks need less brooding than do dabbling ducks. Velvet Scoter ducklings are brooded at night (Koskimies, 1957), as are Eiders at times during the high tide roost

(Mendenhall, 1978), and both are brooded more in bad weather than in good. Koskimies and Lahti (1964) have suggested that sensitive species such as the Mallard minimize their energy requirements as ducklings through intensive brooding but are then at risk of experiencing inadequate feeding time during bad weather. In contrast, cold-hardy species can remain active and forage at low temperatures but must then expend more energy on metabolic activities. Mendenhall (1978) found that Eiders increased their feeding time by 50 per cent in bad weather, leaving the roosts for the feeding grounds earlier than normal. They obtained their usual sleep period during the high tide roost (when food was inaccessible) by cutting down on extraneous activity. Brooding was concentrated into the period at the roost.

NEST SANITATION

Passerine nestlings produce their faeces enclosed in a gelatinous sac. When the young are small these are normally eaten by the adults, but those of older young are normally carried away and dropped at some distance from the nest. In at least some species small young have low digestive efficiency and the adult probably recovers substantial quantities of food in consuming the sacs. Bullfinch nestlings, for instance, digest only a small fraction of the seeds mixed in with their insect diet until after some days of growth (Newton, 1967). The faecal sacs are normally ejected immediately after a feed, the adult waiting until it appears. Boetius (1949) found the modal duration of a feeding visit by an adult Blue Tit nearly doubled (from 4–5 seconds to 8 seconds) when the young produced a sac, but because excretion occurs only in a minority of the visits, the loss in feeding time to such waiting is fairly small.

Nest sanitation is most pronounced in open nests accessible to predators, and the young either adopt special postures signalling the production of a faecal sac or eject it on the rim of the nest, thence to be removed by the adult. Nestlings in safe nests, on the other hand, are often more careless in excretion, the faeces being directed as a watery stream over the edge of the nest. In hole-nesting species the nestlings may eject the faeces in the direction of the tunnel or entrance hole.

Retention of the faeces may be advantageous in some circumstances. In the Cactus Wren the faecal pellets are left within the nest during the hotter part of the breeding season, apparently because the evaporation of the water they contain serves to cool the nest at a period of otherwise dangerous heat stress (Ricklefs and Hainsworth, 1969).

Nest sanitation is generally of high standard, at least until the young hatch, in species with well-structured nests and in species with cryptic eggs and is of low standard in species with little or no nest structure. Amongst the doves few adults defaecate on the nest once laying and incubation have started, but Mourning Doves and Inca Doves are notable exceptions, apparently thereby strengthening their otherwise flimsy nests (Montevecchi, 1974). In the cliff-nesting Guillemot, the accumulation of guano on the narrow ledges acts to reduce the rolling of eggs accidentally knocked from under the crowded incubators (Tschantz, 1968). Shilov (1973) also notes that certain species nesting in the hotter parts of their range build nests with a particularly open weave through which the air may circulate. Defaecation on such nests may reinforce such air circulation and prevent the nest over-heating. If this is indeed the function of such non-removal of faeces, one would expect to find correlated adaptations to keep the nestlings clean. The chicks of the Blue Waxbill, one of the African estrildine finches, regularly defaecate on the inside walls of the nest, but the adults bring a succession of downy body feathers to the nest as long as it contains nestlings and the droppings adhere to these feathers (Goodwin, 1965). In at least one other species – the Black-cheeked Waxbill – absorptive material in the form of soft feathery grass plumes is similarly present.

PARENTAL TRANSPORT OF YOUNG

Parental carrying of their young is one of the more impressive categories of family care in birds. Johnsgard and Kear (1968) have reviewed the occurrence of this behaviour amongst wild fowl. Parental carrying of young on the backs of swimming adults has been reliably seen in three swan species, in two sheldgeese, and in at least seven species of duck. It has also been established for several grebe species. Parental carrying of young in flight has been reported for at least 16 species of seven of the main wildfowl tribes. The majority of these instances involved carrying the young in the bill. The advantages of the adults carrying the young whilst swimming presumably lie in removing the young from predation risks and in reducing thermal stress from cold water, though in the case of the Coot young hatchlings can stay afloat on their own only if their plumage has been oiled by the adult. These benefits would accrue to most waterfowl species, so why do more species not carry young? Johnsgard and Kear suggest that the three most important requirements for the evolution of parental carrying are that the adults must be large, the brood must be small enough to be accommodated, and the adults must not need to fly much. The evolution of the carrying trait has probably been driven by predation pressures, for the trait is most prevalent amongst temperate species experiencing daily several hours of darkness. The young are most vulnerable at night and need protection from predators. Carrying is largely absent amongst Arctic species experiencing continuous (or nearly continuous) daylight. However, Arctic cygnets also grow faster than do those of temperate species and they perhaps too quickly become too large for easy carrying.

Carrying of young has been reported in a variety of non-waterfowl species, principally in the larger precocial species best able to move whilst burdened (Skutch, 1976). Carriage is usually between the legs (or between the legs and breast) of the adults, but young are occasionally recorded carried in the bill or under the wings of an adult fleeing on foot. The mobility of precocial young must predispose them to selection for parental carrying, to rescue the chicks from trouble, and after waterfowl the behaviour has been most recorded of waders. Carriage of young is unknown amongst passerines.

PARENTAL DEFENCE OF YOUNG

Risk taking by adults

Parental defence of their eggs and young is almost universally an integral feature of avian nesting behaviour. The impact of virtually any mortality factor can be lessened by greater parental care but almost always at the expense of greater risk to the adults. As described elsewhere (p. 178), adults will take greater risks to defend their offspring the greater the cost of replacing them. If Trivers' (1974) theory of parental investment is correct, parents with altricial young should differ from precocial species in the extent and timing of maximum parental care (Barash, 1975). At each stage of the nesting cycle the parents are faced with the choice of investing still further care in a nest threatened in any way (e.g. by the approach of a predator) or abandoning the nest to the threat in favour of starting a replacement nest.

Figure 11.2 sets out the constraints to the choice. The probability of success with a nest attempt (including the current one) in which the adults might invest care will decrease as the breeding season advances, for both altricial and precocial young (curve R). But the probability of survival of the offspring without further input of parental care will differ between the two groups once the eggs hatch. (Before this point the nest will fail both in altricial and in precocial species when the eggs chill.) Precocial young have some finite chance of surviving from shortly after hatching and this chance increases with increasing chick age (curve P). At first, therefore, injecting further parental care increases the probability of success from that given by curve P to that given by curve R for the same date. At some point MP the curve for the probability of the young fledging successfully crosses the curve for the probability of a nest given further care successfully producing young (curve R), and this should be the point of maximum parental care and defence of the young. Thereafter, the parents do better to invest care in future nesting opportunities. Altricial young, on the other hand, have little chance of fledging successfully in the absence of parental aid (curve A) until late in the nestling period (or even later in species requiring post-fledging care), by which time the probability of their re-nesting successfully has so dwindled that their best chance of success is to invest further in the current brood. In the case of predator approach this entails sitting more tightly, engaging in distraction displays more readily, using more conspicuous displays, and settling nearer the predator. In the case of a threat to the current brood from food failure this further investment may take the form of the adults feeding themselves less, of foraging in more dangerous situations, or of spending more time hunting on behalf of their young than their own welfare demands.

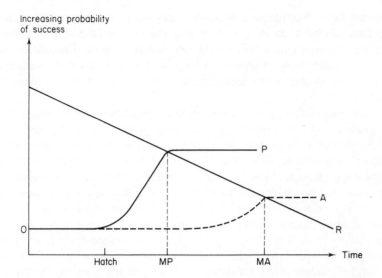

Figure 11.2 Model for the timing of maximum investment of parental defence in precocial (P) and altricial (A) species. Curve R shows a seasonally declining probability of successfully nesting or re-nesting, given parental aid, and curves P and A show the probability of a young precocial or altricial surviving successfully without the parental defence. Points MP and MA indicate the times when maximum parental defence should occur. Redrawn with permission from Barash (1975).

Several studies of distraction displays in altricial and precocial species do indeed show precocial species to have the most intense displays when the young leave the nest (Armstrong, 1958; Jehl, 1973; Barash, 1975). Barash found experimentally that nesting Alpine Accentors flushed at much shorter distances on his approach as the eggs and young aged, with the distraction displays of the female becoming more conspicuous in parallel. Thus, her risk-taking increased as her cumulative investment in the clutch and brood increased and became harder to replace in a further breeding attempt. Similarly, East (1981) found that the rate of alarm calling by European Robins increased more than sevenfold between the end of incubation and the fledging of the young. However, none of these studies completely precludes an alternative explanation suggested by J. H. Marchant (personal communication), that large young are likelier to be hungrier and noisier, so that louder parental alarms are necessary.

Distraction displays can sometimes be avoided if the incubating bird has enough warning. Several species, notably the grebes, cover their eggs with nesting material to camouflage them before leaving the nest. In the South

American Least Seedsnipe this habit is extended to covering the young as well. This can be done in 2–3 seconds if the female (the sole incubator) is disturbed, the adult then leaving the site surreptitiously. Precipitate departure, on the other hand, is always accompanied by distraction displays and the highly precocial chicks can leave the nest even before their plumage is dry (MacLean, 1969).

Similar reasoning applies to nest defence behaviour and is supported by field evidence. Lemmetyinen (1971) found that attacks on model Great Blackbacked Gull 'intruders' by Common and Arctic Tern pairs increased steadily through the nesting cycle, from less than two attacks per minute prior to egg-laying through about 24 attacks per minute in early incubation and 32 per minute in late incubation to 41 per minute during the fledgling period. The violence of these attacks also increased, 72 per cent of the attacks by birds defending chicks yielding strikes on the model against only a 30 per cent strike rate by terns defending eggs. Gottfried (1979) similarly found that the intensity of anti-predator behaviour shown by various songbirds presented with snake models was correlated with the clutch size of each pair, as expected on the investment–replacement cost argument.

Barash's model can be extended in various directions, particularly to predict the incidence of distraction displays in multi-brooded species. If a seasonal decline in nesting success is present the model predicts more intense defence of second than of first broods, but for a species with a seasonal peaking of nesting success the model says that the intensity of brood defence should be correspondingly modified. However, these predictions do not seem to have been tested as yet.

Parental defence and clutch size

The classic explanation for the large clutches of nidifugous species is the reduced need for parental care in such species. Without the need to bear food to the young, so the argument runs, the parents are limited only by the ability of the female to produce eggs and incubate them successfully. Yet Safriel (1975) has shown that the optimal clutch size in Semi-palmated Sandpipers and other nidifugous waders can be limited by considerations of parental care. Precocial young are vulnerable to predation when they move around and may therefore be actively protected by their parents. If the young do not stay together in a group but wander around independently seeking food items, the demands on the parents' attention will be higher the larger the brood size. The demands are also likely to be higher the faster the movement rates of the individual young, which rates are inversely related to the distances between prey items at low prey densities. One might therefore expect clutch size to vary in parallel with densities of prey for the young. But hunting efficiency does not in general increase linearly with food density, since at high densities the chicks spend a greater part of their time actually eating the items encountered and relatively less on searching for the next item. Instead, movement rates level off asymptotically with food density, imposing an upper limit on clutch size. Many Arctic waders have clutches of four eggs, despite their considerable differences in energy needs, thus allowing the possibility of some general limiting phenomenon such as predation. To test this Safriel artificially increased the size of Semi-palmated Sandpiper broods to five. The fledging success of these experimental pairs was lower (average 1.00 young) than for control pairs with broods of four (average 1.74 young), in keeping with this model.

The importance of parental defence limits is also suggested by the behaviour of Sandhill Cranes, in which the chicks from the brood of two are tended separately, one chick to each adult. But the adults maintain contact with each other from their feedng areas 100–150 m apart and unite to defend a threatened young (Harvey *et al.*, 1968). In this species the young are fiercely intolerant of their siblings and will fight furiously if not kept apart. The system of split parental care thus seems to be a compromise between the conflicting interests of adults and young of the type described elsewhere in this book.

Not all species share the defence of the young equally between the two parents, even when both adults attend the young. Evans (1975), studying captive Bewick Swans at the Wildfowl Trust at Slimbridge, found that most behaviours of the cygnets over their first 5 weeks of life were directed to the female alone, with only 5 per cent to the male alone; interactions to both were usually initially directed to the female. The female also led the newly-hatched young from nest to water and made the 'trampling' movements needed to stir the bottom mud to bring food within reach of the young. The male's role was mainly that of guard and protector. Male guarding is similarly marked among geese, but in sexually dimorphic raptors it is often the larger female who provides local defence of the young. For most altricial species with both parents in attendance defence is probably equally shared. In communally nesting species, however, certain individuals may be more intense defenders of the nest than are others (Woolfenden, 1975).

NEST DEPARTURE BEHAVIOUR

The period of nest departure behaviour can be a difficult one for parent birds, for they must tend the nest for the young still there and yet guard and feed the young outside the nest. Altricial young probably fledge spontaneously once ready to do so, most frequently in the early morning and often in the absence of their parents. Where hatching has been more or less synchronous, the young fledge on the same day and the adults can care for the brood as a unit. With asynchronous hatching, however, the young are at different developmental stages when the first young is ready to fledge, and special behaviours may be necessary to compensate. House Sparrows may split their brood temporarily under these conditions, one parent continuing to visit those young still in the nest, the other tending the new fledglings in the nearby undergrowth. Titmouse broods often contain one nestling markedly less developed than its siblings (since incubation commences with the penultimate egg, especially in late nests) and adults have been seen attempting to lure such young from the nest with morsels of food (Walker, 1972). In yet other asynchronously hatching species the young are largely or fully independent of their parents when they leave, so it does not matter if the young depart on very different dates. These are usually species with feeding habits not particularly suitable for feeding flying young.

For the parents there comes a stage when the extra effort needed to feed the young at the nest is more productively expended on either a second brood or on keeping themselves alive until the following year when they can breed again. In keeping with this, the adults may adopt strategems to encourage their young towards independence. Thus, feeding rates to young Bluebirds are reduced in their last day in the nest, most markedly so once one or more young have left the nest: in some cases the young are not fed at all in this period, though they are fed immediately on leaving the nest (Hartshorne, 1962). Ospreys have similarly been seen bringing food to the vicinity of their nest site but withholding it from the young until they have flown from the nest itself (Meinertzhagen, 1954), and a male Redwinged Starling has been observed physically ejecting the young from the nest when they refused to leave to take the food he was offering from a short distance away.

Apart from these fairly clear-cut examples of parental enticement, there are many cases of young receiving few or no feeds in the days prior to fledging, observations therefore interpreted as a 'starvation period' to force the young to leave the nest. In fact, detailed study of many species shows that the reduction of feeding visits is due simply to a reduction in appetite. Table 11.2 shows that a Puffin chick fed *ad libitum* took less food towards fledging

Table 11.2 Changes in the voluntary food intake of an artificially reared Puffin, *Fratercula arctica*, chick. Data from Harris (1976)

Age (days)	Daily food intake (g)
1–5	11.0
5–10	33.6
11–15	65.0
16–20	106.9
21–25	148.7
26–30	110.4
31–35	104.7
36–40	87.7[a]
41–45	51.5
46–49	47.1

[a] Feeding not *ad libitum* on 2 days in this sampling period.

time than previously, even though its weight decreased. Skutch (1976) suggests that the loss of appetite is adaptive, by making the fledglings lighter and more competent in flight.

Altricial nestlings, of course, have had long exposure to their parents by the time of fledging, but in precocial species the nest exodus is usually within a day or two of hatching and is aided by certain parental behaviours, principally vocalizations (Gottlieb, 1965). Both Mallards and Wood Ducks call their broods from the nest, the former usually from a ground (but occasionally a tree-fork) nest, the latter from the base of deep vertical cavities in dead and decaying trees. The Wood Duck female correspondingly vocalizes earlier and more frequently than does the Mallard and at an increasing rate with the passage of time (Figure 11.3). This build-up in calling is also present in Mallards and points to reciprocal stimulative interplay between mother and brood. Unlike the Wood Duck, the Mallard ducklings are usually able to see their mother as she calls them, and experimental testing of the ducklings shows that they are more likely to leave the experimental nest-box if they can connect the calling to a visual stimulus, such as a moving striped box. Parental control of the exodus process in these species is of course linked to the possibility of predation. Before calling the brood out the mother makes a reconnaissance of the area and will delay the exodus if she senses intruders.

Most birds fledge or depart the nest in the early morning, presumably to leave them with daylight for their initial period of independent life. In some

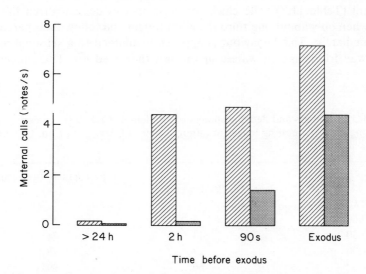

Figure 11.3 The build-up of parental calling by Wood Ducks, *Aix sponsa*, (stippled) and Mallard, *Anas platyrhynchos*, (black) as the time for nest exodus by the brood approaches. Note the earlier and more frequent calling by the hole-nesting Wood Duck. Data from Gottlieb (1965).

species, indeed, the young return to the nest each night: this is particularly true of hirundines and other aerial species which presumably lack an extensive choice of suitable roost sites in the earlier part of the breeding season. Later, adults and juveniles congregate together in reed-beds. In House Martins the young of the first brood may sleep within the nest even with second or third brood nestlings present, but then they also assist their parents with the rearing of these young.

Although departure in early morning provides daylight for the initial explorations of the young, it also increases their risk from predation. Some parents therefore display behaviour patterns assisting the young at the moment of fledging. As already noted, the female Wood Duck reconnoitres the vicinity of the nest before calling her young from the nest cavity (Gottlieb, 1965). To the same end the adults of some altricial species undertake what Skutch (1976) describes as shielding flights. As the young bird descends from its nest it is closely accompanied by a parent flying a little higher than or behind it until it lands and safely reaches cover. Should a raptor pursue the descending pair, the adult would almost certainly be the target so the predator would tend to follow it as it veered to one side at the last moment, thereby giving the fledgling a few seconds grace to reach cover. In colonial species several adults may join in this flight, thereby increasing the protection afforded the young bird.

Some few species fledge in late evening towards dusk, apparently because of the risk of predation by day. Guillemots and Razorbills fledging from a Scottish colony departed at dusk but birds departing around 21.30 were most successful (Table 11.3). The chicks of these species descend from the cliff ledges when only about one-third the adult weight and follow their parents to sea immediately. This behaviour is apparently adapted to a seasonal movement away from inshore waters of the fish they feed on. This movement

Table 11.3 Numbers and fledging success of Guillemot, *Uria aalge*, chicks recorded descending from the nesting ledges at different times of day. Data from Greenwood (1964)

Time	Chicks descending	Percentage successful (%)
19.30–20.00	3	33
20.00–20.30	14	71
20.30–21.00	27	78
21.00–21.30	66	89
21.30–22.00	87	95
22.00–22.30	38	97
22.30–23.00	16	94

necessitates longer and longer feeding flights by adults returning to the colony. An elaborate ceremony of bobbing and bowing by adult and chick on the ledge culminates in the 'water-call' (a loud clear peeping several times repeated) from the young just before it jumps and serves to co-ordinate their activity. The chick, descending on what lift it gets from its coverts and webbed feet, is closely followed by the adult and they re-unite on the water, guided by mutual calling. The whole sequence is most successful when done at the main period of descent, and mutual stimulation of the adults and chicks on each nest ledge serves to synchronize the activity to this period (Table 11.3). Chicks descending early are prone to gull predation, those descending later may be hampered by poor visibility (Greenwood, 1964).

POST-FLEDGING CARE

Parental care continues for a period after fledging in most passerines and in some non-passerines. For Temperate Zone species this period typically lasts for 2–4 weeks, but in larger species it may last months or more (Table 11.4) and in extreme cases young may still be attended by their parents years after leaving the nest (Norton-Griffiths, 1969).

The speed of the transition to independence is linked to the ease or otherwise of juvenile foraging. Parental care is prolonged where a species uses highly skilled feeding methods or where the foods sought are scarce (Ashmole and Tovar, 1968; Fogden, 1972). Fogden observed at least 17 species feeding young 10 weeks after fledging in equatorial forest in Sarawak. Figure 11.4 shows that a slow decline in feeding contributions by the parents was largely independent of species identity and was very much slower than in Temperate Zone passerines (Table 11.4). The majority of these species hunted for insects in foliage and such prey are always relatively sparse and difficult to find there. Moreover, the insect fauna in Sarawak is extremely diverse and has evolved a great variety of protective adaptations, providing substantial security against the development of searching images on the part of the young birds. Fogden suggests that in these conditions great survival value lies not only in ability to learn quickly but also in having the opportunity to learn by experience. This is provided by prolonged association with parents.

Prolonged parental care is also the rule where the young must acquire high skills in prey capture. Ashmole and Tovar (1968) note the occurrence of such cases in such groups as frigate-birds, terns, owls, and kingfishers, all of them species with small clutches and no second broods. Juveniles of these species often spend much time in play activities related to hunting behaviour. Newly-fledged Sandwich Terns begin to develop hunting skills by making shallow plunge dives or dips to the water surface, never catching edible prey but usually picking up pieces of algae or inanimate objects (Dunn, 1972). By the middle of their first winter these young are fishing efficiently on their wintering grounds in Africa, but have still not achieved adult success rates,

Table 11.4   Age of independence of young of various species. Data from various sources

	Age of independence (days)	
	Since fledging[a]	Since hatching
Passerines		
Great Tit, *Parus major*	6–8	27–29
Coal Tit, *Parus ater*	*c.* 14	30–33
Boreal Chickadee, *Parus hudsonicus*	14–21	32–39
Wren, *Troglodytes troglodytes*	15–18	33–36
Spotted Flycatcher, *Muscicapa striata*	17–18	*c.* 30
Bullfinch, *Pyrrhula pyrrhula*	18–19	*c.* 35
Song Sparrow, *Melospiza melodia*	18–20	28–30
Chaffinch, *Fringilla coelebs*	18–24	32–37
Crested Tit, *Parus cristatus*	21–28	38–49
Prairie Warbler, *Dendroica discolor*	31–32	41
Wilson's Warbler, *Wilsonia pusilla*	*c.* 35	*c.* 45
Non-passerines		
Oystercatcher, *Haematopus ostralegus*	14–21	42–49
White Stork, *Ciconia ciconia*	14	67–69
Ringed Plover, *Charadrius hiaticula*	—	*c.* 25
Sparrowhawk, *Accipiter niscus*	*c.* 27	*c.* 59
Lapwing, *Vanellus vanellus*	—	*c.* 33
Wigeon, *Anas penelope*	*c.* 42	—
Common Tern, *Sterna hirundo*	63+	—
Crane, *Grus grus*	—	*c.* 70
Mute Swan, *Cygnus olor*	—	*c.* 120
Black Vulture, *Coragyps atratus*	180+	—

[a] Nest departure for nidicolous species; first flight for nidifuges.

especially when the fish are located well below the surface. Dunn estimated that adults were catching 14 fish for every 10 caught by first-winter birds. Other species of tern are even worse off: Royal Terns and Caspian Terns are still dependent on their parents when on their wintering grounds 6 months after fledging (Ashmole and Tovar, 1968; Recher and Recher, 1969; Dunn, 1972). Similar inefficiency is found amongst Little Blue Herons and in immature Brown Pelicans. At least five other species of seabird and shorebird show age-related differences of this type (Burger, 1980) and it is worth remarking that the majority of these species show deferred reproduction and

185

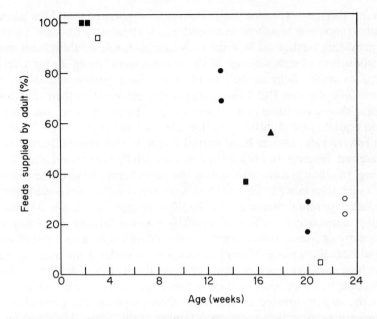

Figure 11.4 The extent and duration of parental feeding after fledging in some tropical species. The vertical axis shows the percentage of feeds supplied by the adult. The species are *Harpactes duvauceli* (▲), *Eurylaimus javanicus* (●), *Criniger phaeocephalus* (■), *Stachyris poliocephala* (○) and *S. erythroptera* (□). From data in Fogden (1972).

smaller clutches in first-time breeders. For many species with such specialist feeding techniques, additional experience in foraging may be vital to success in breeding (Lack, 1954).

TRANSITION TO INDEPENDENCE

What determines the timing of break-up of families? The theory developed by Trivers (1974) is relevant here. Briefly, a parent can either invest care in the current young or can abandon that young in favour of a new breeding attempt. (Included here is the possibility of the adult starting to recuperate in preparation for an intervening winter.) The parent is equally related both to present and to any future offspring. On the other hand, each offspring shares only half its genes with young in the new brood, so the return (in reproductive terms) it needs before it benefits by the diversion of parental investment to the production of siblings has to be double the return acceptable to the parents. In this way parents and offspring can come into conflict about the amount and timing of parental care. The parent should want to terminate care earlier than the young is prepared to accept.

To test this idea experimentally, Davies (1978) argued that if the young was already competent to accept independence it should try to 'trick' its parents into providing further for it, until such time as it too would benefit more by the production of new siblings. If the parents were 'mean' in their response to, for example, begging behaviour, the young bird would do better to become independent. But if the parents were 'generous' in their response, the fledgling should continue in its dependence on them. In effect, the young bird can do equally well in either situation and switches between them according to its reward rate. Davies hand-reared Great Tits on three different regimes of 'parental' feeding. In each he at first fed each fledgling soon after it started begging, but then gradually increased the delay before he gave the food to the chick (such reluctance to deliver food is observed in the wild – see below). He introduced 'parental meanness' to the three groups at 1, 2, and 3 weeks after fledging respectively, with food available for self-feeding at all times. The chicks were of course rather inept at self-feeding but by 8 days out of the nest all could obtain a piece of food in this way in under 1 minute. Figure 11.5 shows how chicks on each regime switched sharply to self-feeding at the time self-feeding becamer more profitable than continued begging on their regime. Since the switch differed between the three regimes, the possibility of the young turning to self-feeding on attaining some particular developmental state can be discounted. The transition to independence is thus an active one,

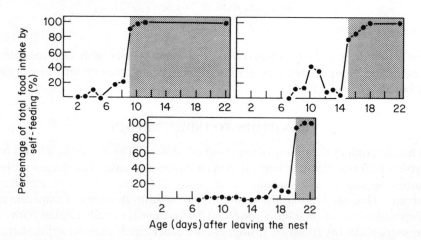

Figure 11.5 The incidence of self-feeding by fledgling Great Tits, *Parus major*, in relation to the timing of parental reluctance to respond to begging by the fledglings. In each graph young were fed promptly at ages to the left of the stippled area but only after a delay at ages within the stippled area. Representative results redrawn from Davies (1978) by permission of The British Ornithologists' Union.

determined by the young on the basis of how quickly it gets fed by begging or by self-feeding.

What happens in the wild? Much the same, according to a field study of Spotted Flycatchers conducted by Davies (1976). These birds feed almost entirely on flying insects, mainly Diptera, by taking them in mid-air sallies from a perch. The young leave the nest at 12–13 days of age when only half-grown and spend their first week hidden in the tree canopy near the nest, being fed there by their parents whenever they beg. In their second week they move about more and in the third week come down to low perches and become independent at 17–18 days. In their first 9 days the parents bring food whether or not the young are calling, but after 10 days do so only if the young beg. Between 10 and 15 days there is a sharp increase in the number of flights the young make after the parents and in the proportion of such flights in which the young goes unfed. The initiative for feeding by the parents thus passes from the parents to the young. Once self-feeding became more profitable than begging, the young became independent and had stopped all begging within 1–2 days.

Parental feeding in Spotted Flycatchers thus lasted until the young were able to feed themselves efficiently and the actual changeover was triggered by begging becoming unprofitable relative to self-feeding. A point of interest here is that both Spotted Flycatcher and Great Tit chicks show a sharp increase in begging frequency just prior to independence. The chicks seem to be testing the relative values of continued begging and of independent foraging.

No studies have yet been conducted as to how the decreases in parental willingness to feed young are governed. Norton-Griffiths (1969) found that such decline occurred also in Oystercatchers. Where the adult would wait on a 1 week old chick for 12 seconds before itself consuming a prepared food item, fledged chicks were allowed a median time of only 1 second to take the item. Norton-Griffiths was able to show that in the Oystercatcher these changes have two components. The first is a motivational change occurring about the third week, leading to a reduction in the parents' role in initiating trains of feeding behaviour. The second is dependent on for how long the young demands food from the parent and is correlated with the ease or difficulty the chick has in self-feeding.

As the young develop in independence within the family the guidance of the parents becomes less pronounced, and the initiatives of the young more so. Thus in Bewick's Swans the proportion of movements initiated by the cygnets, with the parents following, increased from 52 per cent in the first week through 66, 77, and 87 per cent respectively in each of the following 3 weeks (Evans, 1975). The begging rates of Black-capped Chickadees similarly decreased with age as the young began to find food on their own; the parents began to ignore begging by young following them after 8–10 days (Figure 11.6). The frequency of aggressive interactions between adult and young also increased with age (Figure 11.6), as did fighting among the siblings them-

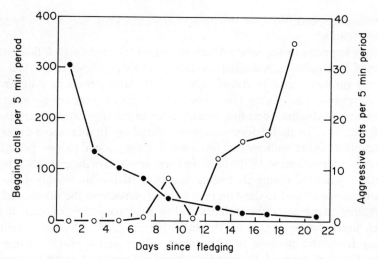

Figure 11.6   Changes in begging behaviour (●) and in frequency of parent–offspring aggression (○) in Blackcapped Chickadee, *Parus atricapillus*, families in relation to fledgling age. Drawn from graphical data in Holleback (1974).

selves. The average distance between family members consequently increased from a metre or so just after fledging to about 30 metres near the time of dispersal at 20 days (Holleback, 1974). Hori (1964, 1969) found that brood attacks in which both parents drive their young away were one of the basic methods by which family ties in the Shelduck were broken. For most young the tie lasted only a day or two after reaching the nursery water.

CRECHES

Chicks of several species of colonially nesting penguins customarily leave their nests during their development period and huddle together in groups or 'creches', so called because early observers concluded the chicks were coaxed or driven into these formations and tended there by a few adult 'nurses'. Later studies have shown such creching behaviour occurs in many other groups, including the Sandwich Tern, pelicans, flamingos, and several waterfowl species (Gorman and Milne, 1972). The details of the behaviour differ between species, even within a single group: thus, all the chicks of a King Penguin colony or of an Emperor Penguin colony mass together in one giant creche, but other penguin species have several creches per colony, each containing 12–20 young (Rockhopper Penguins) or 20–30 young (Gentoos and Adelies) (Pettingill, 1960). Again, waterfowl differ from the other groups mentioned in having self-feeding young, so that their creches must be mobile within the feeding areas.

Adult Eiders and Shelduck guarding creches of young actively defend the creche by giving alarm calls on the approach of gulls or other predators; the young then bunch in response (Gorman and Milne, 1972; Williams, 1974). Rockhopper Penguin chicks unguarded by parents were frequent prey to Antarctic Skuas if solitary but were never taken if in creches (Pettingill, 1960). Feeding considerations may underlie such guarding being of creches rather than of broods. Female Eiders feed little or not at all whilst incubating, instead burning body fat and protein until they can lead their brood to the shore. With their reserves depleted, newly-arrived females are in pressing need of food, but their main prey, the edible mussel *Mytilus edulis*, is spatially remote from the areas of the shore supporting the populations of *Corophium* (an amphipod), *Hydrobia* (a gastropod mollusc), and *Littorina* (a winkle) on which their ducklings can feed. Gorman and Milne (1972) found that individually marked females averaged only 4 days in a creche with their broods before leaving to replenish their depleted reserves. Interestingly, on rocky coasts females and their young remained together without forming creches, but the young now fed on gammarids and small crustaceans from amongst the seaweeds present while the females took *Mytilus* and littorinids from the rocks. For presumably similar reasons creche formation is rare amongst dabbling ducks, in which adults and young feed together by surface sifting, and is commoner in diving species, in which adults and young feed on different sizes (and even taxa) of prey.

Asynchronous hatching regulates sibling rivalry

Excessive rivalry and fighting between siblings could endanger the success of the brood as a whole. Parent birds with only two offspring can rear them separately but others must resort to alternative behaviour. As discussed in detail in Chapter 1, asynchronous hatching within clutches is commonly present in species which need to regulate the effects of sibling rivalry when food is scarce (Lack, 1947, 1968). But a competitive hierarchy amongst siblings can also improve parental reproductive success by reducing sibling rivalry (Hamilton, 1964). If the outcome of a contest between siblings can be predicted, there is little point in engaging in the contest (Maynard Smith and Parker, 1976). Each chick therefore wastes less energy in contest with its siblings if within a hierarchy than if without one, and the efficiency with which the brood uses the food gathered by the adults should increase. Lack's explanation does not predict such an increase. Hahn (1981) actually found the predicted effect amongst Laughing Gulls: 54.2 per cent of the asynchronous broods fledged all three young whilst only 23.1 per cent of the experimental synchronously hatching clutches did so, the additional failures being due to starvation. Her findings thus reveal a very interesting point about the costs of sibling rivalry.

SUMMARY

All young birds except megapode chicks receive some parental care after hatching. Such care may include assistance with hatching, egg shell disposal, nest sanitation, brooding, feeding, and even transport. Chick recognition by parents develops most rapidly in species with mobile chicks and least rapidly in altricial species and is, in general, timed to the mobility of young. Parental brooding is geared to the needs of the young and is most frequent with small young, but in some species the brooding phase is of relatively fixed duration whilst in others it is determined by stimuli provided by the young. The extent to which parents undertake active defence of their young varies with the current value of the offspring to the parents: where the prospects of successful replacement of threatened young are high, defence is weak, and conversely. In Arctic waders and other precocial species the need to guard young may limit the size of clutch that can profitably be laid. Family relations are governed by kin selection and this can lead to conflict between members of the family, especially at nest departure and at the transition of young to independence. Parental care of fledglings is particularly prolonged in species with specialist feeding skills. Creche formation offers an alternative pathway to adult care but the processes behind such behaviour are not fully understood. Asynchrony in hatching reduces the overall cost associated with the sibling rivalry over food.

CHAPTER 12

Mortality of eggs and young

Nesting mortality in birds has been extensively reviewed and analysed by Ricklefs (1969a), concentrating particularly on Temperate Zone passerines. Two broad categories of mortality are distinguished in his analysis: (1) factors associated with events during nesting, such as fertility, laying, hatching, and fledging, and (2) factors which may operate at any time and whose expectation changes with time. The former are qualitative factors: for example, an egg either hatches or it does not hatch. The latter are quantitative and can be attributed 'rates': for example, predation losses can be frequent, generating a high rate of loss, or rare, generating a low rate of loss.

ESTIMATING MORTALITY RATES

If the probability of egg or nestling dying at any moment is constant throughout the nest cycle, the number of eggs or nestlings alive at time t is given by

$$N = N_0 \exp(-mt), \tag{12.1}$$

where N_0 is the number of eggs laid at $t = 0$ and m is the mortality rate. In most studies, nests are visited almost daily and m is usually expressed as a daily mortality rate, as per cent loss per day. For many species it has been shown that mortality differs between different stages of the nest cycle (Ricklefs, 1969a) and separate values of m are needed within the egg stage and within the nestling stage. There has been little empirical investigation of the constancy of m for each.

An important practical bias in the study of egg and nestling losses has been pointed out by Mayfield (1961, 1975). Nests found late in the nest cycle are necessarily more successful than is typical of the species concerned. Estimates of the proportion of eggs or of young that were successful therefore underestimate true mortality, unless a correction is introduced for nests which failed before they could be discovered by the investigator. Many of the early studies of nesting success thus understate the extent of nest failure. Mayfield (1975) and Johnson (1979) describe a reliable and robust method of analysing incomplete nest histories to yield more realistic estimates of mortality.

Partial and whole brood loss

A helpful distinction in considering mortality is one between the loss of the clutch or brood as a whole, as when a nest is deserted or predated, and the loss of only part of the clutch or brood, as when some but not all of the nestlings perish through starvation (Ricklefs, 1969a). Some of the mortality factors considered by Ricklefs are mutually correlated across species, so the patterns apparent in individual factors are not always independent of each other. Use of multivariate analysis to allow for the cross-correlations between nest, egg, and nestling mortalities in small land-birds shows that variation in the general level of mortality – mortality falling more or less evenly on nests, on eggs, and on young – accounts for rather more than half the variance in mortalities (R. J. O'Connor, unpublished). Independent of this variation in general mortality are two other factors: (1) the relative importance of mortality rates of eggs and of nestlings, and (2) the incidence of partial loss (of clutches and of broods) within individual nests.

On general theoretical grounds one would expect selection to operate strongly against any risk of mortality late in the development period. Birds which lose their nest at nestling stage have lost a larger investment – extra incubation and extra feeding effort – than have birds losing their investment early on, and they have less chance of successfully re-investing. Selection will therefore favour genes opposing such losses. Figure 12.1 shows that the expected pattern was present in the species analysed by Ricklefs (1969a). Note in particular the fourfold reduction in desertion rate between egg and nestling periods and the 10-fold reduction in losses to Cowbirds. The effects of weather are unpredictable, by definition, and thus not open to selection, and the same differential is not apparent in weather-related mortality of eggs and young. Hence nest sites should be selected to reduce the risk of their discovery when they contain young (i.e. late in the nest cycle) to a greater extent than they are selected to reduce the risk of their discovery when they contain eggs early on; genotypic failures should be expressed early, as hatching failures, rather than later, as nestling failures; parental defence of a nest when nestlings are present should increase over the level provided for eggs; and parental willingness to desert a nest should be lower for nestlings than for eggs. In these ways the loss of time and efort expended on the failed nests is minimized.

This type of difference between mortality of eggs and of nestlings accounted for 30 per cent of the variance in mortality patterns in my multivariate analysis (R. J. O'Connor, unpublished). The remaining factor, partial loss, is relatively uncommon, accounting for 13 per cent of the variance, and largely reflects the occurrence of nestling starvation. As discussed elsewhere in more detail, certain species are limited in their breeding success by unpredictable variation in the level of food supplies available for their young. These birds therefore maximize their production of offspring by laying excess eggs and subsequently selectively feeding the strongest of their broods to the limit imposed by food availability; the neglected hatchlings quickly die (Ricklefs, 1965; O'Connor, 1978a,c). These

193

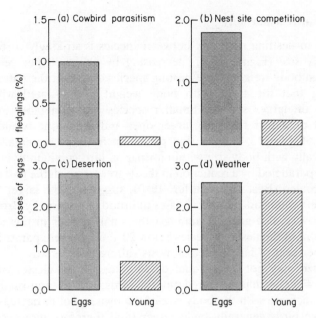

Figure 12.1 Percentage losses of eggs (solid) and nestlings (shaded) to various sources of mortality, showing the reduction (relative to egg losses) of nestling losses to biological factors. Note the different scales. From data for various North American passerines in Ricklefs (1969a).

tactics work best in species for which egg production and hatching are relatively easy, so it is not surprising that they are most prominent in species with low proclivities for egg loss (Figure 12.2).

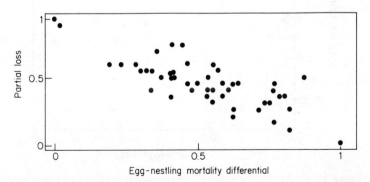

Figure 12.2 Relationship between partial losses of nest contents (largely due to nestling starvation) and egg–nestling differences in mortality in 48 species of small land-birds. Each axis is based on scores obtained in principal component analyses and the units are therefore arbitrary.

194

Adult size affects mortality

Variation in nestling mortality between species is strikingly correlated with adult body size (Figure 12.3), decreasing by about 30 per cent for each doubling in body weight. Hole-nesting species do not fit the pattern, a point suggesting that the trend with body weight is associated with reduced predation on larger species. Smaller species are vulnerable to a greater variety of predators than are larger ones and this alone would explain a reduction. Figure 12.4 shows that predation by large mammals decreased systematically with body size within a riparian bird community in Iowa. The trend was paralleled by a reduction in the degree to which the nests concerned were concealed (Best and Stauffer, 1980), suggesting the larger birds could defend their young adequately. Losses to brood parasitism by Brown-headed Cowbirds were also greatest amongst the small species in this community: 15.4 per cent of the nests of species below 20 g weight were parasitized, whilst only 2.4 per cent of larger species' nests suffered in this way.

Body size may not offer equal defence against all classes of predator. Figure 12.4 shows that nest predation by birds, snakes, and small mammals tended to increase with the body size of the owner, not to decrease. For one thing, large birds generally build larger (and therefore more conspicuous)

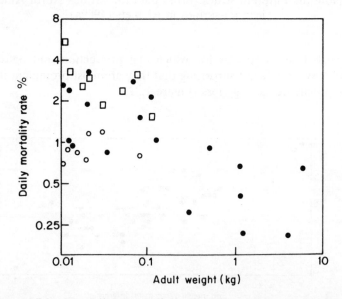

Figure 12.3 Mortality rates of nestlings in relation to adult body weight in Temperate Zone altricial land-birds. *Key*: ○, hole-nesting species; □, ground- and marsh-nesting species; ●, passerines and raptors using open nests above ground level. Redrawn from Ricklefs (1969a).

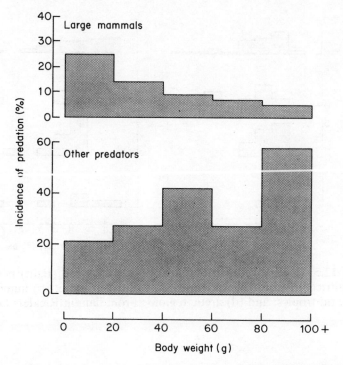

Figure 12.4 Incidence of predation by large mammals (top) and by birds, snakes, and small mammals (bottom) on nests of 15 riparian bird species in Iowa, USA. From data in Best and Stauffer (1980).

nests, thus attracting a greater intensity of attention from those predators that can climb or fly to the nest. Secondly, body size does not defend against venomous reptiles nor against sneak attack in the adult bird's absence.

Larger birds may also enjoy their greater breeding success on account of the greater thermoregulatory abilities conferred on their nestlings by size. Not only do large young cool less rapidly in adverse weather than do small species, they also utilize what energy reserves (notably fat) they possess less rapidly than do small birds with high metabolic rates (Calder, 1974).

REGIONAL VARIATION IN MORTALITY

Amongst small land-birds the 'general' component of mortality is similar in the humid and arid tropics, in Temperate Zone latitudes, and in the Arctic, but the extent of egg–nestling differentials in mortality varies substantially between the four regions (Figure 12.5). Egg mortality is relatively high in Temperate Zone species and in species of the humid tropics, whilst nestling mortality is more prevalent in species of the arid tropics and of the Arctic

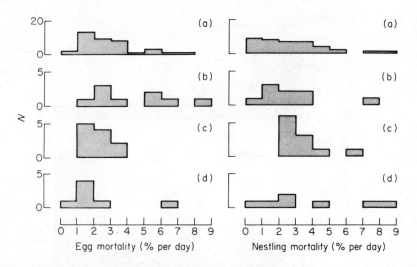

Figure 12.5 Frequency distributions of (left) eggs and (right) nestlings for small altricial land-birds breeding in (a) temperate zone; (b) humid tropics; (c) arid tropics; and (d) arctic regions. From data in Ricklefs (1969a).

(Figure 12.5). When all sources of mortality are considered, birds breeding in tropical woods and scrub are less successful than in Temperate Zone conditions. Birds breeding in tropical forests encounter a very different spectrum of mortality factors than do birds breeding in temperate woodland (Skutch, 1949, 1976). The few arboreal predators of temperate regions – mostly crows and squirrels – are joined in the tropics by specialist nest robbers such as toucans and by a wide variety of monkeys, snakes, and other tree-dwellers, so it is not surprising to find this greater mortality in the tropics. The bulk of this additional mortality must fall on eggs (Figure 12.5) since, as already noted, the 'general mortality' component does not differ between regions.

In arid regions breeding is usually induced by the onset of the rainy season, the birds exploiting the sudden flush of arthropod production brought on by the sprouting of the desert vegetation. However, the ending of the rains is quite unpredictable, particularly in the most arid regions, and may occur before the nest cycle is complete. Nestling mortality through desertion or through starvation is therefore likely to be more severe than with eggs (Figure 12.5).

The nests of Arctic species enjoy greater success than do Temperate Zone species, averaging only two-thirds the mortality of mid-latitude species (Figure 12.5). The difference is more pronouncd for egg survival than for nestling survival. Predation is generally low in the Arctic, at least where small birds are concerned, and indeed needs to be, since the majority of species nest on the ground at these latitudes. Consequently, the relatively poor success of Arctic nestlings must be due either to exposure or to starvation. In a few species predation is nevertheless important: over 72 per cent of eggs and young lost from Lapland Longspur nests near Barrow, Alaska, over a 7 year study (Custer and Pitelka, 1977) fell to predators.

COMPONENTS OF EGG MORTALITY

Hatching failure

Hatching failure can be due to infertility, to death of the embryo, or to death whilst hatching, and in Ricklefs' analysis occurred with 8.1 per cent of eggs. Embryonic abnormalities are responsible for a proportion of hatching failures and are induced by a variety of factors, notably by environmental pollution (below). Some abnormalities are induced by egg chilling prior to incubation. Twinned avian embryos occur only at low frequency – in three cases out of 1153 domestic hen eggs examined by Olsen and Haynes (1948), in only one of 833 waterfowl eggs examined by Kear (1966c), and in none of approximately 2000 unhatched Mallard eggs reported by Batt *et al.* (1975) — but in Mallard eggs held experimentally at low temperatures five cases of twinning occurred amongst 251 hatching failures (Batt *et al.*, 1975). Sturkie (1946) increased the frequency of twinning experimentally by exposing laying domestic hens to temperatures inducing hypothermia. Hatching failures are therefore likely to rise whenever the sitting adult is unable to maintain an adequate incubation regime, particularly in the later stages of incubation.

In a few cases egg mortality in a variety of northern wildfowl and passerines has been linked to high summer temperatures, e.g. Hilden (1964).

Predation

Predation is by far the most important mortality factor for small birds, both for egg losses and for nestling deaths. Ricklefs (1969a) calculated for six common North American passerines that 55 per cent of all eggs and 66 per cent of all nestlings were lost through predation and Lack (1954) estimated that three-quarters of open-nest mortality in Britain is due to such losses. Predation (and the diversity of predators) increases towards the equator and has been vigorously advocated by Skutch (1949) as being the limiting factor to clutch size in tropical regions. Perrins (1977) has reviewed the theoretical basis for this belief. At its simplest, the argument is that large clutches take longer to lay and therefore expose the nest to the risk of predation for longer. In addition, for altricial species the larger brood may require greater frequency of feeding visits by the parents, thus revealing the nest more readily to a watching predator. The idea that the threat of predation limits clutch size also offers an explanation of the latitudinal increase in clutch size observed of nocturnal species. Other explanations based on longer daylength available for foraging at high latitudes fail to account for this.

Competitive egg destruction

Interference competition between species is a significant source of egg loss for some species. When Short-billed Marsh Wrens were experimentally offered nests of a variety of marsh-dwelling species, they responded positively to all conspecific nests and in varying degree to those of the other species (Table 12.1). In 20 of 35 trials they also attacked the eggs, breaking or removing 16 of the 21 eggs offered in the 20 trials. Only unusually large eggs (Teal eggs) offered experimentally had the strength to withstand their attacks. Picman and Picman (1980) suggest that the wrens were removing potential competitors by this behaviour. Predation in its strict sense, i.e. for food, was not the explanation since in only one case were any of the eggs eaten. Moreover, egg attacks were always elicited by conspecific nests (though in a very small sample) and only sometimes by other species' nests (Table 12.1).

Nest desertion

Nest desertion accounts for 0–11 per cent of the egg losses recorded in a variety of studies and constituted between 0 and 29 per cent of these losses (Table 12.2). Desertion of young is much less frequent an occurrence, going unreported altogether in many studies and accounting for a maximum of 8.1 per cent of nests (23.0 per cent of nestling losses) (Table 12.2). Nests are probably most commonly deserted if disturbed by a predator or if the nest has been partially robbed by a predator, because of the probability that the predator will return. But brood desertion may also occur as a response to food shortage (Clark and Gabaldon, 1979). In a year in which Piñon Jays bred unusually early and at low temperatures, nests left with small broods as a result of removal of nest-mates for experiments were deserted by the adults; in normal conditions this does not happen with this species. Raising a small

Table 12.1 Incidence of nest attacks by Short-billed Marsh Wrens, *Cistothorus platensis*, offered strange nests in the vicinity of their courtship centres. After Picman and Picman (1980)

Type of nest	Nests offered	Nests attacked	Percentage attacked (%)
Yellow-headed Blackbird, *Xanthocephalus xanthocephalus*	5	1	20
Long-billed Marsh Wren, *Cistothorus palustris*	5	3	60
Red-winged Blackbird, *Agelaius pheoniceus*	23	19	83
Short-billed Marsh Wren, *Cistothorus platensis*	5	5	100

Table 12.2 Incidence of nest desertion in various passerine species

Species	At egg stage		At nesting stage		Source
	individuals (%)	losses (%)	individuals (%)	losses (%)	
Eastern Bluebird. *Siala sialis*	4.5	19.2	0.0	0.0	Thomas (1946)
Prothonotary Warbler, *Protonotaria citrea*	5.8	9.3	0.0	0.0	Walkinshaw (1953), for Michigan
	11.0	28.6	0.0	0.0	Walkinshaw (1963), for Tennessee
Yellow-headed Blackbird, *Xanthocephalus xanthocephalus*	4.3	8.2	0.0	0.0	H. Young (1963)
Red-winged Blackbird, *Agelaius phoeniceus*	1.4	5.5	0.0	0.0	Smith (1943)
	0.9	2.9	0.0	0.0	Smith (1943)
	3.6	6.9	0.5	0.9	H. Young (1963)
Purple Grackle, *Quiscalus quiscula*	2.1	7.6	8.1	23.0	Peterson and Young (1950)
Song Sparrow, *Melospiza melodia*	0.0	0.0	0.0	0.0	Nice (1937)
Savannah Sparrow, *Passerculus sandwichensis*	8.6	15.1	0.6	2.5	Dixon (1978)

brood may be energetically expensive since the heat conservation properties of a large brood are lost and the young must be brooded to an older age than needed with a brood of normal size (Chapter 8). Yarbrough (1970) found that broods of fewer than three Gray-crowned Rosy Finches never fledged young and pointed out that a lower limit to brood size could be adaptive in cold climates. If, as in the Piñon Jay study, the brood size is reduced suddenly (for whatever reason), it might well be less costly to abandon the current nesting attempt and to expend the reproductive effort otherwise required on a fresh attempt with a larger brood. Indeed, a species suitably equipped can recover part of the energy invested in the brood being abandoned by cannabalizing it. Balda and Bateman (1976) record such behaviour of Piñon Jays whose nesting attempts were overtaken by a severe snowstorm. Nest desertion in the face of declining food supplies has also been documented of hummingbirds (Calder, 1973): live young were abandoned once the flowers in meadows about the nest began to fade, making foraging more difficult.

In Canada Geese nesting at McConnell River, North West Territories, egg losses and desertions were influenced greatly by female body condition. MacInnes et al. (1974) suggest that high Arctic geese populations normally minimize the body reserves held back for incubation in favour of an extra egg in the clutch, producing extra young in most years at the cost of finding themselves unable to complete the incubation process and having to desert in the most severe years.

WEATHER AND BROOD SURVIVAL

In precocial species brood survival is decisively dependent on the weather, either directly through exposure or indirectly through effects on food supply. So marked are these effects that, for example, Arctic Terns in districts with severe weather rarely rear more than one young per brood, even though in the southern parts of their range broods of one and of two young are equally successful (Lemmetyinen, 1972). The youngest broods may be particularly vulnerable. Cold weather intensifies the need for feeding and at the same time reduces the availability of animal food. In diving ducks this requires the young to obtain proportionately more of their food underwater (instead of from vegetation as in fine weather) and results in a greater degree of down soaking and exposure mortality. Amongst seabirds weather conditions may alter the availability of food for plunge-divers, both by altering the flying ability of the adults and by altering sea surface conditions and thus the detectability of prey (E. K. Dunn, 1975). Such species are particularly vulnerable to storms and reproductive success may drop sharply in years with severe storms (E. C. Young, 1963; Wood, 1971). Sladen (1955) indicates that penguin chick mortality increases in periods of snowfall and storms in the Antarctic.

Young Partridge chicks are also critically sensitive to the influences of weather on their insect prey. A warm summer brings the timing of the major sawfly hatch into synchrony with the population peak of young chicks, thus improving chick survival, whilst a cold summer can delay the insect hatch until too late to serve the chicks' needs (Potts, 1973). In this species in Britain, summer temperatures have become exceptionally critical to chick survival over the last two decades because alternative foods – mostly arthropod pests of cereal crops – are largely destroyed by routine crop spraying.

Altricial land-birds are generally not as strongly limited by weather as are precocial species. As already noted in relation to growth rates (Chapter 6), aerial insectivores have been shown to grow more slowly in cold wet weather and suffer higher losses in such conditions. In England annual mortality over a 20 year study of Spotted Flycatchers was high when nestlings were reared in cold cloudy conditions and low in warm sunny summers (O'Connor and Morgan, 1982). Sudden outbreaks of heavy rain can also lead to nest failure in various ways. Many small songbirds previously feeding young started re-nesting after an overnight storm in Southern England in May 1980, apparently because the rain washed practically all available invertebrates from the vegetation in which the birds forage (P. Davis, personal communication). Such storms can also lead to wash-outs of nests of riparian species such as Coot and Moorhen, though species with nests in taller vegetation such as reeds may escape.

Other natural disasters account for a small proportion of nest failures. Some 17 per cent of nests found in forbs in Best and Stauffer's (1980) study failed in this way, largely by being dislodged by growth of the support vegetation. In England a small proportion of Reed Warbler nests in *Phragmites* beds are lost when nests are built attached both to dead and to live reed stems: growth of the latter eventually tips the nest contents into the water below. Other nests may be dislodged by high winds (Stokes, 1950). Nest flooding may also be a significant source of nest mortality for species nesting in low-lying areas (Burger, 1977).

MORTALITY AND AGE

As young birds grow their physical capabilities increase and their risk of death consequently declines. Such improvement in survival partly reflects increased ability to cope with environmental vicissitudes. Thus, during a storm of 3 days duration at a Laughing Gull colony in New Jersey, many more small young died than did large young (Figure 12.6).

PARTIAL LOSS OF BROODS

Broods of altricial young are frequently lost completely, usually to predators but also to inclement weather. Precocial broods, on the other hand, are rarely wholly lost: Eltringham (1974), for example, found that only 26 per cent of Egyptian Goose broods lost all young whilst 23 per cent lost no young. For

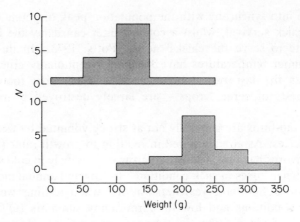

Figure 12.6 Weight distributions of Laughing Gull, *Larus atricilla*, chicks dying in (top) or surviving (bottom) a 3-day storm in a New Jersey colony, showing the greater susceptibility of young chicks. Data from Burger and Shisler (1980).

this species predation by Black Kites was probably responsible and broods generally kept near cover on the shore.

Amongst altricial species partial brood loss is largely (though not entirely (Finch, 1981)) due to starvation. Starvation of nestlings occurs surprisingly frequently among birds. Broadly speaking, two hypotheses are available to account for its occurrence. The first is that growing young impose an increasingly severe strain on their parents in meeting the energy and nutrient needs of the offspring (Chapter 6), and mortality consequently increases with nestling age. Figure 12.7 shows this pattern holds in Red-winged Blackbirds. The second hypothesis (O'Connor, 1978a) is that starvation mortality occurs as an adaptive process by which the adults reduce an over-large brood size to that appropriate to the prevailing food supply and was discussed in detail in Chapter 1.

Density-dependent increase in nestling mortality in part arises through greater competition for the available food supplies. The chances of a young altricial surviving the nest cycle depends, in at least some species, on the level of food available at the time. This has been shown experimentally by supplying additional food to breeding birds. In Sweden such supplementation of Nutcracker foods evoked a larger clutch size than normal, without reducing individual nest success (Lack, 1954). In Scotland Hooded Crows supplied with hen eggs and dead chicks as an additional food supply achieved greater hatching and greater fledging success with constant clutch size (Yom-Tov, 1974). Similarly, Northern Orioles in Manitoba fledged an average of 4.94 young per nest in 1976, the first year of an outbreak of forest tent caterpillars, but only 3.81 young per nest in 1977, when the number of breeding oriole

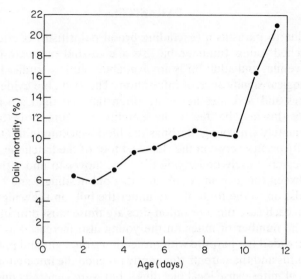

Figure 12.7 Daily mortality in relation to nestling age in Red-winged Blackbirds, *Agelaius phoeniceus*. Based on data in H. Young (1963).

pairs in the area had more than doubled, despite clutch size being similar in both years (Sealy, 1980). For such species territorial interactions at high population densities prevent access to as much food as is available in years of lower population, and reproductive success is consequently lower.

FLEDGLING LOSSES

Starvation of fledglings also occurs in some species, particularly where the method of feeding fledglings demands a greater parental energy expenditure than with nestlings (Royama, 1966a). Young Song Sparrows leave the nest at 9–11 days after hatching but remain within a few metres of the nest for the next few days. Smith (1978) found that feeding rates at this time were some 44 per cent higher than with nestlings 7–8 days old. Similarly, hand-reared Redwing young were found to consume more food once out of the nest than they did as nestlings (Tyrvainen, 1969). Fledglings may be under greater energy stress out of the nest than in because they can no longer save energy by huddling together in the nest (Chapter 8).

Predation of inexperienced young accounts for a significant proportion of post-fledging mortality. Indeed some raptorial species time their breeding season to coincide with the peak of summer fledging in local songbird populations. Thus Sparrowhawks were estimated to take 18 per cent of the juvenile Blue Tits and as much as 28 per cent of the juvenile Great Tits present during a period of 3 months (Perrins, 1979).

PARASITES

Mortality due to parasitism (excluding brood parasitism) is effectively confined to hatched young. Innumerable lists of ecto- and endoparasites found on nestling, juvenile, and adult birds are available, but few studies have focused on the ecological significance of infestation. The available evidence suggests that nestlings and juveniles are more often infested than are adults. With ectoparasites this may be due to the inability of young birds to preen their bodies adequately. In young Starlings the blood-sucking mite *Ornithonyssus bursa* initially occurs between the clenched toes of the hatchlings, feeding on the blood vessels closely underlying the toes, moves to under the wings and bill and between the growing quills in 5 day old nestlings, and finally (on 15 day old birds) on to the folds of skin under the bill, on the wings, and about the body. In each case the infestation sites are those most difficult to preen at the time. The number of mites on the young also increased as the feathers grew and provided the parasite with protection from light and preening. Both the young birds and the parents frequently preened the infected spots – heavy infestations of mites cause local bleeding – but were generally unsuccessful in eliminating the parasites. Heavily infested nestlings were hyperactive, thus losing energy wastefully: rapid quiescence following application of an acaricide showed that the extra activity was directly due to the presence of the mites (Powlesland, 1978).

Endoparasitic infections achieve considerably higher incidence than the 30 per cent found for the mite infestations just discussed and appear to hit juveniles more severely (Table 12.3). In the Starling infestations were most frequent and most intense and involved a greater diversity of parasites in the youngest juveniles (freshly out of the nests in late spring), but declined as the

Table 12.3 Seasonal variation in infestation by helminths in juvenile and adult Starlings, *Sturnus vulgaris*. From Owen and Pemberton (1962)

Incidence measure	April–June	July–September	October–December
Birds infected (%)			
Juveniles	100	100	97.1
Adults	100	93.8	86.8
Birds with three or more species (%)			
Juveniles	75.1	74.6	57.2
Adults	53.9	36.0	16.2
Birds with ten or more helminths (%)			
Juveniles	62.7	44.4	17.2
Adults	53.9	23.5	8.8

season progressed. The same decline was present in adults, showing that the infestation decrease in juveniles was influenced by season rather than by age of bird. However, as rates were always higher in juveniles the younger birds must have been in some ways more susceptible (Owen and Pemberton, 1962).

Infestation by parasites is itself probably only occasionally damaging to growing birds but becomes more harmful when in combination with other factors. Certainly in adult birds parasitism is more often fatal in conditions of stress: of 38 birds found dead and subjected to *post mortem* in two cold spells in Britain no less than 29 were parasitized in some form or another (Ash, 1957). Conversely, parasitism can weaken nestlings and leave them more vulnerable to other factors. In colonies of Purple Martins, adults at nests which were kept free of infestations of the blood-feeding martin mite *Dermanyssus prognephilus* through the application of acaricides reared more and heavier young than did birds in untreated colonies (Moss and Camin, 1970).

Blood-sucking parasites occasionally kill their nestling hosts directly: up to 30 individuals of the Mexican chicken bug *Haematosiphon inodorus* have been recorded on single nestlings of Prairie Falcons, in which whole broods are occasionally lost to the parasite (Platt, 1975).

Infestation of nestlings probably takes place most commonly via the nest material. Some parasites – for example, the hippoboscid lousefly *Craterina pallida* found on European Swifts and on various hirundines – over-winter as eggs, whilst others, such as the mite *Ornithonyssus bursa* already mentioned as a parasite of Starlings, are carried into the nests afresh each year, having over-wintered on the birds themselves (Powlesland, 1978). The incidence of parasitism by parasites residing in nest material can be reduced by parental sanitation activities: larvae artificially added to Great Tit nests were usually removed by the parents over the course of 1–3 days, particularly in smaller nests where the volume of nest material to be kept clean was small. In precocial species, on the other hand, this defence is not available to the birds, and it is notable that the Anseriformes are susceptible to parasitism by a great number of helminth species, with mature birds often bearing very heavy infections of cestodes. High concentrations of geese on the breeding grounds each spring lead to many of the secondary hosts of the cestodes – usually aquatic invertebrates – being infected. When the young geese hatch and move to the ponds to feed, they ingest the parasite in their food. Canada Geese goslings, for example, acquired their parasite burden principally in their first week of life: in one study 46 goslings yielded 14 species of parasite, including five protozoans, four nematodes, and two cestodes (Wehr and Herman, 1954). Once inside the gosling the parasite can breed rapidly: in one case a 2 week old Emperor Goose carried a tapeworm developing new segments at an average rate of two per hour. As a result the tapeworm can produce eggs before the young geese are ready to migrate and thus can increase parasitic egg dispersal (Schiller, 1954).

DISEASE

Little is known of the quantitative impact of disease on wild bird populations but certain forms affect young birds differentially. All bird poxes are host-modified and local bird populations may build up considerable immunity amongst adult birds. Young remain vulnerable until they acquire the immune response, and in particular years may be badly affected. In one study in South Sweden, up to 20–30 per cent of the yearling Woodpigeons examined were afflicted with pigeon pox (Ljunggren, 1971). Death rates were high whenever the lesions proliferated to the bill, damaging it and making feeding impossible.

MORTALITY AND ENVIRONMENTAL CONTAMINATION

Egg mortality is particularly high in areas significantly contaminated with organochlorine pesticide residues, since females feeding in such areas tend to produce unusually thin egg shells which fracture readily during incubation (Ratcliffe, 1967; Newton, 1979). Shell thinning is more marked in species at the top of the food chains than in those lower down, and is more marked the longer the food chain involved. Bird-eating and fish-eating species are more affected than are mammal-eating species. Experimental studies have shown that the degree of shell thinning and the amount of DDE (a residue of DDT) are positively correlated. Because thin shells fracture during incubation breeding success is greatly reduced in contaminated populations, and these in turn become unable to sustain their own numbers (Newton, 1979).

Table 12.4 summarizes other reproductive mortalities associated with organochlorine contamination. Unhatched eggs may be fertile but contain dead embryos killed by inadequate respiration across the affected egg shell or by the toxicity of the residues in the egg. Other eggs have failed to hatch as a result of abnormalities of parental behaviour brought about by the toxicity of the chemicals to the adults. In some cases this extends into the period of nestling care.

Experimental studies of DDE effects in wildfowl have shown that females on a contaminated diet lay eggs less likely to fledge young than do females on an uncontaminated diet (Stickel, 1973). In contrast to wildfowl and raptors, experiments with gallinaceous birds suggest a low susceptibility to shell thinning. Nevertheless, in domestic hens very high doses of pesticides result in fewer eggs, in reduced hatching success, and in poor chick survival (Stickel, 1973).

Nestlings and chicks exposed to environmental pollution show a variety of development abnormalities, often involving bill or limb malformations or the loss of plumage elements. Such casualties are unusual but may locally be significant and traceable to specific contaminants (Gochfeld, 1975). For example, young terns on Long Island showing feather loss had higher levels of mercury in their bodies than had normal chicks of the same age.

Table 12.4 Incidence of reproductive mortalities in captive species[a] fed on diets containing organochlorine compounds.[b] Based on Newton (1979)

	Number of species tested	Percentage showing positive response (%)
Mortality factors		
Infertility	10	50
Reduced egg production	15	88
Egg breakage	4	75
Reduced hatchability	17	82
Increased mortality of young	12	83
Mortality correlates		
Low weight eggs	3	67
Shell thinning	19	53

[a] Species were variously American Kestrel, Screech Owl, Mallard, Domestic Chicken, Pheasant, Bobwhite Quail, Japanese Quail, Partridge, Crowned Guinea Fowl, Ringed Turtle Dove, and Bengalese Finch.
[b] Compounds were variously DDE/DDT, Heod (dieldrin), and PCB (polychlorinated biphenyls)

Juvenile birds accumulate pesticide burdens from their food and these rapidly reach adult levels. In juvenile Herring Gulls fed on polluted fish, the maximum pesticide burdens occurred as the winter fat deposits declined and released the chemicals into the circulation. The lethality of several pesticides is greatest when in circulation, so juveniles with low fat deposits through inexperienced foraging or because of subordinate status are more likely to build up fatal concentrations in their blood than are adults.

SEX DIFFERENCES IN MORTALITY

In adverse conditions sexual differences in survival rates are present even amongst nestlings. Dhondt (1970) found that Great Tit nests in urban habitats in Belgium contained proportionately more (54.6 per cent) males than females 15 days after hatching. The difference was correlated with rearing success, for in suburban and woodland habitats in which rearing success was better (66.9 per cent nestling survival against 57.5 per cent in urban areas) males and females were equally represented among the 15 day old nestlings. Dhondt found similar links between a prepondrance of males and rearing success on a seasonal basis, and concluded that the larger males were able to out-compete small females for food brought to the nest by the adults. Garnett (1981) showed for this species that full-grown males dominate females for food in winter, succeeding by virtue of their larger size. One consequence of this higher female mortality from birth is a relative scarcity of breeding females: at Oxford nearly one-third of the first-year males are unable to find an unmated female (Bulmer and Perrins, 1973).

In the Spanish Sparrow the 'runts' within each brood tend to be female; being thus underweight, females are more likely to die (Gavrilov, 1968). Sexual differences in nestling weight associated with sexual dimorphism in such species as the Hen Harrier (Scharf and Balfour, 1971), the European Sparrowhawk (Newton, 1979), and the Red-winged Blackbird (Cronmiller and Thompson, 1980) are, however, more generally independent of nestling mortality. Indeed, it has been argued that the more rapid growth of the smaller male Sparrowhawks is an adaptation permitting them to fledge before experiencing severe sibling competition from the females (Newton, 1979).

SUMMARY

Interspecific variation in mortality experienced by young as eggs and nestlings originates in differences between species as to (1) the general level of mortality of all types, (2) the relative rates of mortality of eggs and mortality of young, and (3) the incidence of partial loss of clutches and of broods within nests. Mortality decreases with adult body size, probably due to decreased predation on the nests of larger species. Mortality also varies with latitude, principally in relation to egg losses. Predation is the major cause of losses of eggs, of nestlings, and of fledglings, but eggs are also lost to hatching failure (usually associated with inadequacies of incubation or with environmental pollution), nest competition (especially in hole-nesting species), and nest desertion (often following either nest disturbance by predators or energy deficits in sitting females). Nestling mortalities are generally lower than with eggs but partial brood loss, usually due to starvation, is a significant additional cause of death in altricial young. Mortality due to adverse weather falls especially heavily on precocial young who cannot feed and be brooded simultaneously. As birds grow their risk of dying decreases markedly, except where species-specific behaviour exposes them to new risks, but in some species parents may be unable to feed dispersed fledglings as adequately as they can a brood of nestlings. Disease and parasite infections may be more important mortality factors for young birds than is generally realized. Finally, in some species one sex, probably most often the female, may suffer greater mortality than does the other, due to sex differences in behaviour or in physiology.

CHAPTER 13

Instinct and learning

In newly-hatched birds some behaviours appear in what seems to be an almost instinctive manner, without the young having had any opportunity to learn or to practise the behaviour. Thus neonatal altricials will beg, young gull chicks will peck at the adult's bill, and young Kittiwakes and other species with high nests grasp the nest lining when they are tipped or lifted. Such behaviours appear to be innate and clearly must have a genetic base though, as pointed out elsewhere in this book, embryos can be stimulated by their environment even whilst in the egg. Once out of the egg, the continuing development of behaviour depends on the interplay of the genotype with the physical maturation of the young and with the stimulation and opportunities for learning provided by the young bird's environment (including siblings and parents). In this chapter we consider some of the principal developments in nestling and chick behaviour.

PECKING RESPONSES

Many young birds, especially of precocial species, have an innate pecking response which they direct at small objects in the environment and which presumably aids them in the ingestion of food. Studies of this response have been significant in advancing our knowledge of how feeding behaviour develops. Initially this pecking is directed at a variety of objects, but the response to inedible objects gradually decreases with age (Figure 13.1). The response also narrows down to objects in a size range suited to the chicks and their bill size; e.g. Curtius (1954) found that Lapwing chicks took objects of 1 mm size, domestic fowl objects to 2.5 mm, and Turkeys objects of 5 mm, in each case preferentially from a range of sizes.

Solidity cues from targets are likely to be the most general of cues eliciting pecking from a visual omnivorous feeder. Dawkins (1968) offered White Leghorn chicks a choice between a solid hemisphere and a flat disc and found significant preference for the more solid stimulus. Moreover, chicks reared in complete darkness until testing showed the solidity preference immediately, as did chicks with one eye temporarily covered by an eye-patch. The chicks thus had some innate ability to identify the hemispheres as solid. Dawkins was able to show that this ability was to recognize surface lighting cues in

209

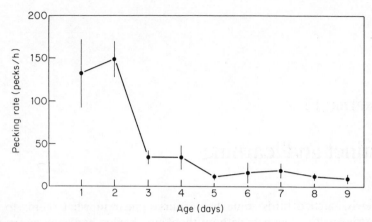

Figure 13.1 Pecking rate of Indian Peafowl, *Pavo cristatus*, chicks during their first 9 days of life, when presented with inedible test objects. Based on data in Dilger and Wallen (1966).

much the same way as humans do: almost all chicks responded to a photograph of a solid (half a ping-pong ball lit from above) when the photograph was oriented 'correctly', i.e. with the lighter half on top. Were this recognition of surface-shading cues learned, chicks reared with under-floor lighting should have treated the inverted photograph as solid: in fact, they made the same choice as normally reared chicks, showing the innate nature of shading recognition.

Newly-hatched Herring Gulls peck at the tip of their parents' bills, apparently directing their pecking at the red spot present on the otherwise yellow bill. This apparently instinctive behaviour has been used in one of the classic studies of what induces young birds to beg (Tinbergen and Perdeck, 1951). The almost mechanical response of the gull chicks to test stimuli (parental head models) allowed Tinbergen to show that the chick behaves as if it perceives only certain aspects of its situation. It ignores, for instance, the details of the head of the parent in the natural situation, and processes only certain highly relevant and normally adequate stimulus characteristics – bill shape, colour, contrast, and motion. The effect of pecking in response to these characteristics is, in the absence of experimenting scientists, to strike the tip of the parent's beak and induce the adult to regurgitate a food mass for the chick.

Tingerben was impressed by the stereotypy of the pecking response and by how it was present in all gull chicks from soon after hatching, with little opportunity for any learning of the behaviour. He therefore regarded it as an example of purely instinctive behaviour. Hailman (1967) re-examined the same behaviour in Laughing Gull chicks and detected significant variability between chicks. These chicks also improved in the accuracy and efficiency of the pecking as they grew and they displayed greater discrimination between parental models with age. Hailman suggested three conclusions. First, older chicks are more accurate in their pecking because they have stronger leg muscles that allow them to stand more steadily. Second, they are more accurate in judging the distance to the parents' bills because their initial pecking pitches them on their faces if delivered from too far away and on their backs if delivered too forcefully from too near, thus creating punishments for pecking incorrectly. Finally, they become more discriminating following learned associations of food and parental features.

FOOD RECOGNITION

How do young birds approach a completely novel situation, such as an encounter with a hitherto un-met food source? Several studies have found that various birds may be very hesitant to sample a new food (Hogan, 1965; Coppinger, 1969), but it appears that the avoidance may be moderated if the food shares characteristics of previously encountered cues. Shettleworth (1972) found that 9–10 day old chicks learned to avoid drinking quinine-flavoured water faster if the water colour was unfamiliar, and learned to consume palatable water faster if its colour was familiar, and conversely.

The foods to which a bird becomes familiar in early life in this way tend to be preferred again when a choice is available in later life (Rabinowitch, 1968, 1969). This is particularly true of species with rather restricted or specialist diets. To test for a finite period over which such preferences are formed Rabinowitch (1969) reared Zebra Finches for various periods of time on one of three diets: (a) white millet, (b) red millet, and (c) canary seed. In the wild, Zebra Finches are reared in the nest for about 14 days, being fed half-ripe and ripe grass seeds by regurgitation from the parent's crop. After fledging they are fed by the parents for a further 14 days. In the aviary experiments to be described, the food preferences developed over this period as fledglings proved to be crucial. In the experiments the captive adults had access only to the training or to the tests seeds for food on which to rear the young under test. Initial testing showed that the millets were preferred to canary seed in the absence of training, with little difference between the two millets. However, if the birds were reared on only one seed – millet or canary seed – for their first 5 weeks, they developed a preference for that seed. If now exposed for a further 16 weeks to a different seed or to a mixture of the three seeds, this preference declined gradually in most individuals, though a few birds retained the initial preference. Since Zebra Finches become sexually mature at 3 months, this is obviously of some importance in determining their adult diet. Rabinowitch finally conducted some shorter term experiments to test whether it was nestling or fledgling experience that determined the observed preferences. When young fledged from each nest, he changed the seeds placed in that cage and tested the young for preferences on becoming independent. In each case the birds preferred the food eaten as fledglings.

Table 13.1   Seed preferences of Zebra Finches, *Taeniopygia castanotis*, reared on one seed as nestlings, a second seed as fledglings, and a third 'neutral' seed for 2 weeks before testing. Data from Rabinowitch (1969)

Rearing regimen[a,b]				Preferences shown were for		
Nestling food	Fledgling food	Pre-test food	N	Nestling food	Fledgling food[d]	Random choice
Canary seed	Millet W	Millet R	4	0	4	0
Millet R	Canary seed	Millet W	5	2*	1	2
Millet W	Canary seed	Millet R	4	2	2	0

[a] R and W indicate red and white millet respectively.
[b] All periods were of 2 weeks duration.
[c] Preferences were individually significant at $P < 0.05$ except for one case (*) significant at $P < 0.01$.
[d] Preferences were individually significant at $P < 0.01$.

However, these foods were also those they had more recently eaten, so he placed the birds on a third seed for a further 2 weeks and re-tested them (Table 13.1). In most cases the birds maintained a preference for their fledgling diet, but a few fed on millet as nestlings reverted to it in the preference tests. The table shows, though, that these birds spent their fifth and sixth weeks (the pre-test period) on millet of the alternative colour, so that these anomalies were probably a 'recency' effect. (The ideal experiment would have included the sequences nestling food red millet, fledgling food white millet, pre-test food canary seed and its interchange white/red/canary seed). The importance of fledgling experience in determining subsequent food preferences suggested by these experiments is confirmed by studies of 11 species of insectivorous birds by Blagosklonov (1976), who found that experience in the transition period between nestling life and independence was critical.

Domestic chicks are especially likely to accept new stimuli as potential sources of food on their third or fourth day, i.e. about the time they first forage after depletion of their yolk sacs. In the experiment summarized in Figure 13.2, chicks of various ages received a food reward in association with an experimental stimulus. When tested 3 days later, but this time without reward for a response to the stimulus, most chicks quickly lost interest in it. However, those that were 3 or 4 days old when they first met the rewarded stimulus were more persistent in their (unrewarded) pecking activity. Thus a lasting preference for cues to potential food sources had formed at a very critical time – that of yolk depletion – for the chicks. Of course, this finding does not preclude incorporation of new items into the diet at later ages, provided that the reward rates for responding to the cues concerned are sustained.

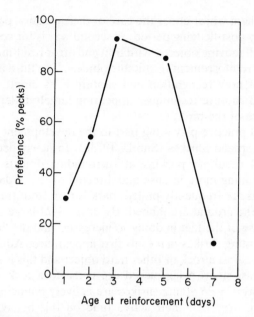

Figure 13.2 Retention by domestic chicks, *Gallus domesticus*, of an unrewarded pecking stimulus preference (established 3 days earlier by reinforcement with food rewards) in relation to the age of the chick at the time the preference was initially established. Based on Hess (1962).

Similar interaction between innate capabilities and learning occurs in respect of food selection in some species, particularly those with specialized foods. Piñon Jays specialize on seeds of the pine *Pinus edulis*, some of whose cones contain little nourishment as a result of disease or insect attack. Time spent opening bad rather than good seeds is thus wasted and Piñon Jays show considerable accuracy in discriminating against the former. Ligon and Martin (1974) investigated the basis of this ability by comparing the behaviour of hand-reared juveniles and yearlings with that of wild-caught adults. Inexperienced birds showed an immediate preference for piñon seeds over other objects, but distinguishing good seeds from bad was a learned ability. The discrimination was eventually made on the basis of a combination of visual, tactile (possibly weight), and auditory cues, the latter through an innate 'bill-clicking' of the seeds. The integration of these cues depended on experience and discrimination therefore improved with increasing practice.

Prey recognition by carnivorous young

Predatory birds taking live prey can come to recognize and kill prey in one of three ways (Mueller, 1974). First, they may learn to identify suitable prey and how to attack them whilst under the care of their parents, particularly during the immediately post-fledging period. A second way is for young predators to pursue any small moving objects, with trial and error resulting in perfection of techniques and reinforcement of suitable objects. The third and final mode of development of prey recognition and capture is as innate behaviour, with recognition and capture techniques appearing largely independently of any prior experience of the bird.

Learning and practice play a big part in the development of prey catching by young Loggerhead Shrikes (Smith, 1973a). In newly-fledged young (age about 18 days) desultory pecking at surrounding objects is apparent immediately, becoming more intense and directed over 3–4 days. At this stage leaves and twigs are frequently pulled, bark is torn from trees, and contrasting objects on the ground are jabbed. By days 23–24 large objects are being manipulated, use of the feet in doing so increases, and the 'aim–flutter–peck' sequence of hunting shrikes makes its first appearance. Adults typically beat captured prey against a rock or other hard object and this is seen in fledglings 23 days old, grass stems being picked up, taken to a rock and beaten, then dropped. By days 25–26 young shrikes are actively pouncing on grasses and sticks, spending over half their active time on this behaviour (which first appeared only 2 days earlier!). Pecking at twigs and leaves becomes scarcer. On day 28 instances of young swooping down from a high perch to attack ground arthropods become common (and even successful) and by day 30 (12 days after fledging) most hunting time is directed at live prey, mainly in typical shrike 'sit and wait' style. The young are still fed by adults but regularly catch grasshoppers and crickets for themselves. Finally, on days 37–38 chasing of small birds appears for the first time (Smith, 1973a).

Catching dangerous prey involves risk to the young bird, so the techniques for doing so are likely to be innately determined. However, even small mammals may bite or scratch an attacking bird. Table 13.2 summarizes a series of experiments with hand-reared American Kestrels, experiments which suggest innate recognition and attack of mice by this species. Only one bird responded to paper models and then only to moving models of mice (one 'played' with the tissue ball). All nine birds attacked the live mice, in two cases within 10 seconds of first sight. These attacks were intense and well oriented, with the mouse being grasped by the head, neck, or thorax and biting directed at the head. All birds achieved expert performance within six trials, suggesting that experience plays only a minor role in recognition and killing of mice by this species (Mueller, 1974).

Comparative studies of prey attack in different species indicate that innate components are present whenever the young bird would be at risk from the prey or where the prey item might regularly escape. In three omnivorous species with 'safe' prey – Blue Jay, Gray Catbird, and Black-capped Chickadee – a plain model in motion was attacked at the trailing edge. In contrast, Loggerhead Shrikes presented with moving models invariably attacked the leading third, normally the head end and therefore that at which the prey is most easily disabled. Young shrikes were even noted stepping

Table 13.2   Respnses by naïve American Kestrels, *Falco sparverius*, to various model and live prey presented to them. Nine individual birds were tested, though not equally with all models. Summarized from Mueller (1974)

Model or prey	Percentage attacked when	
	Stationary	Moving
Tissue ball	2	0
'Tissue' mouse	0	15
Stuffed mouse	13	11
Dead mouse	23	33
Live mouse	—	87

round the moving models presented so as to attack the 'head' end. The shrikes were able to kill live mice efficiently in a precise stereotyped way (attack to the back of the neck) from first exposure, but some maturation of response led to improvement in the orientation of attacks to the head through the birds' fourth week out of the nest (Smith, 1973a). Very similar behaviour was shown by hand-reared Turquoise-browed Motmots, a 'sit and wait' predator of lizards, small snakes, and various insects. Tests with models showed that movement was by far the most highly directing cue used by the motmots, though 'head' markings were also responded to. Again, therefore, attacks were concentrated where they were most likely to disable the prey (Smith, 1976). It is worth noting that lizard prey attacked at the tail could probably escape by shedding the tail, so the young motmots do best, even with harmless species, by attacking the head.

Finally, one other innate response by motmots is remarkable. Naive young presented with various plain-coloured models were not inhibited from attacking them. But when presented with a model with yellow and red rings along it they were much alarmed and flew to the remote parts of their cage, none attempting to attack the model. The pattern closely resembled that of the rather dangerous coral snake and the responses by the naive motmots suggested they possessed an innate recognition of the snake. Coral snakes are rather secretive and during the day might be met partly exposed amongst forest litter. A motmot attacking the exposed part would be in danger of being fatally bitten, most so if the part were not the head (Smith, 1975).

Tutoring by adults

Some observations suggest that adults may actively join in tutoring their young, especially where the techniques involved are difficult to learn. LeCroy (1972) observed adult Common Terns and Roseate Terns fishing in late September, with each accompanied by a juvenile. The latter closely followed the adults in their manœuvres but never plunged into the water. Amongst LeCroy's observations were an adult catching a fish and flying up to its young, to drop the fish in front of it and catch it again before it had fallen more than a few feet, and another sighting of an adult passing a fish to a juvenile under the water surface whilst both swam on the sea. Since tern chicks taken into captivity do not overtly recognize fish swimming in a bowl put before them but will eat dead fish from the bowl, it seems that learning to forage is a two-stage process for the young. First, they must learn to recognize live fish as prey, to which end LeCroy's observations are pertinent. Secondly, they must learn to use the appropriate capture techniques. The high post-fledging mortality of species such as terns and Kittiwake suggests that these birds have severe problems learning how to catch food for themselves, so this parental tutoring is vitally important.

RECOGNITION OF PREDATORS

Recognition of avian predators for what they are is of no small importance to young birds, but the processes involved are not fully understood. Some birds respond to the visual stimuli presented by an owl at first exposure to it and may mob it (Nice and ter Pelkwijk, 1941). Experiments involving the presentation of model owls to naive young birds show that significant characteristics of the model in eliciting mobbing include the general outline and colour pattern and the presence of forward-looking eyes and a beak. Response to these characteristics is obviously adaptive but the results do not prove that owl-like features are crucial *per se*, for the birds tested were reared with siblings present. Thus, they could have become familiar with their own form and attacked the models subsequently in response to divergence by the model from significant cues in recognizing siblings. As we saw earlier, familiarity is a major component of stimulus acceptability for young birds.

Much the same problem arose in experiments on the escape response of birds to a hawk flying overhead. Escape is elicited by model hawks if they possess three shape features: wings, a short neck, and a long tail. Lorenz (1939) found that young geese exhibited escape behaviour when a hawk model was moved across their pen, but showed no concern when the same model was flown backwards over the pen. Lorenz noted the resemblance of the hawk model when flown backwards to a flying goose with long outstretched neck and short tail, and suggested that the young geese were innately equipped with predator recognition ability. Schleidt (1961a,b) resolved the controversy that this suggestion caused. Naive Turkeys exposed to objects of approximately hawk size and rate of movement gave strong escape responses, irrespective of the shape of the model. But birds then repeatedly exposed to these models habituated to them but responded strongly to the presentation of one of the less familiar models. It therefore appears that young birds habituate to other species of birds flying overhead and respond with escape behaviour to the unfamiliarity of their rare encounters with a hawk.

For young birds these responses would be enhanced by the alarm behaviour of nearby adults, not necessarily their parents or even conspecifics. Many species of the African Turdinae are notable for the frequency and accuracy with which they imitate the calls of other species. When potential predators intrude in the vicinity of nestlings or attended fledglings the adults intersperse their own basic notes with the alarm calls of other species, presumably teaching the young to react appropriately to early warning by unrelated species (Oatley, 1971). Predators of birds are relatively numerous in Africa, so that any added awareness of danger could be expected to be beneficial, particularly during the critical early months. Amongst Temperate Zone species the same outcome is thought to have been evolved through convergence of alarm call structure (Marler, 1959): quite unrelated species sharing the same set of predators have alarm and 'mobbing' calls with very similar characteristics (themselves believed to reflect functional requirements such as difficulty of location and ease of location respectively).

Juveniles and fledglings which do not quickly learn their own predators are unlikely to survive, but behaviour in the face of nest predators can become more efficient with age more slowly. Hedgehogs, for example, do not kill full-grown gulls but they are a threat to eggs and to small young, e.g. in Black-headed Gulls (Kruuk, 1964). Most first-time breeders are therefore unlikely to have experienced hedgehogs as a threat and have to learn of them as such. One way to reduce the cost of acquiring such experience is to make use of the experiences of conspecifics. When a predator such as a stoat enters a breeding colony of gulls or a hedgehog approaches a gull nest, many birds may gather above the predator without attacking it. Kruuk (1976) has analysed this behaviour experimentally in gulls, using model predators to elicit the behaviour. His results (Figure 13.3) show that gulls seeing a dead conspecific apparently taken by a model stoat displayed greater wariness than was apparent on sighting the stoat alone: after such a 'kill' the birds continued to avoid the predator to a greater extent than earlier. Young inexperienced birds can thus learn by observation the threat constituted by the predator.

LEARNING TO BUILD NESTS

In general, the more complex a pattern of avian behaviour the more likely it is that its development involves significant learning by experience. The Village Weaverbird builds an elaborate domed nest with discrete stages of construction on each of which the successful completion of later stages depend. Nests built by first-year birds are often crude in appearance due to lack of practice in selecting and preparing nest materials, but getting even that far depends on obtaining practice in weaving (Collias and Collias, 1973). Complete deprivation of nest materials for the first year of life in a captive colony significantly retarded the development of weaving ability by yearlings, but prolonged deprivation during adult life had no effect on males who had learned to weave in their first or second years. Much practice with nest materials takes place in

218

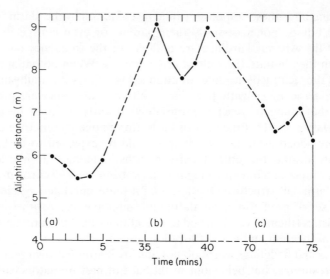

Figure 13.3 Alighting distances of breeding Herring Gulls, *Larus argentatus*, and Lesser Black-backed Gulls, *Larus fuscus*, in relation to models of stoats presented (a) alone, (b) with a dead gull beside it, and (c) alone but following exposure of the gulls to the 'stoat plus gull' situation. Redrawn from Kruuk (1976).

the first 7 weeks of a male's life, when the juveniles play with and mandibulate such materials. The effects of this practice were linked to the young bird's position in the dominance hierarchy prevailing. Dominant birds in the aviary often inhibited attempts to weave nests by subordinates, either chasing them from good sources of nest material or from good nest sites, so that the amount of practice a young bird got depended on its social rank.

The extent to which nest-building behaviour is learned or inherited has been particularly well studied in the lovebirds of the genus *Agapornis* (Dilger, 1962). The nest-building behaviour differs between species. Female *A. roseicollis* carry strips of suitable material (paper, bark, leaves) tucked amongst the feathers of the lower back or rump, carrying several strips there at one time. Female *A. fischeri* use similar materials supplemented by twigs but carry each item one at a time in the bill. Hybrids of the two species almost invariably attempt to tuck material into the feathers but fail frequently: the female fails to let go of the piece just tucked, holds the piece at an inappropriate position, attempts to tuck the material elsewhere than on back or rump, switches from tucking to preening in mid-stream, or, most comically, attempts to get its bill near its rump by running backwards! The hybrids are successful only if they carry the material in the bill. Dilger's work thus shows that the various behaviours needed to collect nesting material were under genetic control but their integration was undone by the hybridization.

With prolonged practice the hybrids eventually learned to carry the material only in the bill, but even then the flight to the nest site was often preceded by the bird turning its head towards its rump.

ENVIRONMENTAL MODIFICATION OF INNATE ABILITIES

Although chicks are innately able to perform elementary behaviours (such as distress calling in precocial young or gaping in altricial young) from birth, the details of their development turn on the subsequent experiences of the young birds. Hoffman *et al.* (1966) devised an experiment to demonstrate such effects in distress calling by ducklings. The experimental arrangement presented 5 second exposures of a mother-surrogate to two ducklings simultaneously but in response to calls of only one of them. In this way, both ducklings received identical patterns of stimulus presentation. By monitoring the calling rates of both birds it was possible to show a greater frequency of distress calling on the part of the bird whose calling actually determined the presentation schedule. This bird had therefore learned by experience of its ability to obtain feedback from the mother-surrogate.

A similar effect influences gaping by young (altricial) thrushes (Tinbergen, 1973). Young songbirds will gape readily when stimulated by the return of their parent or by any of the stimuli associated with such return, but older nestlings are more selective as to what they will respond to. The initial stimuli are mechanical in nature (a slight jarring of the nest, for example, suffices) and are oriented to gravity: that is, the nestling gapes vertically irrespective of the position of the parent. Older chicks begin to orient towards the adult's head, doing so 1–3 days after their eyes have opened. Tinbergen and Kuenen (1939) discovered a curious transitional phase at this stage of development in Blackbirds and Song Thrushes: nestlings whose eyes were open would gape on seeing a model but directed their begging vertically and not at the model itself. There must therefore be two 'control systems' for begging behaviour, one for its release, the other for its orientation, with the two not coming together until some time after each is independently functional.

COLOUR PREFERENCES

Experiments in which young birds are induced to peck at coloured spots show that some colours are preferred over others. This result might at first sight be related to the colour of the natural foods of precocial young, or to the colour of the parent's beak where young are fed. In fact, the colour preferences shown by chicks of closely related species prove to be similar, even when their foods are quite different (Kear, 1964). Amongst 41 species of the Anatidae tested, a preference for green was general, with some species and individuals pecking also at yellow. The tribes differed, however, in the colours they avoided, some ignoring blue, others red and orange. Pheasants also preferred green, but chicks of other species which peck the parent's beak in feeding – Herring and Lesser Black-backed Gulls, Moorhen, and Coot – all preferred

red or orange. Except for the Coot (which has a white bill), this is the colour on the beak of the adults. However, other anomalies for a parental feeding explanation also exist, for in the Wideawake Tern (which has a black bill) the preference for red is also apparent in the chicks (Cullen, 1962), and in the Magpie Goose the adult has a red beak and feeds its young but the latter prefer green (Kear, 1964).

A chick's preferences are determined in part by its embryonic environment, once the appropriate sense organs are functional. Wada and his colleagues have investigated experimentally the effects of colour exposure during incubation on the colour preferences of White Leghorn chicks (Wada et al., 1979). Eggs were incubated in darkness or under either red or green illumination for a particular part of the incubation period. Chicks incubated in darkness subsequently rested more readily (68 per cent of cases) under red than under green illumination, thus displaying a preference for the former. Chicks reared under red illumination subsequently showed an even stronger preference for red (78 per cent of cases) whilst chicks exposed to green lights during incubation later showed a bias away from red (only 42 per cent of cases) and towards green. Chick preferences were therefore modifiable by manipulating the embryo's environment. However, the timing of this manipulation was significant. Anatomical development of the visual system of the embryo is not complete until day 17 of incubation (Schifferli, 1948). Exposure of embryos to green light up to but not beyond day 18 had no effect on colour preferences, but if exposure continued to day 20 or was confined to a 24 hour period on days 19–20 the colour preference was shifted in favour of green. Thus the embryonic adjustment of colour preference depended on visual stimulation received once the visual system was complete.

AVIAN PLAY

In animals play can prepare the young for adult life in at least three important ways (Ficken, 1977): (1) it enhances muscular development; (2) it results in discoveries and finding out the consequences of the animal's own actions; and (3) it can adjust the social relationships needed later in life in a social species. Behavioural sequences during play are often broken off before completion or are low intensity actions. That this is so is especially important in aggressive and in sexual play. In play with objects, however, this is less important, the main function being to provide practice in areas requiring dexterity or judgement. Playing with objects is widespread amongst raptors, cormorants, pelicans, and many fish-eating birds, species with skilled foraging techniques (p. 183). Bildstein (1980) has shown that juvenile Hen Harriers 'play' with objects (corn cobs) of the same size spectrum as their usual vole prey, ignoring larger and smaller objects also available. Practice thus accrues to the muscles later used in manipulating captured voles.

Play is particularly well developed in young Ravens, a species noted for its initiative and learning powers. This is probably associated with it being the extreme of non-specialists: individual birds are highly variable in many behavioural patterns, with learning through play being at least partially responsible for their detailed adjustment to the requirements of their surroundings. Although Ravens spend a long time – 5 months – with their parents (followed by a period with other groups of young until the end of their third year), this play appears not to be of a social nature but relates to imitation of skills needed in non-social contexts, such as feeding (Ficken, 1977).

Other categories of play suggested to occur in birds are locomotory play and acoustic play. Young birds of various species have been recorded running or flying about their cages in ways reminiscent of predator escape reactions but without there being an overt stimulus for doing so. However, the true motivation of such behaviour is unknown. Similarly, it is questionable whether the 'practice' subsong of young birds is truly play (Ficken, 1977).

LEARNING FROM PEERS

When young birds spend their early life together they can learn from each other (Turner, 1964). Young from different broods may have different learning experiences – about food, about predators, etc. – so their collective knowledge is greater than that of any one individual. Hence membership of a family group or of a peer group may serve to modify a chick's behaviour to stimuli outside its own range of experience or preference. Experimental tests of imprinting (Chapter 10) show that individual ducklings respond in widely differing ways to presentations of a mother surrogate, some approaching but others ignoring the model. Tested together, however, the brood shows considerable behavioural synchrony, the group matching to those individuals with the strongest response tendencies (Klopfer and Gottlieb, 1962). This may extend subsequently into adult life, where social facilitation can be particularly influential in such activities as choice of feeding area and of food type, of roosting site, of breeding ground and of the timing of breeding, and even of migratory direction! Other experiments by Hailman (1967) on gull chicks suggest that naive chicks may learn to take new foods by responding to the feeding actions of their siblings. Indeed, Storey and Shapiro (1979) suggest that ducklings separated as a brood from the mother remain together rather than disperse to find the missing mother, though if separated as an individual each duckling attempts to locate the source of any maternal calls heard.

Tool-using is relatively rare in birds but one case provides valuable insight into the ontogeny of behaviour. On the Galapagos Islands juvenile *Cactospiza pallida* placed in captivity were provided with a variety of tools (twigs, tooth-picks, etc.) and situations in which to use them. Most of the birds hunted prey visually and resorted to tools once prey (or a likely feeding hole) was found to be beyond the reach of the beak. But one individual never probed with a tool in captivity and this individual was the only one of the six captives to have been isolated in the first 3 months after capture. This suggests that learning to use tools was at least partly observational, possibly with a critical period constraint also present (Millikan and Bowman, 1967).

HABITAT SELECTION

Although habitat selection by birds is obvious in the wild (Lack, 1971), little is known of the factors involved nor of the role of experience in determining the choice made by the young. In an early set of experiments Chipping Sparrows were hand-reared and subsequently tested for perch preferences (Klopfer, 1963). These young showed a preference for perching on pine leaves rather than on oak leaves, a preference also shown by wild-caught birds. But other young reared in the presence of oak foliage showed a reduced preference for pine. Thus a preference exists from hatching but it can be modified by experience. Other experiments (Klopfer, 1965, 1967b) showed that the extent of such preferences and the degree to which they might be modified by experience were variable from species to species.

Comparison of habitat selection by Blue Tits and Coal Tits has revealed some of the factors behind such stereotypy (Partridge, 1974, 1976b, 1979). Table 13.3 shows that naive Coal Tits had strong preferences for pine needles against oak leaves whilst naive Blue Tits had preferences for the oak foliage. These choices match those seen in the wild, Blue Tits living mainly in broad-leaved woods, Coal Tits in coniferous woods, and each species survives rather better in its preferred wood than in the other type. The two species have different feeding behaviours: the Blue Tit spends a lot of time using feeding techniques involving hammering, hacking, and tearing at food items, e.g. for insects in cones, whilst the Coal Tit feeds more by minute examination and gentle probing, using its feet to manipulate objects to a lesser extent. This is partly determined by morphological differences between the two species (Partridge, 1976a). It is thus possible that these titmice choose their

Table 13.3 Perch choice by Blue Tits, *Parus caeruleus*, and Coal Tits, *Parus ater*, in respect of Oak and Scots Pine. Compiled from Partridge (1974)

	Percentage on pine rather than on oak	
Substratum type	Blue Tit	Coal Tit
Whole branches	25.8	59.0
Branches only	45.0	62.3
Leaves only[a]	43.4	93.4

[a] Pegged in bundles on wooden perches.

habitat in relation to the physical constraints of their build – if you like, to where they feel most comfortable. To test this possibility, naive birds of both species were trained to hunt for food in the test devices illustrated in Figure 13.4. The various devices require different skills used in the wild but, unlike the natural situation, the possible reward rate could be made constant for

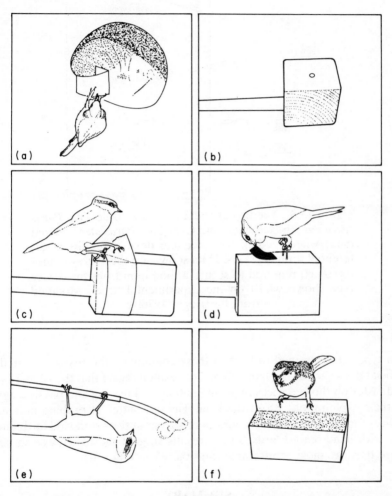

Figure 13.4 Devices used by Partridge (1976b) in testing feeding behaviour preferences of titmice: (a) coky, half-coconut shell suspended with mealworm prey hidden under masking tape in the interior; (b) proby, wooden block with hole drilled to receive mealworm quarter under a cotton-wool cap; (c) hacky, wooden block with prey in large hole covered by brown gummed parcel tape; (d) pully, as hacky but covered with tear-resistant though peelable plastic tape; (e) springy, flexible spring with rubber tube at end to contain prey; (f) hoppy, box containers with prey hidden in debris. From Partridge (1976b).

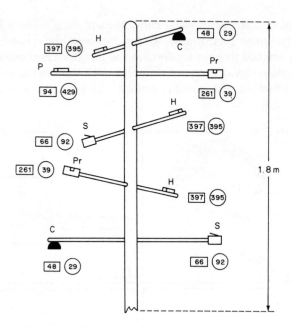

Figure 13.5   Feeding efficiency of naïve Blue Tits, *Parus caeruleus*, (boxed data) and Coal Tits, *Parus ater*, (circled data) when feeding from feeding devices requiring differential foraging skills. Data shown are the average times (seconds) required to secure a food item from the device type shown whilst on the experimental 'tree'. Modified from Partridge (1976b).

each type. Once trained to hunt in these containers, the tits were tested on artificial trees such as in Figure 13.5. The result indicate that the Blue Tits got food faster on those devices requiring pulling and tearing and physical agility for their exploitation, whilst the Coal Tits did better at probing into small crevices for their food. Differences were similar in tests with wild birds and in tests with hand-reared birds, so the results suggest the tits choose to perch where they are most comfortable (and therefore efficient).

SUMMARY

Newly-hatched young already have well-developed feeding behaviours but these improve with physical development of the young and with improvements in discrimination of the eliciting stimuli. When young fledge at periods of food abundance, opportunities for practice may be abundant and, where complex feeding skills are involved, may be aided by adult tutoring or by extended periods of play in which the appropriate capabilities are developed. Young of species specializing on dangerous prey show innate killing behaviours and innate responses to prey characteristics, thereby minimizing the risk to themselves in attacking. Anti-predator responses of young birds are probably based on their avoidance of unfamiliar stimuli coupled with learned responses to the alarm calls of other species. The development of appropriate behaviours in areas of such complexity as nest-building and habitat recognition depends on the interaction of genetically controlled behaviour and morphology with learning and experiential feedback.

CHAPTER 14

The development of song

The development of the young bird's ability to produce the calls and song of its species deserves separate consideration from other forms of behavioural development. Song development has not only been particularly well studied but is extremely diverse amongst birds, ranging from the largely innate vocalizations of pigeons and doves through to the learned mimicry of mynahs and mockingbirds. Furthermore, various features of the different development modes adopted can be shown to be correlated with selective pressures generated by the ecological situation of each species. This chapter therefore examines how these factors interact to shape the development of calls and song in young birds.

CALLS AND OTHER NOTES

Precocial young

Precocial chicks with their demanding life-style have a wider repertoire of calls than have altricial young. Kear (1968) has studied the calls of many of the waterfowl reared in captivity at the Wildfowl Trust at Slimbridge, England. Calls common to many species include contact or contentment calls, essentially a social call given during exploration and feeding and preening; a greeting call; a distress call, given when separated from parents or siblings; and a sleepy call, usually a soft trill given before resting periods and perhaps conducive to brood assembly for maternal brooding. Semi-precocial and semi-altricial species have additional calls associated with feeding by the adults or – in, for example, grebes – the demands of the young for transporting on the parent's back. Examples of these development patterns in gulls are discussed in Chapter 5.

Altricial nestlings

Newly-hatched altricial songbirds produce only quiet cheeps, giving way in older chicks to loud and characteristic begging calls which may persist past fledging. These notes are typically sharp repetitive notes with a wide frequency range and thus easily localized. Whilst clearly advantageous after

226

fledging, in that the parents can relocate their offspring on returning with food, these calls can frequently draw the attention of predators to hungry broods of nestlings. In the Wytham Woods study area of Oxford University weasels are thought to detect nest-boxes in this manner (Dunn, 1977). It has therefore been argued that these noisy calls have evolved as a form of 'blackmail' of the adults by the young, each of whom endangers the whole brood by calling and thus forces the parents to work harder on behalf of that young than they might otherwise do (p. 139). Altricial species lack the 'distress' and 'contentment' calls of precocial chicks at hatching, though a distress call may be produced in extreme discomfort. Fledglings may quickly develop a full repertoire of non-breeding calls: European Blackbirds, for example, have their 12 calls developed by 3 weeks of age (Messmer and Messmer, 1956). Many of these calls can be traced back to one of the juvenile calls (Lanyon, 1960). Thus the location call of European Blackbird fledglings develops into the adult alarm call but also gives rise to social calls.

ANATOMICAL CONSIDERATIONS

Avian vocalizations are produced by the flow of air past the elastic membranes of the syrinx, with changes in the tension of these membranes determining the changes in pitch. In the adult bird the syrinx is little affected by environment (Nottebohm, 1970) but its early growth imposes some constraint on the sounds a young bird can produce (Abs, 1970, 1974). In domestic ducks the pitch of utterances produced by ducklings is determined by the resonant frequency of the trachea: as this lengthens with growth it supports longer wavelengths, i.e. lower and lower pitch (Figure 14.1). The voice of the drake therefore decreases from about 3.5 kHz to 1.5 kHz, mostly between the third and sixth week. The voice eventually 'breaks' as the trachea nears its final length, taking on a wide range of frequencies as the simple physical resonance of the trachea ends (Figure 14.1). In the domestic pigeon this break coincides with sharp increase in the size of the male gonads (Abs, 1974) and in both species its timing can be modified by treatment with steroid hormones or by castration. In cockerels precocious crowing can be elicited by testosterone injections, an effect largely independent of age among birds ranging from 1 to 12 weeks old (Marler et al., 1962a). In detail, however, there are parallels between the frequency spectra of the crowing calls produced precociously and the sizes of the external tympaniform membrane (area) and of the trachea (length) of the bird concerned, at least over the age span 4–12 weeks (Abs, 1975). Bird song is thus constrained initially by developmental processes in the morphology of the sound production system and then by elements of the hormone system. The latter is particularly relevant to the time course of song learning (p. 235).

Call notes are inherently simpler to produce than is full song and are therefore less likely to require training or practice. Experimental deafening of young birds has been shown to have little effect on the subsequent develop-

228

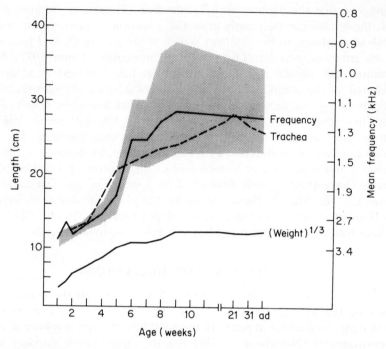

Figure 14.1   Mean (plus standard deviation stippled) frequency of utterances by young Mallard, *Anas platyrhynchos*, in relation to age and growth of the trachea (dotted line) and changes in body weight (below). Modified with permission from Abs (1970).

ment of calls in both the altricial Ringed Turtle Dove (Nottebohm and Nottebohm, 1971) and the precocial Domestic Fowl (Konishi, 1963). Also, basic call notes requiring little modulation develop normally in the absence of full neuromuscular control but more complex song requires a degree of training or practice under auditory feedback (Nottebohm, 1970; Lemon, 1973). Lemon (1975) offers the analogy of a violin string: a novice can get the natural sound of the string, but only a trained musician can achieve controlled harmonic vibrations over any length of time.

THE ONTOGENY OF SONG

Song learning in European Chaffinches

The development of song in the young Chaffinch has been studied by Thorpe (1958). The begging calls of the nestling develop in fledglings into an early subsong, a quiet rather unstructured sequence of notes trailing unevenly over a wide frequency range. This is probably a practice phase of the young and persists through late summer and early autumn (when the adults often show a resurgence of song), then ceases for the winter. Subsong is resumed in the following spring and develops into a period of plastic song in which the chirps and rattles of early subsong are mostly absent: these songs are rather longer and less stereotyped than is full song. Full song consists of three phrases lasting 1.5–2.5 seconds, occasionally to 3 seconds. Phrase 1 consists of between four and 14 notes, usually crescendo and decreasing in mean frequency between its two parts (from 4.5 to 3.0 kHz). Phrase 2 is usually distinct from phrase 1 and consists of between two and eight notes of 2.0–2.5 kHz. The concluding part is a phrase of between one and five notes at about 3.5 kHz rising into a complex terminal flourish. No two individual birds sing identical songs but within any region certain features are common to a local dialect (Thorpe, 1958). Moreover, each individual has several songs which it uses, and these become rather stereotyped at the time the young bird completes its first breeding season, i.e. when about 14 months old.

Thorpe caught young Chaffinches in the autumn of their first year, i.e. when they had had opportunities to hear adult birds in song and to themselves engage in some subsong, and kept them in groups in a soundproofed room where they heard each other but not older birds. These captives eventually produced adult-type songs slightly less elaborate than normal but definitely trisyllabic and with a terminal flourish. Hand-reared birds removed from the nest at 5 days and kept in individual auditory isolation developed quite differently, however. In the songs of these birds there was only a trace of division into three phrases and, in most cases, no terminal flourish. These basic songs resembled the normal ones only in being about the right length, in having about the right number of notes, and in being of normal tonal quality. Such birds received auditory feedback only through hearing themselves singing. When this feedback is increased, by rearing similar birds in groups rather than alone, the songs that developed were of better structure and phrasing but were extremely alike within each group of birds. These experiments thus suggested that young Chaffinches learned from hearing conspecifics singing and subsequently modified their own songs in the light of what they heard. Since the autumn-caught birds developed almost normal song, these characteristics of the species song must have been learned as juveniles.

Given that young Chaffinches learn a model of adult song in the way just described, how do they subsequently match their own song to that model? One possibility is that they constantly monitor their own song production to keep it in line with the model. Another is that they transfer their memory model to proprioceptive patterns (in the same way as humans can learn a skill such as music or typing and subsequently perform 'mechanically'). Nottebohm (1968, 1970) deafened young Chaffinches of various ages to remove the auditory feedback they obtained from hearing themselves sing. Birds deafened as adults, after they had acquired their (normally stable) repertoire of individually characteristic songs, maintained those songs more or less intact following deafening. Birds deafened in spring after varying degrees of practice with subsong developed songs whose quality was correlated with the extent of that experience. Finally, a bird deafened after 3 months in auditory isolation since being taken as a nestling (about 5 days old) showed a pattern described as a continuous screech.

Song development in White-crowned Sparrows

Figure 14.2 sets out schematically the ontogeny of song in the White-crowned Sparrow in relation to the experimental evidence for the sequence, as revealed by the now classic studies of Marler and Tamura (1964) and Konishi (1965). Unlike the Chaffinch, each White-crown male sings only one song which is shared by all the local population. The salient features of the natural situation are that young White-crowns experience a sensitive period for song learning whilst between about 10 and 50 days old. During this period they learn only from conspecifics and can then develop a normal song at 8 months of age, even if they have not heard an adult male singing in the interim – but only if they can hear themselves singing. Finally, once they have developed their song at this age they retain the ability to produce it thereafter, even if experimentally deafened.

Figure 14.2 shows the effects of depriving young White-crowns of normal experiences at various ages. Males reared in complete isolation develop a song pattern unlike that of wild birds. If reared in groups with other isolates they do not, like the Chaffinch, produce a common community song: the group members show no systematic differences of song structure from individual isolates. Males exposed to tutors prior to the sensitive period produce songs unlike the tutors. Birds isolated after conspecific tutoring during the sensitive period develop the song pattern of the tutor, irrespective of subsequent isolation and of subsequent exposure to alternative tutors; and birds exposed during the sensitive period only to other sparrow species develop neither the normal song patterns of their home area nor the song patterns of the tutor species: they develop as if in acoustic isolation (Marler and Tamura, 1964). Exposure to conspecific song during a restricted sensitive period is thus essential to young White-crowns if they are to develop normal song later. This development depends on good auditory memory of what was heard, for the songs are produced at 8 months without further tutoring. Konishi's (1965) experiments showed, however, that the young bird merely learns an auditory model during the sensitive period. It needs to hear itself sing when first coming into song if it is to match the model: birds deafened between sensitive period and first song produce distorted versions lacking the pure whistles and trills of typical White-crown song. That is, deafening the trained bird before it can match its own performance against the song model makes the training unusable. Once the period of song crystallization is complete, however, the song pattern is subsequently immune to surgical deafening. The bird continues to recreate the learned song despite its inability to monitor its performance by ear. Control of song production must therefore

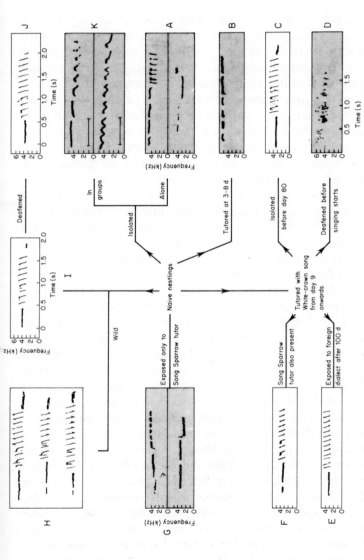

Figure 14.2 Outline of experiments demonstrating the constraints on song learning in young White-crowned Sparrows, *Zonotrichia leucophrys*, showing that (a) hearing adult song is essential (compare box A and box C); (b) this learning must take place within a critical period between 9 and 80 days after fledging (compare boxes B, C and E); (c) this learning is selectively restricted to White-crown song (boxes F and G); (d) this learning must be followed by opportunities for the young to hear itself sing (compare box D with box C); and (e) once adult song is learned fully it can persist without further auditory feedback (box I and box J). Lightly stippled boxes A, B, D, G and K indicate atypical song patterns. All vertical axes are frequencies (in kilohertz); all horizontal axes are time (in seconds). Based on Brown (1975), from whom sonographs have been redrawn or modified.

have passed to proprioceptive or motor memories, in much the same way as a human typist or musician can perform 'mechanically' and without concentration.

Song learning in other species

Other studies have revealed that song learning may affect only syllable structure or only syllable sequencing, and may act independently on the size of the song repertoire developed by the young bird. In addition, invention of entirely new song components may occur. When Mulligan (1966) fostered Song Sparrow eggs and young under Canaries in a soundproof room, he found that the songs the young produced were similar in patterning to those of wild birds but the repertoire of each bird was smaller than in wild birds, both in the number of syllables and in the number of song types. In this way the absence of a tutor depressed the variety of song output by isolate young.

Much the same holds for the Cardinal but the details are better known (Dittus and Lemon, 1969; Lemon, 1975). Each individual has several distinctive songs, with birds in any one locality being very similar in song type and birds from different localities being obviously different. Young Cardinals disperse from their natal area, normally by less than 30 miles (50 km), although occasionally to over 500 miles (800 km), so learning is possible either in the home area or in the settlement area. By tutoring groups of young taken from nests in Ontario at 5–8 days of age on different regimes of song exposure, various points were established. First, tutoring from between 1 and 5 weeks until 15 weeks with tape-recordings of home and foreign songs resulted in some of both being copied; moreover, home songs not on the tape were never produced. Second, a group exposed in their first autumn (at ages 2–7 weeks) to one tape-recording and in their first spring (at ages 6–8 months) to another tape learned songs from both tapes, indicating that copying of both home and settlement area dialects was possible on the part of dispersing young. Third, all young produced songs which were not on the tape-recordings but which were invented. Such songs were also produced by untutored isolates and these differed from the parental songs. Finally, the syllable repertoires of tutored birds were about the same size as those of wild birds but those of isolates were smaller. These results are particularly important in showing that repertoires are determined only partly by copying of birds heard after fledging and on spring settlement. In principle, of course, copying cannot persist indefinitely as the sole process for song development. For many females, courtship song is the only evidence they have of a male's fitness, with song repertoire an important avenue of variation between males. Hence sexual selection favours variability in song and some source of variation is necessary to prevent a population becoming completely fixed in respect of song types. The Cardinal studies show that invention (rather than, say, error in copying) is a significant source for such variation. European Blackbirds show similar improvization (Thielcke-Poltz and Thielcke, 1960).

Yet another pattern of song development has been established for juncos (Marler, 1967, 1975). Arizona and Oregon Juncos have songs with complex structuring of the component syllables (Figure 14.3). Birds raised as social isolates do not develop this complexity but sing a simpler song which is at times merely a trill of a single note (Figure 14.3). By hand-rearing Oregon Juncos in two groups, one in soundproof boxes, the other where they could hear not only their own songs but those of other species in the laboratory, Marler et al. (1962b) showed that both groups developed stable songs. Those reared in auditory contact with many other birds developed the greater diversity of song independent of imitations. Both species of junco thus seem to respond to diversity of song experience rather than by imitation of the specific tutor model. This is quite different from the situation with Chaffinches, which learn the tutor model quite successfully (Thorpe, 1958).

Selectivity in song acquisition

In species in which the song is learned from other individuals, there must be some restriction as to what the young bird will copy: otherwise, it may learn a quite inappropriate song. In some species this takes the form of a predisposition to learn species-specific song patterns. Chaffinches reared in isolation and exposed to tape-recordings of normal song in the course of their first autumn subsequently produced almost normal songs. But when presented with artificial songs of various types, they accepted only those of Chaffinch quality, e.g. Chaffinch songs played backwards (Thorpe, 1958). Such restrictions are even tighter in birds which have been exposed to conspecific song: juvenile Chaffinches caught by Thorpe (1958) after experience of adult songs in the wild did not respond to tutoring with re-articulated Chaffinch recordings.

Other species have a less clear avoidance of alien song. Red-winged Blackbirds presented with conspecific song and with Baltimore Oriole songs over a loudspeaker incorporated songs of both types into their repertoire (Marler, 1975). This does not happen in nature, so some other factor must be present to ensure selective learning of conspecific song. A link to social display and recognition of plumage is one obvious possibility.

An alternative pathway to selectivity in song learning is for young birds to accept tutoring only from known individuals. Young Bullfinches, for example, will learn new sounds only from their father (Nicolai, 1959). Young Crested Larks will learn from a territorial neighbour (Tretzel, 1965, cited by Immelmann, 1975b). Young Zebra Finches are selectively responsive to individuals with whom they have a strong personal bond (Immelmann, 1975b). When fostered under Bengalese Finches but within sight and sound of breeding pairs of their own conspecifics, the young Zebra Finches developed songs resembling those of their foster parents. This and other experiments indicate that the young have a hierarchy of preferences as to tutor: they prefer first to imitate a male with which they have a personal bond, but if they have

234

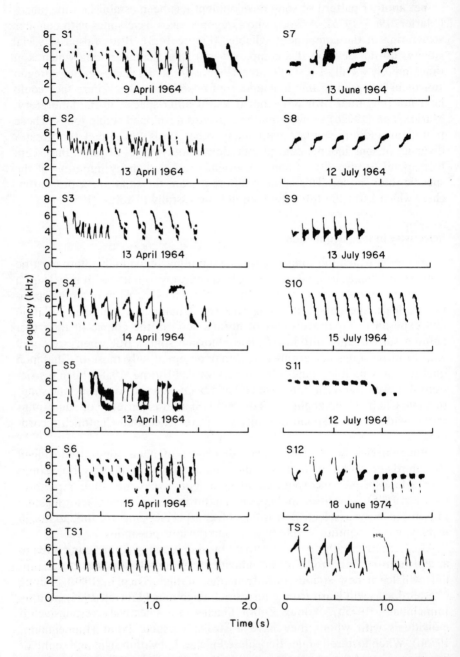

Figure 14.3 Normal song in an Arizona Junco, *Junco phaeonotus*, (bottom right) and that produced by six birds raised in auditory isolation (right-hand column S7-S12) and by six birds tutored with playback of recordings both of normal conspecific song (TS2) and of normal Oregon Junco, *Junco oreganus*, (TS1) song (left-hand column S1–S6). From Marler (1975).

two such bonds (or none) selection for their specific song type takes place.

Mimics are a particularly fascinating class of song learners, incorporating as they do the songs or calls of other species into their own song. Well-known groups with this behaviour are starlings, corvids, and mockingbirds, though other species also do so. In these groups the adaptive significance of imitation is unresolved, in contrast to the obviousness of host mimicry by brood parasites. However, even with perfect mimicry of the target songs or calls the starling or mockingbird is immediately recognizable by the presence of its own calls or repetition pattern in the song sequence as a whole, and the mimicry is probably acquired in ontogeny in the same way as other song learning (Dobkin, 1979).

The influence of hormonal status

The timing of the sensitive period for song learning is affected by the hormonal state of the young bird. In young male Zebra Finches singing can be induced earlier than normal by injections of testosterone proprionate (Sossinka *et al.*, 1975). When the injection was on day 16, juvenile song normally occurring on days 34–37 could be shifted forward to day 18 or 19, but both earlier and later timing of the injection delayed singing until days 25–35. Thus there seems to be a point prior to which juvenile song cannot be induced, perhaps due to the maturation of central nervous system components. The advanced juvenile song in these experiments lacked the maturity of rhythm of the normally timed song.

In another study Nottebohm (1969) found that a young male Chaffinch castrated in January of its first year failed to sing until implanted with testosterone 2 years later. Under tape-recorded tutelage it abandoned its own developing theme for that of the 'tutor' and retained this theme in the face of further tutelage with a new theme the following year. These findings show that the sensitive period is not age-dependent but follows the crystallization of song, whenever that happens. Such an end to the sensitive period could come about through a loss of motor plasticity to develop new songs (i.e. becoming physically unable to perform novel actions), or through a loss of the ability to acquire new auditory templates. Young Meadowlarks have been shown to have a limited period in which to acquire the motor patterns for song expression (Lanyon, 1957). In the Chaffinch the sensitive period closes permanently at the end of the bird's first spring, but in at least some species – Red-backed Shrike and Canary – a sensitive period re-opens seasonally (Poulsen, 1959; Blase, 1960).

Song learning and social dominance flight calls amongst social finches are not restricted by sensitive periods but are added to by imitation as individual relationships form within winter flocks and between breeding pairs (Mundinger, 1970). In playback experiments, incubating American Goldfinch females were found to respond to recordings of their mate's flight call but not to recordings of other males' calls, i.e. these calls served for individual recognition. Mundinger further showed that flight calls amongst finches can be acquired by imitation and that in winter flocks mutual imitation by male European Siskins was related to their relative dominance. Where calls are used for such social relationships within flocks or pairs whose membership changes from year to year, a sensitive period against the incorporation of new calls could be maladaptive.

Environmental control of song learning

Further evidence that song learning is not rigidly constrained to an age-dependent sensitive period has recently been obtained for the Marsh Wren (Kroodsma and Pickert, 1980). Marsh Wrens have an extended breeding season along the Hudson River, with young fledging throughout the period from June to August. Those juveniles hatched early in the year hear many adult songs during a period of long daylength. Those hatched late in the year, on the other hand, hear few (if any) adult songs during the short days then prevailing. The early fledglings seemed to have a sensitive period for song learning confined to ages of 60–80 days. Kroodsma and Pickert therefore devised the experiment summarized in Table 14.1, to investigate the relative importance of photoperiod and opportunity to learn from adults. Nestling males were taken from July nests and isolated in soundproofed chambers artificially lit with either a June or an August photoperiod which was subsequently varied in line with that of the home latitude. Within each photoperiod group, various young were tutored with 0, 16, or 32 songs in autumn, and all received tutoring with an additional 12 songs in the following spring. The results showed a major effect of photoperiod on the development

Table 14.1 Results of an experiment to investigate environmental influences on song learning by young Marsh Wrens, *Cistothorus palustris*. Based on Kroodsma and Pickert (1980)

Song tutoring (number of songs)		Songs learned by bird in photoperiod of			
		June type		August type	
Autumn exposure	Spring exposure	Autumn songs learned	Spring songs learned	Autumn songs learned	Spring songs learned
32	12	Many	None	Appreciable	None
16	12	Several	None	Fewer	Some
0	12	None	None	—	Some

" Exposure between days 20 and 100.

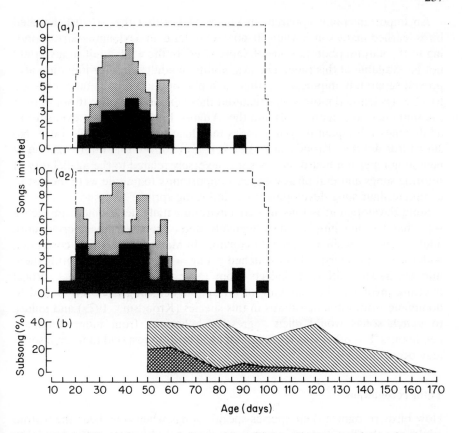

Figure 14.4 The influence of time of hatching and opportunity to learn on song development in Marsh Wrens, *Cistothorus palustris*. Young were taken from natural nests and reared on experimental photoperiods corresponding to (a₁) August hatching or (a₂) June hatching and were tutored either with 16 (black) or with 32 (light stipple) songs over the period indicated by hatched lines, at the rate of one (or two) songs per 5 day period. (a₁) and (a₂) Song learning was most frequent between 30 and 60 days in age and this sensitive period was extended in the less tutored groups and in June birds. (b) Subsong (measured as percentage of 1 min sample periods recording subsong) was more prolonged in June photoperiod birds (◩) than in August birds (▨). After Kroodsma and Pickert (1980).

of subsong: birds on June photoperiods continued to learn songs when August photoperiod birds had ceased to do so (Figure 14.4(a)); they spent more time in subsong than did August photoperiod males of the same age (median times in subsong of 56 minutes and 11 minutes respectively; Figure 14.4(b)); and they developed subsong which was advanced and stable (whilst 'August' birds never progressed beyond rudimentary early stage subsong).

An important point apparent in Figure 14.4 is that the 'August' hatched birds learned many songs when 30–60 days old, i.e. in daylengths corresponding to the natural photoperiods of September. In the wild, adult songs would not be available at this time. The availability of adult songs for imitation also proved separately important: within each photoperiod groups birds exposed to 32 songs learned more songs than did those exposed to only 16 songs over the same period (Figure 14.4) and the 'August' birds given least exposure to adult songs subsequently copied songs they heard in spring (Table 14.1). No June birds did this. Kroodsma and Pickert also noticed that improvization of new song types not heard on tapes was inversely related to the availability of tutoring songs and that birds with least opportunity to imitate were the last to complete their song development the following spring.

Song development is thus in part environmentally determined, in such a way that late-hatching young effectively defer their sensitive period until adult songs are again available for copying. In Marsh Wrens this is correlated with the great dispersal of late-hatched young, a dispersal likely to bring them into areas with different dialects from those of their natal areas. Vocal duetting involving the matching of an opponent's song is closely linked to dominance–subordinance status in this species (Kroodsma, 1979) and failure to match songs would elicit aggressive behaviour from more dominant neighbours. For other species failure of local birds to respond to foreign songs may occur (p. 239).

Tutor recognition

How birds recognize their species-specific songs when they hear them from adults for the first time is not fully understood. Young Chaffinches learn re-articulated songs played back to them but reject alien songs (Thorpe, 1958), yet young Red-winged Blackbirds learn Baltimore Oriole song in playback experiments (Marler, 1975). In one of the most thorough investigations of this problem to date, Marler and Peters (1977) have identified the structure of individual song elements as a critical factor in learning by young Swamp Sparrows. They constructed artificial songs from Swamp Sparrow and Song Sparrow elements, some being a regular slow trill as used by adult Swamp Sparrows, others being the alternation of fast-slow elements as used by Song Sparrows. When exposed to these artificial songs young Swamp Sparrows, isolated even as embryos (by fostering under Canaries) from any exposure to adult songs, developed only those songs with their own specific note structure. This knowledge thus appears to be innate.

In complete contrast, young Zebra Finches learn the songs of their foster parents (p. 233), precluding innate recognition of species song. The explanation may lie with the accelerated reproductive development of these grass finches (young Zebra Finches may be sexually mature at 6 months of age) in mixed colonies or whilst feeding in mixed species feeding flocks. The opportunities for erroneous copying from a closely related species could select for exclusive learning of parental song, in much the same way as with other imprinting (Chapter 10).

Culture and song learning

Song patterns modified during ontogeny by learning of cultural traditions may be preserved over many generations, even when isolated from the ancestral population. Extant songs of 11 species introduced via wild-caught birds to New Zealand more than 100 years ago are not distinguishable by ear from extant songs of their German or Middle European conspecifics (Thielcke, 1974). Four species tested by sound spectrograph in both regions showed no structural distinctions. Whilst this could be explained if the songs were under simple genetic control, at least two of the species thus tested, Blackbird and Chaffinch, are known to modify and transmit their songs (at least in part) by learning. One powerful reason why young birds should be traditional in their learning is that songs are signals for intraspecific communication as well as for sexual isolation. Hence a song pattern once achieved should stay within the normal range of intraspecific variability: too foreign a song will not be recognized by conspecifics. This can happen even within the variation provided by the evolution of local dialects. Song Sparrows, for example, showed significantly less response to playback of conspecific song recorded only a few miles away than they did to songs recorded in their own locality (Harris and Lemon, 1974).

Experimental evidence for total failure of recognition of altered conspecific song was obtained by Thielcke (1973). Short-toed Treecreepers in Morocco sing very differently from their European conspecifics and when the latter were presented with tape-recorder playback of the Moroccan songs they ignored the songs completely. European recordings, by contrast, evoked intense territorial responses. Moroccan treecreepers have thus diverged too far from the ancestral song for communication with European populations of their species.

Intraspecific communication is less likely to be lost if song modification occurs by the addition of extra notes. In the early part of this century Yellowthroats in New England were generally described as singing three-note phrases, yet 60 years later about two-thirds of tape-recorded Yellowthroats in the region had four- and five-note phrases instead (Borror, 1967). Such changes can only evolve if they affect elements of the song not critical for species recognition.

SUMMARY

Call-notes are simpler to produce than is full song and their development proceeds in most species largely independently of training or practice. Precocial young have a more developed repertoire of calls than have altricial young. Song learning is a more complex process, in general involving exposure of the young bird to one or more tutors and a period of song practice in which the young compares his own production of notes against song models obtained by hearing the tutors. In general young birds will copy only species-specific song elements and in some species copying may be further restricted to notes produced by individuals related or otherwise known to the young bird. The effectiveness of tutoring and of practice may be restricted to particular sensitive periods in the young bird's life, with this timing under hormonal control. In at least some species late-hatched young can defer their period of song learning until the following spring, when adult song is again common enough to provide adequate tutoring. Song elements can be transmitted culturally as well as genetically, but extreme variations in song types learned by cultural transmission may be restricted by the need for potential mates to recognize the songs produced.

CHAPTER 15

Migration and dispersal

As young birds become capable of locomotion they begin to move from the confines of the nest. Baker (1978) argues that this is the beginning of a step by step expansion of the young bird's 'familiar area', within which it will eventually move in the certain knowledge of where it is and of its location with respect to any desired destination. However, within any such general process are three features of developmental interest: (1) the existence of differential selective pressures for dispersal between species of different developmental modes; (2) the acquisition of navigational ability by the young bird; and (3) the existence of narrow and developmental 'windows' within which the young bird decides permanently on its subsequent home area or areas. These features of the young bird's development are the subject of this chapter.

SOME PATTERNS OF POST-FLEDGING DISPERSAL

The dispersion of semi-precocial chicks is closely linked to predation pressure. Many gull chicks, for example, face major predation (often cannibalistic) only during their first 7 days, and parental guarding can see them through this period. Thereafter, the chicks remain close to the nest, near which they usually have individual hiding places when threatened (Tinbergen, 1963). In several skua species nesting in more open tundra habitat, on the other hand, the young disperse rapidly from the nest (E. C. Young, 1963; Andersson, 1971). Long-tailed Skua chicks, for example, remain on the bare nesting mounds only for a few days, dispersing 200–300 metres from them within only 2 weeks into areas of dwarf birch and willow brush where they are more difficult to find (Andersson, 1971). Moreover, the two young normally stay 50–100 metres from each other, as in other skuas. This may reflect sibling competition, since they will attack each other if placed together, but it certainly reduces the chances of both being taken by a predator. Again, Sabine's Gull chicks are initially reared on exposed tundra and likewise leave the nest within a few days and go to pools of water where they are safer (Brown *et al.*, 1967). Arctic Tern chicks in Sweden similarly resort to tundra pools when harassed by foxes (Andersson, 1971).

241

In other precocial species the broods may remain intact but are significantly over-dispersed (i.e. systematically spread out) with respect to each other, in so far as the habitat permits this. Female Blue Grouse accompanied by very young chicks avoided aggregating during the early brood season in Washington State but eventually clumped together in thickets in July and August, irrespective of whether they were accompanied by chicks. Such clumping was associated with over-grazing of grasslands in the vicinity (Zwickel, 1967).

The dispersal of ducklings on nest departure depends on the availability of suitable habitat (Bengston, 1971b). In Iceland, rearing areas for diving species are predominantly on open areas, with bare lava gravel shores used as loafing areas. Dabbling species, by contrast, use areas closer to aquatic vegetation for both feeding and loafing, though older ducklings will move towards more open water. Most newly-hatched broods travelled 250–300 metres on their first day but Gadwall broods averaged 365 metres, Wigeon broods 390 metres, and Mallard 480 metres. Once on a rearing area, the dabbling species showed a greater degree of site tenacity than did diving species, taking 26–28 days for 90 per cent of the broods to move further as against the 9–21 days observed of diving species. The one-brood territorial species, Barrow's Goldeneye, dispersed from the territories slowly, with 50 per cent of the young moving over about 3 weeks and 90 per cent by 30 days.

Such relatively stereotyped use of habitat during initial brood dispersal may also underlie the annual passage of migratory waterfowl along chains of traditional stopping places (Hochbaum, 1955). Waterfowl typically travel in family groups within which knowledge of suitable stopping places for habitat-limited species may be passed from generation to generation. Even in species not dependent on limited habitats, such as the more ubiquitous gulls, young dispersing in the company of their parents do better than if alone. Fledgling Franklin's Gulls in Minnesota survived well if their parents accompanied them and fed them through the first phase of the dispersal (Burger, 1972). Nisbet and Drury (1972) also considered parental care in the post-fledging dispersal period to be of particular significance to juvenile Herring Gulls.

Guidance along a dispersal or migration route need not always come from the young birds' parents, particularly in social species. European populations of the White Stork comprise a western population which crosses the Mediterranean at Gibraltar and an eastern population which migrates down the eastern shore over the Bosphorus. Western birds breeding in the Rhineland thus migrate in a generally south-westerly direction; those breeding in East Prussia, on the other hand, generally head south-east or south-south-east (Schüz, 1963). Eastern nestlings transferred experimentally to the Rhineland were recovered along the south-western route if released at the time local birds departed on migration. If they were held until later, when the local birds had already departed, they took the south-eastern route (Schüz, 1951). Young storks normally depart before their parents, so these results suggest that the eastern birds responded socially to their western peers, though it is always possible that some adults – perhaps failed breeders – were present in the juvenile flocks.

Prior experience of a migration route confers some hidden advantages. Among species lacking parental guidance young birds may be more prone to occurring in 'falls' at coastal concentration points along their migration route, simply because they come on them unexpectedly. Adult birds which have made the journey on a previous occasion halt well in advance of an extensive sea-crossing so that they can feed away from the concentration of competitors (and predators) along the coastal belt (Alerstam, 1978).

MECHANISMS OF NAVIGATION

Compass navigation by young Starlings

At some point in their lives young birds have to navigate for themselves. The processes involved in such navigation by young Starlings have been analysed experimentally by Perdeck (1958, 1967a, 1974). Starlings caught on autumn migration near The Hague in the Netherlands are from breeding areas around the southern part of the Baltic (Denmark, southern Sweden and Finland, Russia, Poland, and northern Germany). They pass though Holland on a west-south-westerly or westerly heading, to winter in the western parts of Belgium, north-western France, southern England, and Ireland. The Dutch breeding birds move out in autumn and are replaced for the winter by more eastern populations. Perdeck (1958) trapped some 11000 migrant Starlings at The Hague and flew them to Switzerland for release there. The subsequent recoveries of these birds show striking differences between juvenile and adult birds (Figure 15.1). Juveniles released in Switzerland continued to migrate on more or less the same west-south-westerly heading and over about the same distance as they would have flown without displacement. Their winter distribution thus lay south of the normal winter range for the species. In contrast, the adults corrected for the displacement and completed their migration in a north-westerly direction, bringing them back to their normal wintering range (Figure 15.1). No differences were found between juveniles released alone and those released in flocks with adults present.

This experiment shows that juvenile Starlings on their first migration fly in a standard direction and are unable to detect experimental displacement. Adult birds, returning to previously used winter quarters, show goal orientation: they know their intended destination from previous experience and navigate there by taking the appropriate (and not a fixed) direction. The later recoveries of experimentally displaced juveniles fitted this idea, for most returned in their second and subsequent winters to the new winter grounds in France and Iberia. A few, however, appeared in later years on the normal wintering grounds. Spaans (1977) has subsequently suggested that Starlings in practice navigate directly to the wintering region and then respond to local conditions (which vary from year to year) found there.

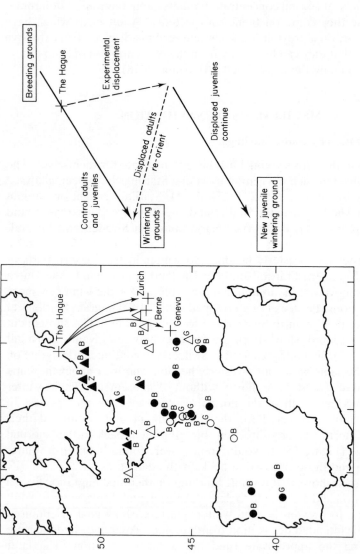

Figure 15.1 Recoveries of ringed Starlings, *Sturnus vulgaris*, trapped on south-westward migration through The Hague, Holland, and flown to release points at Zurich (Z), Berne (B) or Geneva (G), with a schematic interpretation of the distributions. The experienced adults recognized their displacement and re-oriented but naïve juveniles continued on their original bearing. *Key:* △, adults in October–November; ▲, adults in December–February; ○, juveniles in October–November; ●, juveniles in December–February. After Perdeck (1958) by permission of The

Ontogeny of a stellar compass

How do young birds acquire the ability to maintain the fixed compass heading revealed by the displacement experiments just described? Emlen (1972) examined this question by experimentally depriving hand-reared Indigo Buntings of celestial information normally available to them. This species is a nocturnal migrant which apparently uses the night sky as a navigational aid. Emlen therefore used a small planetarium at Cornell University to study the use made by naïve birds of the different elements visible overhead. He found, first, that young birds which had been deprived of experience with day and night skies oriented less well than birds previously exposed to these cues. Emlen then asked whether the birds had some genetically inherited know-ledge of star patterns to which to respond, of whether their exposure to celestial rotation through the night was involved. Adults are able to orient under a stationary planetarium sky (Emlen, 1967) but when juveniles without prior sight of the night sky were tested under these conditions, each individual's movements were randomly spread around the cage (Figure 15.2(a)). In contrast, another group given experience of the rotating night sky were able to orient under a stationary planetarium sky (Figure 15.2(b)). A third group were trained under a planetarium sky modified to rotate around the bright star Betelgeuse instead of about the Pole Star. These birds then oriented correctly if their directions were computed by reference to Be-telgeuse rather than to the Pole Star (Figure 15.2(c) (d)). Further experiments removing parts of the night sky map displayed within the planetarium showed that removal of a 35° sector around the Pole Star hindered orientation by normally trained birds. Birds trained with respect to Betelgeuse retained their orientation and lost it only if a 35° sector around Betelgeuse was removed. Emlen's experiments thus preclude young birds possessing an innate star map. Instead they show that the young learn that the night sky rotates and identify the axis of rotation as 'north'.

The timing of this learning was the subject of a final experiment by Emlen (1972). He retained the Betelgeuse-trained birds in captivity until spring, keeping them in a windowless aviary lit to the day–night schedule of the birds' wintering grounds. The birds were then held in an outdoor aviary with sight of the night sky. But when tested in the following autumn, the birds continued to orient incorrectly with respect to the Pole Star and correctly with respect to Betelgeuse. Their late summer training in their first year had thus remained dominant into adult life, despite considerable subsequent experience of the correct sky patterns. Their compass is thus 'set' during a sensitive period prior to their first autumn migration.

Emlen's experiments show that young birds use the apparent rotation of the night sky to identify an axis of rotation to be treated as a north–south axis. Once they have learned the location of this axis they transfer the information to a star map which then provides faster indication of where 'north' lies. Adults can thus orient even under a stationary planetarium sky (Emlen, 1967).

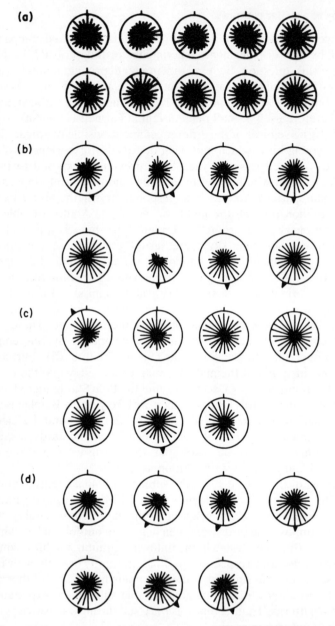

Figure 15.2 Orientation of young Indigo Buntings, *Passerina cyanea*, prevented from viewing natural celestial cues: (a) birds tested under a stationary planetarium sky; (b) birds exposed to normal rotating planetarium sky in training but tested under a stationary sky; (c) birds trained and tested under a planetarium sky rotating about an axis through the position of Betelgeuse, with orientation plotted relative to Polaris; (d) the same data plotted relative to Betelgeuse – arrowheads indicate significant orientation in that direction by the bird concerned. See text for interpretation. From Emlen (1972).

Magnetic compass in gull chicks

Young Ring-billed Gulls have been shown to possess orienting abilities linked to the geomagnetic field at their colony, and thus presumably based on a magnetic sense (Southern, 1969, 1972). Chicks aged 2–20 days placed in the centre of circular orientation cages showed pronounced tendencies to walk to the south-east. This is the direction of autumn banding recoveries from the Lake Huron (Michigan) study colony, and is close to the east-south-easterly bearing taken by juveniles captured within the colony and released at various sites 11 km south, 18–19 km north, and 29 km west of it (Southern, 1969). The chicks thus displayed from a very early age a spontaneous preference for their subsequent dispersal bearing.

Southern (1969) found that this preference was apparent both in sunny and in overcast conditions, but it was less marked during periods of magnetic disturbance at the colony. The preference deteriorated at high levels of sunspot activity and disappeared altogether at still higher levels (constituting magnetic storm conditions). Southern's work therefore shows that even very young chicks can detect some form of magnetic cue (or a factor correlated with it).

Solar compasses and the integration of navigational information

Diurnal migrants often make use of the Sun for navigation and may show a degree of confusion when flying under totally overcast conditions (Keeton, 1979). Even so, many experienced adults can maintain a constant bearing in such conditions. Pigeons equipped with bar magnets lose this ability, however, implicating a magnetic compass as well as a solar one. A number of experiments indicate that young birds have to learn the integration of several such navigational systems. Young pigeons making their first homing flight need both solar and magnetic cues (Keeton, 1979). If they are released under overcast conditions or if they are released with bar magnets attached to their backs, they depart in random directions, even though adult pigeons can navigate successfully in either of these conditions. A crucial experiment indicates that it is the problem of integrating cues that underlies the young birds' difficulties. When young pigeons were trained exclusively in days of total overcast, so that they never saw the Sun, they oriented perfectly when released for homing for the first time. Never having learned about the Sun, these birds had not incorporated it into their navigational system and did not miss it when absent.

Another experiment indicates that the sun-compass has to be calibrated by young birds. Normally-reared birds treat the morning sun as rising in the east and make a time correction for its progression across the sky through the day. If placed in an aviary with a light–dark cycle some hours out of phase with the outdoors, the birds adjust their internal clocks to the new regime. When tested outdoors immediately after removal from the artificial light–dark cycle, the birds orient incorrectly, being mis-directed by 15° for each hour of time-shifting. Wiltschko et al. (1976) showed that this link between time and direction was learned rather than inherited. They found that young pigeons under a light regime which was permanently 6 hours slow oriented correctly when released to home, whereas normal birds placed in such conditions would have been 90° out. When Wiltschko *et al.* placed their experimental birds on the natural light regime the birds oriented incorrectly by 90°. The young birds must therefore have calibrated their knowledge of 'north' against some more fundamental reference whose nature is currently unknown (Keeton, 1979).

Physiological control of compass directions

The experiments described above account only for the acquisition of compass ability by young birds. They do not explain why Indigo Buntings orient south-south-eastwards or American Robins orient southwards, in each case to the species' normal migratory heading in autumn. The direction concerned is probably innate. This direction varies from season to season, however, for birds heading south in autumn need a northerly bearing the following spring. Emlen (1969) raised two groups of male Indigo Buntings on opposite light regimes such that, at the same time, one group was in spring condition, the other in autumn condition. When the two groups were tested under the same planetarium sky, the group anticipating a spring migration oriented predominantly north-north-eastwards whilst that anticipating an autumn migration oriented south-south-eastwards. Since birds differ between spring and autumn as to their hormone level, Emlen suggested that the changes of direction between the two groups might be linked to changing endocrine status. This suggestion was subsequently confirmed experimentally by injecting White-throated Sparrows with prolactin and corticosterone and obtaining a reversal of polarity linked to the time course of the injections (Martin and Meier, 1973).

Evolutionary significance of learned compasses

Why should the navigational ability of young birds be determined by such a mixture of innate and learned components? One possibility, akin to that suggested for imprinting processes in Chapter 10, is that stellar cues are unreliable on the time scale needed for evolutionary change. The Earth's axis is tilted slightly with respect to its annual plane of motion but the tilt is not constant. Instead, the Earth oscillates slightly, such that every 25 800 years the axis has described a cone of radius 23.5° (Emlen, 1972). The star patterns seen from Earth therefore vary on this time scale, the spring stars of today becoming the autumn stars of 13 000 years hence. Similarly, the Pole Star will alter to a new declination of 43°N, and so on. Emlen points out that a genetic star map would have to change sufficiently rapidly to compensate for these changes, were young birds to depend on innate cues. A system by which the axis of celestial rotation is first determined, thus automatically determining the direction of north–south, and only then transferred to the (observed) star map avoids this problem.

LEARNING THE NATAL AREA

In the Starling displacement experiments performed by Perdeck (1958, 1967a), the young birds spent their first winter in unfamiliar surroundings away from their natal areas. A further question of interest is how the young birds established their breeding grounds in their first summer. When birds traped on spring migration in Holland were displaced to Switzerland and released, the majority were subsequently recovered within the normal breeding range, though a few were found in areas close to the release point, mostly in France (Perdeck, 1974). The experiment thus showed the majority of the first-year Starlings had a definite goal to their return journey and were not simply retracing their autumn route. Young Starlings already knew their natal area as they left on autumn migration.

At what stage does a young bird thus learn its natal area? Young Collared Flycatchers do so during the period between fledging and their pre-migratory moult (Löhrl, 1959). Hand-reared birds released in three groups at a point 90 km south of their rearing area differed in return rates observed the following spring. In one group released soon after fledging and in another released 2 weeks before migration, 18–19 per cent of the males reappeared the following spring. In a group released after moult and the start of normal migration no males reappeared. In this species, therefore, the males establish their natal area characteristics during their immediate post-fledging wanderings. Similar results were obtained for the Pied Flycatcher in Lower Saxony, Germany, by Berndt and Winkel (1979). They interchanged eggs and nestlings between two breeding areas 250 km apart and in each case found that the birds returned to the area in which they fledged. Using data in the Dutch ringing scheme, van Balen (1979) showed that all recoveries of Pied Flycatchers under 45 days old (assuming all were ringed at age 10 days) were within 10 km of the ringing site. The majority of birds 45–65 days old had moved between 70 and 200 km. Thus, as with the Collared Flycatcher, this species has a very narrow time period within which juveniles can disperse and imprint on new breeding areas before leaving on migration to Central Africa. Young Reed Warblers similarly decide on their breeding habitat on the basis of post-fledging feeding experience, even if this is in a habitat different from their natal one (Catchpole, 1974).

Breeding site selection by young Lesser Snow Geese is on the basis of experience of feeding areas acquired by females as yearlings rather than as goslings. (Male Snow Geese pair with a female on the wintering grounds and then accompany her to the area of her choice.) In their first summer, yearlings return to the Arctic with their parents and feed in the vicinity of their parents' new nest site, in most cases (87 per cent) close to the very areas on which they themselves were reared the previous summer. They then return to its vicinity to breed for the first time when 2 or 3 years old. A small number of successful parents do, however, move their nest site and the yearlings accompanying them therefore feed in an area different from that they knew as goslings. These young subsequently breed near their yearling feeding area rather than near their gosling feeding grounds (Cooke and Abraham, 1980).

...AND WHERE TO WINTER?

Just as Collared Flycatchers decide on their breeding grounds only within a narrow time slot before autumn migration, so too do birds settle on their wintering grounds. Ralph and Mewaldt (1975, 1976) experimentally displaced adult and sub-adult sparrows (Emberizidae) of several species between a number of Californian ringing sites. They found that adults never settled in the sites to which they were displaced but returned to the home site which they already knew. Young birds gave quite different results (Table 15.1). If

Table 15.1 Numbers of sub-adult *Zonotrichia* sparrows of various species displaced in one winter and returning to the home site or to the new site the following winter. Data from Ralph and Mewaldt (1975)

	Date of displacement					
Site of recapture	28–29 Nov.	20 Dec.	9 Jan.	30 Jan.	13 Feb.	25 Feb.–6 Mar.
New site	6.4	1.0	3.8	0.0	0.0	0.0
Home site	0.8	4.5	14.3	18.4	16.1	12.0

displaced before mid-January, they were likely to settle in the new site for the rest of the season and to return to the new site the following winter. If displaced after mid-January, they did not settle at the new station (though some spent the remainder of the displacement winter there) but returned to the home station the following season.

Petersen (1953) displaced Black-headed Gulls wintering in Copenhagen parks to various sites in Denmark, Sweden, and Holland, and found that adult birds were more likely to return the same winter than were first-year birds. Schwartz (1963) has also found Northern Waterthrushes to show age differences in displacement, with young birds more likely to settle in new areas.

The Starling experiments of Perdeck (1958, 1967a) showed that young birds experimentally displaced during autumn migration eventually wintered in locations more or less equivalently displaced from the adult wintering grounds. A question of considerable interest is how these first-time migrants knew where to stop for the winter. Two explanations are possible. First, the young Starlings might respond to external factors, such as characteristics of the landscapes through which they pass. Second, they might respond to internal factors, such as the time elapsed since the start of migration. Perdeck (1964) tested between these two alternatives by a simple displacement experiment. If external factors are important, catching birds on migration and transporting them to suitable wintering grounds will terminate their journeying. If internal factors are important, on the other hand, the birds will continue to travel further. Perdeck trapped Starlings in Holland in autumn and flew them to Barcelona in Spain. The nearby Ebro valley provided suitable wintering grounds for such Starlings as chose to stop migrating. The results were also compared with those obtained in experimental displacements to Switzerland, where Starlings can rarely survive the winter. Table 15.2 shows his findings, which proved to depend to some extent on the time of migration. Early birds (still in the first half of their migration) released in suitable habitat continued to migrate and reached the south-western corner of Spain. Late birds (in the latter half of their migration) released in Spain halted in the Ebro valley. Birds released in Switzerland (unsuitable for overwintering) always moved on (Table 15.2). Thus, although external factors cannot induce a premature termination of the journey they can, when unfavourable, prolong it.

Young birds of other species have been shown to have endogenously set levels of migratory drive, varying from species to species. This is as one would expect from Perdeck's results on the ending of travel by juvenile Starlings. Berthold (1973) hand-reared nestlings of six species of the warbler genus *Sylvia*, and studied their nocturnal activity in cages with microswitch-equipped perches. Migrants held captive during their migratory period showed marked nocturnal restlessness or *Zugenruhe*. Berthold measured the intensity of this behaviour throughout the first autumn of his hand-reared birds. When these data are set alongside the average distance travelled during

Table 15.2   Mean distance (in kilometres) travelled by juvenile Starlings, *Sturnus vulgaris*, trapped on autumn migration in the Netherlands and displaced to suitable (Spain) and unsuitable (Switzerland) winter quarters. After Perdeck (1964)

	Release site		
Time of trapping	Netherlands	Spain	Switzerland
Early (1–23 October)[a]	450	480	700
Late (24 October–8 November)[b]	290	80	720

[a] 'Early' birds originate in Holland, Denmark, and Germany and normally winter in the west of England or in Ireland.
[b] 'Late' birds breed in Sweden, Finland, Poland, and Russia and normally winter in Holland, Belgium, and north-western France.

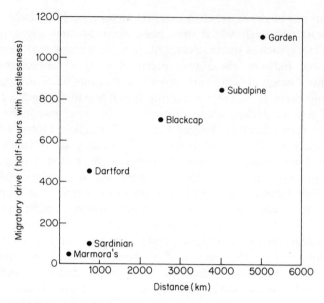

Figure 15.3 Migratory drive (number of experimental half-hours with migratory restlessness or *Zugenruhe*) in relation to average journey length on migration in six species of *Sylvia* warblers. Drawn from data in Berthold (1973). Species are: Garden Warbler, *Sylvia borin*; Subalpine warbler, *S. cantillans*; Blackcap, *S. atricapilla*; Dartford Warbler, *S. undata*; Sardinian Warbler, *S. melanocephala*; and Marmora's Warbler, *S. sarda*.

migration by each species, a striking correlation is apparent (Figure 15.3). Young of the long-distance migrants showed more migratory activity than did short-distance species, presumably driving them a greater distance from their natal areas before winter-ground characteristics can act to halt them.

DISPERSAL PATTERNS AND BREEDING SITE FIDELITY

Offspring seeking to breed for the first time must choose between moving from the natal area in the expectation of breeding successfully elsewhere sooner and waiting in the natal area for death of their parents or their peers to create a breeding vacancy. Baker (1978) suggests that there result from such choices three patterns of variation in the dispersion of young birds between breeding sites. In the first category, exemplified by the Pied Flycatcher, the males establish themselves in a territory before females select a mate and the male undertakes both territorial defence and a degree of courtship feeding. Here, juveniles are very likely to abandon their natal area for the chance of a better area elsewhere.

A second category contains species inhabiting irregularly arid regions, species feeding on the seed crops of trees, and species feeding on cyclic populations of rodents. For such birds, food abundance is synchronized over large areas separated from each other by areas of scarcity and, whilst locally unpredictable, is unlikely to be high for 2 years in succession. In such conditions selection for adults to move on to new areas at any stage in their lives is high, and differentials in movement tendency between adults and juveniles are low. Thus, seed-eaters, rodent predators, and rain-dependent desert species all show nomadic breeding. Nomadism is favoured in species with increased clutch size, high juvenile survival, and low adult survival (Andersson, 1980). With high juvenile survival the relative effect on future populations of breeding early is large, so juveniles should seek new sites allowing early breeding rather than await the next flush of food in the natal area.

Baker's third category covers colonially nesting species, especially seabirds. Here a prolonged period of pre-reproductive life is often accompanied by much movement and a low to medium probability of settling outside the natal colony. Once settled, though, such species are very faithful to the colony, probably because a considerable social investment is involved. If positions within the colony vary in value to the birds, individuals with greatest resource-holding power will get the best sites, and those with least ability will get the poorest sites or be excluded altogether. In such circumstances long-lived birds may do best to return to the colony for which they have already learned their own abilities relative to those of their neighbours. For the same reasons these species are usually characterized by persistent, often lifelong, pair-bonds (see Chapter 16).

Greenwood et al. (1979) investigated natal dispersal in the Great Tit population of Wytham Wood, Oxford. Great Tits are more markedly site-faithful as juveniles than are many other passerines. Nearly 25 per cent of the males but only 10 per cent of the females established a territory close to that in which they were born. Great Tit broods remain with their parents for 1–2 weeks after leaving the nest, dispersing in late summer. By September the surviving young are usually 700–1000 metres from their birthplace and by October the sex difference is already apparent. This period of dispersal coincides with heavy juvenile mortality and in the Wytham population the annual variation in this mortality is the critical factor causing the substantial variations in breeding density from year to year. In a similar study in Swedish oakwood Dhondt (1979) found that only 22 per cent of the young were still alive at the beginning of September.

Is Great Tit dispersal driven by food shortage in summer? When the family parties break up the juveniles join larger flocks, which usually contain a few adults and young from a number of broods. Within these flocks chases are frequent and in interactions at feeding stations a dominance hierarchy is evident. Hence it might pay subordinate juveniles to leave the breeding areas, perhaps to go to places where food exists but where breeding is prevented by the absence of nest-holes. Dhondt (1979) installed sunflower seed dispensers in his study woods at the beginning of August. In areas with large populations of birds using nest-boxes, these foods were immediately taken by Great Tits,

Marsh Tits, and Nuthatches, though not by Blue Tits. In an area without nest-boxes, sunflower seeds were not taken until the beginning of September. These results imply that summer foods can become scarce when large numbers of juveniles are about. Kluyver (1971) found in Holland that experimental removal of first broods led to unusually high survival and persistence of juveniles from the second broods.

In various communally breeding species young males do not disperse from their natal territory. Instead, they stay to help defend the family territory, which may increase in size as a result. In the Florida Scrub Jay, young males can then either inherit their father's position on his death or may be able to bud off a portion of the family land for their own use (Woolfenden and Fitzpatrick, 1978). Females must disperse to seek a breeding vacancy elsewhere and they suffer greater mortality whilst away from the safety of a 'home' group.

SEX DIFFERENCE IN DISPERSAL

Several studies, of a variety of species, have demonstrated sexual differences in the dispersal of fledglings. In a study of Scottish Hen Harriers Picozzi (1978) found that six of 11 nestling females tagged were seen on the study area in subsequent seasons, with the only other female recovery found just 10 km away; young males, however, were recovered at an average distance of 273 km. For Manx Shearwaters Brooke (1978) estimated that half the females reared on Skokholm Island off the coast of Wales emigrate while few, if any, males are likely to emigrate. Amongst gulls, males of the Red-billed Gull (Mills, 1973) and of the Herring Gull (Chabryzk and Coulson, 1976) show greater fidelity to the natal colony than do females, which disperse more. Other examples were cited in the preceding section.

INHERITANCE OF DISPERSAL TENDENCIES

Apart from these sex-related differences, is there an innate tendency for some young birds to disperse more than others? The genetic component of dispersal has been estimated for Great Tits by Greenwood et al. (1979), by comparing the distances moved between natal site and place of subsequent breeding for parents against the distances subsequently moved by their offspring. Because the dispersal distances varied annually the analyses were conducted on distances transformed to zero mean and unit variance, thus standardizing each bird against the population's dispersal pattern for the year (Figure 15.4). The distances moved by males were correlated both with the distances previously moved by their fathers and with the distances moved by their mothers. Daughter dispersal was correlated with that of the mother; a father–daughter correlation was present but was not statistically significant. Overall, some 56–62 per cent of the variance in offspring dispersal was attributable to a genetic component. Thus, males moving short distances tend

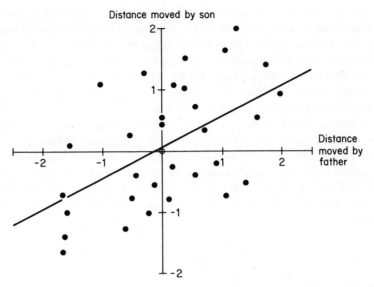

Figure 15.4 Correlation between distance moved by a young male Great Tit, *Parus major*, between birth and first breeding and the equivalent distance moved by its father in its pre-breeding period. Distances expressed as standard deviations from the mean, to eliminate non-hereditary factors. Redrawn with permission from Greenwood *et al.* (1979).

to produce sons who settle close to their natal areas, and males moving afar equally tend to produce some likely to disperse long distances. However, it is not possible to attribute this correlation to an innate predisposition to disperse because the birds concerned could conceivably share some other trait linked to mobility. Dhondt (1979) also found that siblings were more alike in dispersal distance than were non-siblings.

SUMMARY

Precocial young at first disperse from the nest sites to avoid predation, but their later distribution is determined, as for altricial species, by the availability of suitable feeding areas. Among waterfowl, and in species with dependent fledglings, the young are usually guided to these areas by adults. In migratory species young birds may possess an innate compass ability which they calibrate by reference to the learned axis of rotation of the night sky. They may additionally possess a magnetic compass and the integration of magnetic and stellar information is learned by experience. Young birds can migrate in the correct direction in autumn and can reverse it in spring, and this setting of the compass is under hormonal control. Natal areas are learned during a sensitive period in the birds' first autumn, and the young in general subsequently return to these areas each summer. Wintering areas are determined differently, with the young possessing some form of endogenous setting of its minimum migratory travel along its set compass course, though travel is continued if the area thus arrived at proves unsuitable, as shown by displacement experiments. Once the wintering and breeding areas are thus established, displaced birds will correctly compensate for natural or experimental displacement and will return to the home (winter or summer) area. Juveniles are most likely to move away from their natal areas if food is scarce or if population pressure is high there, particularly if juvenile survival is high. In communal and in colonial species, however, kin selection or knowledge of its actual social status relative to its colony mates may result in a young bird staying within the natal area, even if initially as a non-breeder. Dispersal tendencies are inherited, at least in some species, and may also be sex-linked.

CHAPTER 16

Growing up and growing old

In this chapter are considered various topics impinging on the transition of young birds from juvenile status to experienced adulthood. Certain biases may be present in any study of age effects in birds. One is that it may be only the more efficient birds which survive into the older age classes. Such a survival pattern would generate a correlation between age and breeding success in cross-sectional studies, even were there no biological dependence of success upon age. Only longitudinal studies – the same sample of birds examined at each age – can preclude such a bias. Secondly, older birds in general have had more experience of wintering, of breeding, or whatever, so that it is often impossible to distinguish effects of age from those of experience. For many purposes, though, the different mechanisms do not matter.

ENTERING THE WINTER FLOCK

Getting itself established within the wintering flock is probably the earliest test of a juvenile's individual fitness. In many flocking species, access to food is regulated by the formation of a dominance hierarchy within the group. In its simplest form a hierarchy is linear: one bird, the alpha-individual, dominates all other members of the group and can successfully supplant them for food, roosting site, or the like, as it will. A second individual, the beta-bird, is dominant over all but the alpha-individual, a third dominates all but the alpha- and beta-individuals, and so on down to an individual subordinate to all others in the group. The hierarchy initially develops through overt aggression and fighting between individuals, and it is at this stage that the young bird can win for itself a high or a low position in the hierarchy. Figure 16.1 shows how the incidence of fighting in young geese changes with age, eventually giving way to more stereotyped rank associated ceremonies. The advantages of the flock hierarchy to high-ranking individuals take the form of improved food intake as a result of supplanting subordinates who have located food. The advantages of a hierarchy to the subordinate individuals are less obvious and must lie with the ecological benefits of flock-living. Unless these outweigh the disadvantages of losing food to more dominant birds, subordinate individuals should opt for solitary lives, at least for the winter.

258

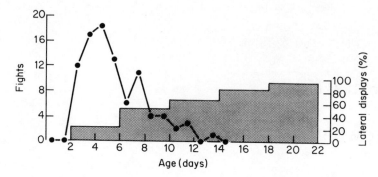

Figure 16.1 Changes in behaviour with age, from fighting (solid line) to lateral cackling displays (histogram), in resolving dominance disputes among Grey Lag, *Anser anser*, goslings. After Radesäter (1974).

Large broods tend to suffer proportionately greater mortality than small ones. Even so, a large family size may serve to provide a few members of the brood with the social experience and aggressive drive that they need to rise to high status in the world at large. In grouse and pheasants, individuals typically flock together for at least part of their annual cycle (usually the winter period), with social relationships within the flocks organized in a dominance hierarchy. Boag and Alway (1980) found that experience acquired during the brood period appears to influence the rank that an individual acquires when the broods have broken up and the juveniles come into the winter flocks (Figure 16.2). The more male siblings a male has, the more aggressive or the more effective in coping with aggression he becomes, and the higher the

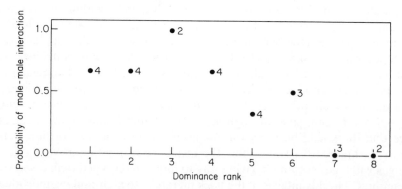

Figure 16.2 The relationship between the eventual dominance rank of young male Spruce Grouse, *Canachites canadensis*, in a winter flock, and the size and sex composition of their natal brood, showing how the more aggressive and dominant birds came from larger broods. Birds of rank 1, 2, and 4 were from the same brood. Figure by each point gives the number of young in the brood. Data from Boag and Alway (1980).

status he can acquire in the establishment of the winter flocks. This will happen even if individual mortality in large broods is disproportionately severe when compared with that within small broods. Effectively, what Boag and Alway suggest is that a large brood provides a small number of its members with the social experience needed to get to such high positions in later life that their reproductive success there more than compensates for the excessive mortality of their siblings as juveniles. This kin selection explanation is exactly analogous to that advanced by O'Connor (1978c) in relation to altricial survival in brood reduction, but the benefit is one of social experience, not a greater share of the available food.

Kin selection likewise underlies the integration of yearlings into family flocks in some communal breeders. Young Piñon Jays are recorded a privileged status at feeding stations as they commence independent feeding and are given way to by older and stronger birds even when the yearlings are aggressive (Balda, 1975). This permissive acceptance of the yearlings is, however, founded on an earlier period of communal care and feeding.

Recognition of social rank

The maintenance of a social hierarchy depends on some form of recognition of the ranks of interacting individuals. This may involve individual recognition, with each bird in the flock known to all other members and their relative rank determined by initial aggression. A young bird entering a hierarchical flock establishes its own status through a series of disputes, until mutual recognition and knowledge of the relative strength of the two opponents as a result of past aggressive encounters between them allow prediction of the outcome of further disputes. In this way a stable social hierarchy reduces the value of overt aggression between flock members.

Recognition of opponents as individuals is not strictly essential for the maintenance of a hierarchy. Just as army sergeants are recognizable as superior in rank to corporals by their wearing three rather than two stripes on their arms, so may birds bear some external evidence of their ability. Such evidence need not be as stereotyped as the sergeant's stripes: few people would willingly start a fight with a man with the build of a heavyweight wrestler. Where individual recognition is not the basis of the hierarchy then, young birds have to learn to discriminate rank.

In at least one species, the Harris Sparrow, social status has been shown to be linked with age-related plumage markings (Rohwer, 1977). Individuals of this species differ in the amount of black feathering on their crowns and throats, the dominant birds having most black. In the Harris Sparrow old males are more heavily marked than old females and these in turn are more heavily marked than young males and then young females, so that plumage marking is normally broadly in line with experience and fighting ability.

In other species again, social rank is indicated not by relatively permanent age-related plumage markings but by the adoption of special signal postures communicating the willingness or otherwise of the bird concerned to fight or otherwise react to a conspecific. Evidence is available to show that juvenile birds may be less adept at interpreting and responding to the signals of older birds (e.g. Groves, 1978). In the Ruddy Turnstone a 'tail-depressed' posture is shown by foraging birds in aggressive state. Individuals so displaying are particularly likely to attack and chase any other Turnstone approaching them during feeding. Observations of migrating Turnstones on the Massachusetts coast in autumn showed that juveniles were still in the process of learning the meaning of this posture. Aggressive interactions were seen in 24 out of 37 flocks of mixed age and were significantly more likely to involve an interaction between an adult and a juvenile than one between two adults. But why should there be this lag in learning that birds in 'tail-depressed' posture are being aggressive? Perhaps because the young were escorted by their parents throughout the pre-fledging period on the breeding grounds without aggression, they are slow to learn that the context of their interactions with adults on the feeding grounds is very different (Groves, 1978).

Social rank, size, and experience

Young birds are usually subordinate to older adults, partly because the latter are often slightly larger in size but also because the adults have greater experience of aggressive encounters (Groves, 1978). Among Rooks, for example, the adults secure the best roosting places when these are scarce and first-year birds suffer accordingly (Swingland, 1977). Even within juvenile peer groups body size and experience may determine social rank. Garnett (1981) hand-reared Great Tit nestlings and tested their success in interactions with other juveniles when they fledged. In the first week of testing, dominance within the aviary lay with young from early broods, but in the second week this dominance gave way to one based on tarsus size: larger juveniles dominated smaller ones. Seasonal variation in growth rates thus affects fledgling survival in two ways. Early broods have time to develop and to learn to look after themselves before late broods fledge and can therefore dominate the latter when they first leave their nests. Within a short while, though, those late fledglings which survive are both skilled enough and large enough to dominate small young from early broods, and dominance and survival become size-related.

What would happen to the young of late broods were there no first brood young around to dominate them in the immediate post-fledging period? Kluyver (1966) experimentally removed 60 per cent of eggs or young from Great Tit pairs nesting on the Dutch island of Vlieland, repeating the experiment for several years after an initial control period of 4 years with no removals. Following this drastic reduction of fledgling numbers, the mean adult survival more than doubled and juvenile survival almost doubled. When this type of experiment was repeated with only young of the first broods removed, the young of the second broods subsequently showed a much higher local survival than normal (Kluyver, 1971). These experiments point strongly to inter-fledgling competition as a major determinant of the survival prospects of young of different body size.

AGE OF FIRST BREEDING

In many species young birds are physiologically capable of breeding long before they actually do so. However, attempting to breed involves a risk to the bird's own survival, particularly so when it is inexperienced. Most deferred breeders are long-lived species, so in many of them more can be gained by acquiring further non-breeding experience than is lost in foregoing early breeding opportunities. When it eventually breeds, the older bird has a better chance of surviving its first nesting attempts and of going onto many further years of breeding.

The age of a breeding bird affects its success most in species with 'difficult' food resources and least in species with readily available prey. House Martins in Scotland feed on aerial insects and have difficulty in catching enough to initiate breeding when they first return each season (Bryant, 1975). Successful completion of nesting attempts was commoner for older parents, and Bryant's (1979) analysis of contributory factors shows that practically every component was better in the case of older birds, for males and females alike (Table 16.1). Several of these findings are also true of Arctic Terns which, in hunting

Table 16.1 Components of breeding success amongst House Martins, *Delichon urbica*, in relation to age. Modified from Bryant (1979)

	Males		Females	
	First year	Older	First year	Older
Nesting attempts yielding one or more young (%)	91.1	96.0	90.2	98.0
Clutches with interrupted laying (%)				
First clutches	31.5	25.0	25.7	28.6
Second clutches	7.4	0.0	8.3	0.0
Duration of laying interruptions (days)	4.6	2.2	2.9	3.6
Time to re-lay following clutch loss (days)	8.0	7.9	10.5	9.6
Eggs lost (%)				
First clutches	7.6	10.0	10.5	9.6
Second clutches	5.1	4.1	5.1	4.4
Young lost (%)				
First broods	3.1	4.4	4.7	3.9
Second broods	1.7	7.0	5.1	6.2
Duration of nest cycle (days)	50.3	50.9	50.1	50.0

pelagically for small fish or plankton, also experience feeding difficulties (Coulson and Horobin, 1976). Deferred reproduction is also common amongst species in which nest sites or mates are in short supply, so that only the most experienced adults are successful.

Young birds may breed unusually early whenever environmental changes reduce the breeding population of older adults and their competition for breeding resources. In extreme cases young may be able to breed whilst still in immature plumages! Immature Brown Pelicans have been found breeding in California where the species experienced a population decline, but few or none have been detected in the very stable colonies of this species in Florida and North Carolina (Blus and Keahey, 1978). Similarly, the culling of large numbers of Herring Gulls from the Isle of May in Scotland as part of the conservation management programme there resulted in a doubling of the number of young birds acquiring a territory (Chabrzyk and Coulson, 1976).

Expanding colonies likewise often offer young birds the opportunity to breed sooner than they could within an established colony. Newly established colonies of the Kittiwake contain a high percentage of inexperienced breeders (Coulson and White, 1958) and, indeed, the proportion of first-time breeders even in established colonies of this species has been increased by creating new breeding sites in the form of artificial ledges.

Unusually high proportions of young breeders (again with some in sub-adult plumage) also occur in populations of Pomarine Skuas in years of lemming plagues (Pitelka et al., 1955). Young breeders have also been noted in populations of Australian Magpies and among young penguins in years of abundant resources (Lack, 1968). With an abundance of food available, first-time breeders can attempt a brood without endangering their own survival and future breeding prospects.

These three lines of evidence – from natural and experimental population crashes, from expanding colonies, and from cases of unusually abundant food – show that the young birds are prevented from earlier breeding by environmental restriction rather than by physiological inadequacies.

Age of first breeding and sex ratio

An imbalance in the ratio of males to females entering the breeding population can cause sexual differences in the age distribution of first-time breeders. In the Great Tit study at Oxford, about one-third of the first-year males fail to find mates and breed for the first time at 2 years, whilst most females breed in their first year (Perrins, 1979). Here a slight differential in sex ratio amongst first-year birds is magnified by mortality amongst the females: annual survival by females is 48 per cent compared with 56 per cent in males, the difference due largely to predation of sitting females by weasels. Females are thus a scarce resource and can choose to pair with older, more experienced males. In other species the sex ratio is biased in favour of males, who then breed at lower ages than do females: amongst the Red-billed Gulls of New Zealand females are present in surplus numbers but form only 19 per cent of the breeding 2-year-olds, most deferring first reproduction until their third, fourth, or even fifth year (Mills, 1973).

AGE AND REPRODUCTIVE COMMITMENT

In long-lived species a young bird in its first breeding season can expect to see several future breeding attempts before it dies. Consequently, its conduct of its early breeding attempts should be moderated by the need to avoid risking these future reproductive opportunities. But, in addition to such a trade-off of present versus future reproductive effort, young birds trying to breed for the first time are also vulnerable to reduced reproductive success at three points. First, their clutch size may be lower because they are inexperienced at finding food for egg laying and are less able to feed young (Lack, 1968). Secondly, their ability to acquire a nest site, territory, or other resource for breeding may be limited by adverse social relationships with adults and with other first-time breeders (Wynne-Edwards, 1962). Finally, their ability to recognize and respond to potential predation may be inferior to that of more experienced adults. Nor, indeed, need these various factors be independent of each other. We saw earlier that laying females may have difficulty accumulating enough food for the formation of the eggs. For many species this results not only in later laying by the younger females but also in reduced egg size (Figure 16.3). As the females gain breeding experience they form steadily

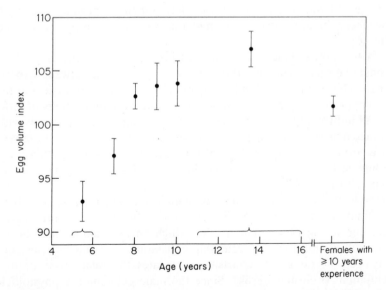

Figure 16.3 Egg size in relation to female age in the Manx Shearwater, *Puffinus puffinus*. Redrawn with permission from Brooke (1978).

larger eggs, at least to some maximum size, an adaptive trend by which nestling growth or survival are correlated with egg weight or volume (Chapter 3). The egg size of known age (or experience) females has been determined over long age spans for very few species, but in several cases where this has been done – for the Ruff (Andersen, 1951), Herring Gull (Davis, 1975), and Manx Shearwater (Brooke, 1978) – evidence has been obtained for a decrease in egg size with still greater age.

Age versus experience

Some aspects of breeding performance improve with age alone, independently of the experience which is normally gained with increasing age, and some aspects improve with the greater experience of breeding acquired with age. Finney and Cooke (1978) compared clutch sizes of different age groups of Lesser Snow Geese breeding in Manitoba. In 1974, but not in 1973, a large proportion of 3 year old females had bred the previous year as 2-year-olds. In 1973 most 3-year-olds were therefore without breeding experience and had an average clutch size of 83 per cent of the adult average for the year whilst in 1974, with the benefit of one year's breeding experience, the 3-year-olds averaged 94 per cent of adult clutch size. Experience thus provided the 1974 3-year-olds with the ability to increase their clutch size by 11 percentage points. In addition, an 18 percentage point increment is attributable to age change alone: in 1973, as 2-year-olds, the same birds had averaged 65 per cent of the adult clutch size, against the 83 per cent of adult values averaged in 1974. The experienced birds also bred earlier than inexperienced birds of the same age. As clutch size is the major determinant of productivity in Lesser Snow Geese, breeding experience had a direct effect on reproductive success in this study.

Data from other studies confirm that breeding experience has effects on reproductive success beyond those due to a correlation between experience and age. The Arctic Skua tends to form pairs in which both male and female have similar breeding experience but many pairs contain partners of different ages and individual breeding experience. Breeding success increases with the combined experience of the pair, even when the two birds have not previously bred together (Table 16.2). Part of this success is due to an ability to start breeding earlier in the season, since the earliest birds are the most successful (Davis, 1976).

In the Red-billed Gull male age and female age contribute separately to the determination of clutch size (Figure 16.4). This is partly correlated with earlier egg laying by older pairs, for which an explanation may lie in the mutual stimulation found in courtship behaviour. For some species, such as the Ringed Turtle Dove, courtship stimulates the final phase of ovarian development (Lehrman, 1964). Since the male gulls mature physiologically progressively earlier with age and season, they may stimulate the completion of ovarian development in their partner correspondingly earlier by their own more rapid initiation of courtship.

Table 16.2 Breeding success of Arctic Skua, *Stercorarius parasiticus*, breeding together for the first time, in relation to the previous experience of the birds forming the pair. From Davis (1976)

Previous experience of male plus that of female (years)	Chicks fledged		
	Mean	Standard error	N
0	0.65	0.07	60
1	0.94	0.10	33
2	1.10	0.15	20
3–5	0.87	0.11	23
6+	1.38	0.11	24

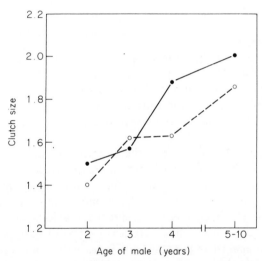

Figure 16.4 Clutch size of Red-billed Gulls, *Larus novaehollandiae*, in relation to the age of male and female partner. 'Young' females (O---O) are those 2–4 years in age, 'old' females (●—●) are 5–10 years in age. Data from Mills (1973).

PAIR-BOND DURATION, AGE, AND DIVORCE

Long-lived birds also develop in reproductive success with increase in experience of the same breeding partner. Arctic Skuas bred earlier and more successfully when both birds of the pair had previously bred together than when the pair was a new one (even though between two experienced birds) (Davis, 1976). Fulmars showed a drop in breeding success on taking a new

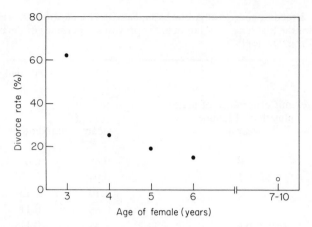

Figure 16.5 Divorce rate (changes of mate whilst original partner was still alive) in relation to age of female in Red-billed Gulls, *Larus novaehollandiae*. Data from Mills (1973).

mate following the disappearance (presumably through death) of their previous mate, from 70 per cent success with the old partner to 50 per cent success with their new one, and the breeding success of established pairs was significantly correlated with the duration of their pair-bond (Ollason and Dunnet, 1978). Very similar results were obtained for the Kittiwake (Coulson, 1966).

In several species birds 'divorce' their partner if their reproduction has been poor: Kittiwakes are three times more likely to separate following hatching failure in a breeding attempt than following a success (Coulson, 1966). Similarly, in Red-billed Gulls those individuals who changed mates between seasons had been less successful the previous summer than were birds retaining their mates. They were less successful both in hatching their eggs (64 per cent against 78 per cent) and in fledging young once hatched (69 per cent against 84) (Mills, 1973). Since birds changing their partner lose the benefits of pair-bond duration already mentioned and suffer reduced success for the first year or two with their new mates, changes of partner should be most frequent early in life (Figure 16.5). Divorce seems to be an option open to young birds seeking a compatible partner to share what can be expected to be a long life together. In Fulmars the divorced birds show greater success (75 per cent) with their new partners than they did immediately prior to divorce (64 per cent).

AGE AND BREEDING SEASON

Within a breeding season the younger birds often breed later than do older adults, a difference reflected in the gonadal maturation of the different age-groups (Figure 16.6): both older males and older females in a Red-billed Gull population showed greater enlargement of the gonads the older they were (Mills, 1973). What is uncertain with such data, however, is whether a young bird is prevented from breeding earlier by physiological limitations within its endocrine system or whether the changes in maturation state are merely correlates of the overall changes in breeding season.

Because young birds breed rather later each season than do older ones, partners tend to be similar in age (except where, as discussed earlier, a sex differential in survival exists). The older birds arrive early and pair with each other, leaving later arrivals few opportunities of pairing with other than another young bird. For example, 26 per cent of 212 Red-billed Gull pairs had partners of equal age and a further 42 per cent were only 1 year apart (Mills, 1973). Corresponding percentages for Yellow-crowned Penguins in New Zealand were 16 per cent and 22 per cent respectively (Richdale, 1957) and for Arctic Skuas in the Orkney Islands 42 per cent and 25 per cent (equating breeding experience to age) (Davis, 1976). Thus, although a young male entering the breeding population should ideally mate with an older female, in

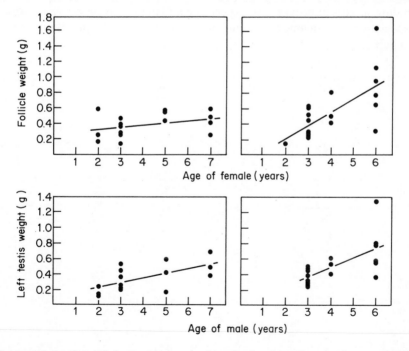

Figure 16.6 Gonadal maturation in female (top) and male (bottom) Red-billed Gulls, *Larus novaehollandiae*, in relation to age. Left-hand diagrams: samples taken 24 days prior to egg-laying starting in the colony. Right-hand diagrams: samples taken 2 days before egg-laying started. Redrawn with permission from Mills (1973).

practice he must, in the majority of cases, mate with a bird of similar age. Young females are likewise constrained. Once mated, the advantages of mate-retention favours the maintenance of the age similarity.

SUMMARY

Young birds have to establish themselves within the adult population and this may depend on their social dominance. Juvenile birds have to learn the significance of signal markings and postures used in social interactions. In some species a very few young from larger broods may obtain especially high status in the winter population as a result of the experience in dominance interactions which they acquire within the brood. In other species their eventual standing may depend on their physical size and strength. The age at which birds first attempt to breed varies with the ease of breeding, being late wherever the task of securing a nest site and territory or food is especially difficult, and conversely. This may differ between the sexes. Young birds generally have a low commitment to their earliest breeding attempts, at least in long-lived species, and may have special difficulties in relation to nest predators, territory quality, egg production, and the feeding of offspring. Breeding success improves with experience and in long-lived species 'divorce' is especially common amongst young breeders whenever the pair is unsuccessful together.

References

Abs, M., 1970. The breaking of the voice in domestic ducks. *Ostrich* Suppl. **8**: 77–83.

Abs, M., 1974. Zur Entwickling der Lautausserungen bei der Haustaube (*Columbia livia domestica*). *Verh. dt. Zool. Ges.* **1974**: 347–350.

Abs, M., 1975. Uber entwicklingsbedingte Anderungen des Krahrufes vom Haushahn. *Arch. Geflügelk.* **1**: 29–36.

Alexander, R. D., 1974. The evolution of social behavior. *A. Rev. Ecol. Syst.* **5**: 325–383.

Alerstam, T., 1978. Re-oriented bird migration in coastal areas: dispersal to suitable resting grounds? *Oikos* **30**: 405–408.

Andersen, F. S., 1951. Contributions to the biology of the Ruff (*Philomachus pugnax* (L.)). III. *Dansk. orn. Foren. Tidsskr.* **45**: 145–173.

Andersson, M., 1971. Breeding behaviour of the Long-tailed Skua *Stercorarius longicaudus* (Viellot). *Ornis scand.* **2**: 35–54.

Andersson, M., 1976. Population ecology of the long-tailed skua (*Stercorarius longicaudus* Vieill.). *J. Anim. Ecol.* **45**: 537–560.

Andersson, M., 1980. Nomadism and site tenacity as alternative reproductive tactics in birds. *J. Anim. Ecol.* **49**: 175–184.

Andrew, R. J., 1956. Normal and irrelevant toilet behaviour in *Emberiza* spp. *Br. J. Anim. Behav.* **4**: 85–91.

Ar, A. and Y. Yom-Tov, 1979. The evolution of parental care in birds. *Evolution* **32**: 655–668.

Armstrong, E. A., 1958. Distraction display and the human predator. *Ibis* **98**: 641–654.

Aschoff, J., and H. Pohl, 1970. Der Ruheumsatz von Vögeln als Funktion der Tagezeit und der Körpergrösse. *J. Orn., Lpz.* **111**: 38–47.

Ash, J. S., 1957. Post-mortem examinations of birds found dead during the cold spells of 1954 and 1956. *Bird Study* **4**: 159–166.

Ashmole, N. P., 1971. Seabird ecology and the marine environment. In D. S. Farner and J. R. King (eds), *Avian Biology*, Vol. I, Academic Press, New York, pp. 222–286.

Ashmole, N. P. and H Tovar, 1968. Prolonged parental care in Royal Terns and other birds. *Auk* **85**: 90–100.

Aulie, A., 1976. The pectoral muscles and the development of thermoregulation in chicks of the Willow Ptarmigan (*Lagopus lagopus*). *Comp. Biochem. Physiol.* **53A**: 343–346.

Aulie, A., and P. Moen, 1975. Metabolic thermoregulatory responses in eggs and chicks of Willow Ptarmigan *Lagopus lagopus*. *Comp. Biochem. Physiol.* **51A**: 605–609.

Austin, G. T., 1974. Nesting success of the Cactus Wren in relation to nest orientation. *Condor* **76**: 216–217.

Austin, G. T., and R. E. Ricklefs, 1977. Growth and development of the Rufous-winged Sparrow (*Aimophila carpalis*). *Condor* **79**: 37–50.

Baeyens, G., 1981. Magpie breeding success and Carrion Crow interference. *Ardea* **69**: 125–139.

Baker, R. R., 1978. *The Evolutionary Ecology of Animal Migration*, Hodder and Stoughton, London.

Balda, R. P., 1975. Care of young Piñon Jays and their integration into the flock. *Emu* **74**: 306.

Balda, R. P., and G. C. Bateman, 1976. Cannabalism in the Piñon Jay. *Condor* **78**: 562–564.

Baldwin, S. P., and S. C. Kendeigh, 1932. Physiology of the temperature of birds. *Scient. Publ. Cleveland Mus. nat. Hist.* **3**: 1–196.

Balen, J. H. Van, 1973. A comparative study of the breeding ecology of the Great Tit *Parus major* in different habitats. *Ardea* **61**: 193.

Balen, J. H. van, 1979. Observations on the post-fledging dispersal of the Pied Flycatcher, *Ficedula hypoleuca*. *Ardea* **67**: 134–137.

Balen, J. H. van, and A. J. Cavé, 1970. Survival and weight loss of nestling Great Tits, *Parus major*, in relation to brood size and air temperature. *Neth. J. Zool.* **20**: 464–474.

Balph, M. H., 1975. Development of young Brewer's Blackbirds. *Wilson Bull.* **87**: 207–230.

Baltin, S., 1969. Zur Biologie und Ethologie des Talegellahuhns (*Alectura lathami*) unter besonderer Berücksichtigung des Verhaltens während der Brutperiod. *Z. Tierpsychol.* **26**: 524–572.

Barash, D. P., 1975. Evolutionary aspects of parental behavior: distraction behavior of the Alpine Accentor. *Wilson Bull.* **87**: 367–373.

Barth, E. K., 1951. Kroppstemperatur hos måkeunger. *Nytt Mag. Naturvid.* **88**: 213–245.

Bartholomew, G. A., and T. R. Howell, 1964. Experiments on nesting behaviour of Laysan and Black-footed Albatrosses. *Anim. Behav.* **12**: 549–559.

Bartonek, J. C., and J. J. Hickey, 1969. Food habits of Canvasbacks, Redheads and Lesser Scaup in Manitoba. *Condor* **71**: 280–290.

Bartov, J., and S. Bornstein, 1976. Effects of degree of fatness in broilers on other carcass characteristics: relationship between fatness and the composition of carcass fat. *Br. Poult. Sci.* **17**: 17–27.

Bateson, P. P. G., 1966. The characteristics and context of imprinting. *Biol. Rev.* **41**: 177–210.

Bateson, P. P. G., 1973. The imprinting of birds. In S. G. Barnett (ed.), *Ethology and Development*, Heinemann Medical Books, London, pp. 1–15.

Bateson, P. P. G., 1974. Length of training, opportunities for comparison and imprinting in chicks. *J. comp. physiol. Psychol.* **86**: 586–589.

Bateson, P. P. G., 1978. Sexual imprinting and optimal outbreeding. *Nature, Lond.* **273**: 659–660.

Bateson, P. P. G., 1979. How do sensitive periods arise and what are they for? *Anim. Behav.* **27**: 470–486.

Bateson, P. P. G., and G. Seaburne-May, 1973. Effects of prior exposure to light on chicks' behaviour in the imprinting situation. *Anim. Behav.* **21**: 720–725.

Bateson, P. P. G., and A. A. P. Wainwright, 1972. The effects of exposure to light on the imprinting process in domestic chicks. *Behaviour* **42**: 279–290.

Bateson, P. P. G., G. Horn, and S. P. R. Rose, 1969. Effects of an imprinting procedure on regional incorporation of tritiated lysine into protein of chick brain. *Nature, Lond.* **223**: 534–535.

Bateson, P. P. G., S. P. R. Rose, and G. Horn, 1973. Imprinting: lasting effects on uracil incorporation into chick brain. *Science, N.Y.* **181**: 576–578.

Bateson, P. P. G., G. Horn, and S. P. R. Rose, 1975. Imprinting: correlations between behaviour and incorporation of [^{14}C]uracil into chick brain. *Brain Res.* **84**: 207–220.

Batt, B. D., J. A. Cooper, and G. W. Cornwell, 1975. The occurrence of twin waterfowl embryos. *Condor* **77**: 214.

Beer, C. G., 1970. On the responses of Laughing Gull chicks (*Larus atricilla*) to the calls of adults. I. Recognition of the voices of the parents. *Anim. Behav.* **18**: 652–660.

Bengtson, S. A., 1971a. Food and feeding of diving ducks breeding at Lake Myvatn, Iceland. *Ornis fenn.* **48**: 77–92.

Bengtson, S. A., 1971b. Habitat selection of duck broods in Lake Myvatn area, North-East Iceland. *Ornis scand.* **2**: 17–26.

Berndt, R. and W. Winkel, 1979. Verfrachtungs-Experimente zur Frage der Geburtsortprägung beim Trauerschnäpper (*Ficedula hypoleuca*). *J. Orn., Lpz.* **120**: 41–53.

Berthold, P., 1973. Relationships between migratory restlessness and migration distance in six *Sylvia* species. *Ibis* **115**: 594–599.

Bertram, B. G., 1980. Breeding system and strategies of Ostriches. *Proc. 17th Int. Orn. Congr.*: 890–894.

Best, L. B., 1977. Nestling biology of the Field Sparrow. *Auk* **94**: 308–319.

Best, L. B., and D. F. Stauffer, 1980. Factors affecting nesting success in riparian bird communities. *Condor* **82**: 149–158.

Bilby, L. W., and E. M. Widdowson, 1971. Chemical composition of growth in nestling Blackbirds and thrushes. *Br. J. Nutr.* **25**: 127–134.

Bildstein, K. L., 1980. Corn cob manipulation in Northern Harriers. *Wilson Bull.* **92**: 128–130.

Birkhead, T. R., 1976. Effects of sea conditions on rates at which Guillemots feed chicks. *Br. Birds* **69**: 490–492.

Blagosklonov, K., 1976. [Fixation of food selection in behaviour development of insectivorous birds.] *Biol. Nauki* **1976** (2): 87–92 (in Russian).

Blaker, D., 1969. Behaviour of the Cattle Egret *Ardeola ibis*. *Ostrich* **40**: 75–129.

Blase, B., 1960. Lautäusserungen des Neuentöters (*Lanius c. collurio* L.). *Z. Tierpsychol.* **17**: 293–344.

Blem, C. R., 1973. Laboratory measurements of metabolized energy in some passerine nestlings. *Auk* **90**: 895–897.

Blem, C. R., 1975. Energetics of nestling House Sparrows, *Passer domesticus*. *Comp. Biochem. Physiol.* **52A**: 305–312.

Blus, L. J., and J. A. Keahy, 1978. Variation in reproductivity with age in the Brown Pelican. *Auk* **95**: 128–134.

Boag, D. A., and J. W. Alway, 1980. Effect of social environment within the brood on dominance rank in gallinaceous birds (Tetraonidae and Phasianidae). *Can. J. Zool.* **58**: 44–49.

Bochenski, Z., 1961. Nesting biology of the Black-necked Grebe. *Bird Study* **8**: 6–15.

Boecker, M., 1967. Vergleichende Untersuchungen zur Nahrungs- und Nistökologie der Flusseeschwalbe (*Sterna hirundo* L.) und der Küstenseeschwalbe (*Sterna paradisaea* Pont.). *Bonn. zool. Beitr.* **18**: 15–126.

Boetius, J., 1949. Feeding-activity in some insectivorous birds. *Dansk. orn. Foren. Tidsskr.* **43**: 45–59.

Borror, D. J., 1967. Songs of the Yellowthroat. *Living Bird* **6**: 141–161.

Brenner, F. J., 1964. Growth, fat deposition and development of endothermy in nestling Red-winged Blackbirds. *J. Sci. Lab. Denison Univ.* **46**: 81–89.

Brindley, L. D., and S. Prior, 1968. Effects of age on taste discrimination in the Bobwhite Quail. *Anim. Behav.* **16**: 304–307.

Brisbin, J. L., 1965. A quantitative analysis of ecological growth efficiency in the

Herring Gull. MS thesis, University of Georgia.

Brisbin, J. L., 1969. Bioenergetics of the breeding cycle of the Ring Dove. *Auk* **86**: 54–74.

Brisbin, J. L., and L. J. Tally, 1973. Age-specific changes in the major body components and calorific value of growing Japanese Quail. *Auk* **90**: 624–635.

Brody, S., 1945. *Bioenergetics and Growth*, Reinhold, New York.

Brooke, M. de L., 1978. Some factors affecting the laying date, incubation and breeding success of the Manx Shearwater, *Puffinus puffinus*. *J. Anim. Ecol.* **47**: 477–496.

Broom, D. M., 1969. Behaviour of undisturbed 1- to 10-day-old chicks in different rearing conditions. *Devl. Psychobiol.* **1**: 287–295.

Brosset, A., 1971. L'‘imprinting’, chez les Columbidés – étude des modifications comportementales au cours du vieillissement. *Z. Tierpsychol.* **29**: 279–300.

Brown, J. L., 1972. Communal feeding of nestlings in the Mexican Jay (*Alphelecoma ultramarina*): interflock comparisons. *Anim. Behav.* **20**: 395–403.

Brown, J. L., 1975. *The Evolution of Behavior*, Norton, New York.

Brown, L. H., and E. K. Urban, 1969. The breeding biology of the Great White Pelican *Pelecanus onocrotalus roseus* at Lake Shala, Ethiopia. *Ibis* **111**: 199–237.

Brown, R. G. B., N. G. Blurton-Jones, and D. J. T. Hussell, 1967. The breeding behaviour of Sabine's Gull (*Xema sabini*). *Behaviour* **28**: 110–140.

Bryant, D. M., 1973. The factors influencing the selection of food by the House Martin (*Delichon urbica* (L.)). *J. Anim. Ecol.* **42**: 539–564.

Bryant, D. M., 1975. Breeding biology of House Martins *Delichon urbica* in relation to aerial insect abundance. *Ibis* **117**: 180–216.

Bryant, D. M., 1979. Reproductive costs in the House Martin (*Delichon urbica*). *J. Anim. Ecol.* **49**: 655–675.

Bryant, D. M., and A. Gardiner, 1979. Energetics of growth in House Martins (*Delichon urbica*). *J. Zool., Lond.* **189**: 275–304.

Buckley, P. A., and F. G. Buckley, 1972. Individual egg and chick recognition by adult Royal Terns (*Sterna maxima maxima*). *Anim. Behav.* **20**: 457–462.

Bulmer, M. G., and C. M. Perrins, 1973. Mortality in the Great Tit *Parus major*. *Ibis* **115**: 277–279.

Burger, J., 1972. Dispersal and post-fledging survival of Franklin's Gulls. *Bird-Banding* **43**: 267–275.

Burger, J., 1974. Breeding adaptations of Franklin's Gull (*Larus pipixcan*) to a marsh habitat. *Anim. Behav.* **22**: 521–567.

Burger, J., 1977. Nesting behaviour of Herring Gulls: invasion into *Spartina* salt marsh areas of New Jersey. *Condor* **79**: 162–169.

Burger, J., 1980. Age differences in foraging Blacknecked Stilts in Texas. *Auk* **97**: 633–630.

Burger, J., and J. Shisler, 1980. Colony and nest site selection in Laughing Gulls in response to tidal flooding. *Condor* **82**: 251–258.

Cain, B. W., 1976. Energetics of growth for Black-bellied Tree Ducks. *Condor* **78**: 124–128.

Calder, W. A., 1973. The timing of maternal behaviour of the Broadtailed Humming bird preceding nest failure. *Wilson Bull.* **85**: 283–290.

Calder, W. A., 1974. Consequences of body size for avian energetics. In R. A. Paynter (ed.), *Avian Energetics*, Nuttall Ornithology Club, No. 15., pp. 86–144.

Calder, W. A., and J. R. King, 1974. Thermal and caloric relations of birds. In D. S. Farner and J. R. King (eds), *Avian Biology*, Vol. IV, Academic Press, New York, pp. 260–413.

Carew, C., H. Rahn, and P. Parisi, 1980. Calories, water, lipid, and yolk in avian eggs. *Condor* **82**: 335–343.

Case, T. J., 1978. On the evolution and adaptive significance of postnatal growth rates in the terrestrial vertebrates. *Quart. Rev. Biol.* **55**: 243–282.

273

Catchpole, C. K., 1974. Habitat selection and breeding success in the Reed Warbler (*Acrocephalus scirpaceus*). *J. Anim. Ecol.* **44**: 363–380.

Chabrzyk, G., and J. C. Coulson, 1976. Survival and recruitment in the Herring Gull *Larus argentatus*. *J. Anim. Ecol.* **45**: 187–203.

Chamberlin, M. L., 1977. Relationships between egg pigmentation and hatching sequence in the Herring Gull. *Auk* **94**: 363–365.

Chura, N. G., 1963. Diurnal feeding periodicity of juvenile Mallards. *Wilson Bull.* **75**: 90.

Clark, G. A., 1961. Occurrence and timing of egg teeth in birds. *Wilson Bull.* **73**: 268–278.

Clark, G. A., 1964. Life histories and the evolution of megapodes. *Living Bird* **3**: 149–167.

Clark, G. A., 1970. Avian bill-wiping. *Wilson Bull.* **82**: 279–288.

Clark, L., and D. J. Gabaldon, 1979. Nest desertion by the Piñon Jay. *Auk* **96**: 796–798.

Clay, D. L., J. L. Brisbin, and K. A. Youngstrom, 1979. Age-specific changes in the major body components and caloric values of growing Wood Ducks. *Auk* **96**: 296–305.

Collias, E. C., and N. E. Collias, 1973. Further studies on development of nest-building behaviour in a weaverbird (*Ploceus cucullatus*). *Anim. Behav.* **21**: 371–382.

Collias, N. E., and E. C. Collias, 1963. Selective feeding by wild ducklings of different species. *Wilson Bull.* **75**: 6–14.

Cooke, F., 1978. Early learning and its effect on population structure. Studies of a wild population of Snow Geese. *Z. Tierpsychol.* **46**: 344–358.

Cooke, F., and K. F. Abraham, 1980. Habitat and locality selection in Lesser Snow Geese: the role of previous experience. *Proc. 17th int. Orn. Congr.*: 998–1004.

Cooke, F., and C. M. McNally, 1975. Mate selection and colour preferences in Lesser Snow Geese. *Behaviour* **53**: 151–170.

Cooke, F., P. J. Mursky, and M. B. Seiger, 1972. Colour preferences in the Lesser Snow Goose and their possible role in mate selection. *Can. J. Zool.* **50**: 529–536.

Cooke, F., G. H. Finney, and R. F. Rockwell, 1976. Assortative mating in Lesser Snow Geese (*Anser caerulescens*). *Behav. Genet.* **6**: 127–140.

Coppinger, R. P., 1969. The effects of experience and novelty on avian feeding behaviour with reference to the evolution of warning colouration in butterflies. I. Reactions of wild-caught adult Blue Jays to novel insects. *Behaviour* **35**: 45–60.

Coulson, J. C., 1966. The influence of the pair bond and age on the breeding biology of the Kittiwake Gull *Rissa tridactyla*. *J. Anim. Ecol.* **35**: 269–279.

Coulson, J. C., and J. Horobin, 1976. The influence of age on the breeding biology and survival of the Arctic Tern *Sterna paradisea*. *J. Zool., Lond.* **178**: 247–260.

Coulson, J. C., and E. White, 1958. The effect of age on the breeding biology of the Kittiwake *Rissa tridactyla*. *Ibis* **100**: 40–51.

Coulson, J. C., and R. D. Wooler, 1976. Differential survival rates among breeding Kittiwake Gulls. *J. Anim. Ecol.* **45**: 205–213.

Craig, J. L., 1975. Co-operative breeding of Pukeko. *Emu* **74**: 308.

Cronmiller, J. R., and C. F. Thompson, 1980. Experimental manipulation of brood size in Red-winged Blackbirds. *Auk* **97**: 559–565.

Crook, D., 1975. Chipping Sparrows feeding grit to offspring. *Wilson Bull.* **87**: 552.

Cullen, E., 1957. Adaptations in the Kittiwake to cliff-nesting. *Ibis* **99**: 275–302.

Cullen, J. M., 1962. The pecking response of young Wideawake Terns *Sterna fuscata*. *Ibis* **103b**: 162–170.

Curtius, A., 1954. Ueber angeborene Verhaltenweisen bei Vögeln, insbesondere bei Hühnerkücken. *Z. Tierpsychol.* **11**: 94–109.

Custer, T. W., and F. A. Pitelka, 1977. Demographic features of a Lapland Longspur population near Barrow, Alaska. *Auk* **94**: 505–525.

Davidson, J., W. R. Hepburn, J. Matheson, and J. D. Pullar, 1968. Comparisons of heat loss from young cockerels by direct measurement and by indirect assessment involving body analysis. *Br. Poult. Sci.* **9**: 93–109.

Davies, N. B., 1976. Parental care and the transition to independent feeding in the young Spotted Flycatcher (*Muscicapa striata*). *Behaviour* **54**: 280–295.

Davies, N. B., 1978. Parental meaness and offspring independence: an experiment with hand-reared Great Tits *Parus major*. *Ibis* **120**: 509–514.

Davies, N. B., and R. E. Green, 1976. The development and ecological significance of feeding techniques in the Reed Warbler (*Acrocephalus scirpaceus*). *Anim. Behav.* **24**: 213–229.

Davies, S. J. J. F., and R. Carrick, 1962. On the ability of Crested Terns, *Sterna bergii*, to recognize their own chicks. *Austral. J. Zool.* **10**: 171–177.

Davis, J. W. F., 1975. Age, egg-size and breeding success in the Herring Gull *Larus argentatus*. *Ibis* **117**: 460–473.

Davis, J. W. F., 1976. Breeding successes and experience in the Arctic Skua *Stercorarius parasiticus* (L). *J. Anim. Ecol.* **45**: 531–536.

Dawkins, R., 1968. The ontogeny of a pecking preference in domestic chicks. *Z. Tierpsychol.* **25**: 170–186.

Dawkins, R., 1976. *The Selfish Gene*, Oxford University Press, Oxford.

Dawson, D. G., 1972. The breeding ecology of the House Sparrow. DPhil thesis, Oxford University.

Dawson, W. R., and M. J. Allen, 1960. Thyroid activity in nestling Vesper Sparrows. *Condor* **62**: 403–405.

Dawson, W. R., and F. C. Evans, 1960. Relation of growth and development to temperature regulation in nestling Vesper Sparrows. *Condor* **62**: 329–340.

Delius, J. D., and G. Thompson, 1970. Brightness dependence of colour preferences in Herring Gull chicks. *Z. Tierpsychol.* **27**: 842–849.

Dhondt, A. A., 1970. The sex ratio of nestling Great Tits. *Bird Study*: 282–286.

Dhondt, A. A., 1979. Summer dispersal and survival of juvenile Great Tits in southern Sweden. *Oecologia* **42**: 139–158.

Diamond, A. W. 1973. Notes on the breeding biology and behavior of the Magnificent Frigatebird. *Condor* **75**: 200–209.

Diehl, B., 1971. Productivity investigations of two types of meadows in the Vistula valley. XII. Energy requirement in nestling and fledgling Red-backed Shrikes (*Lanius collurio*). *Ekol. Polsk* **19**: 235–248.

Diehl, B., C. Kurowski, and A. Myrcha, 1972. Changes in the gross chemical composition and energy content of nestling Red-backed Shrikes (*Lanius collurio* L.). *Bull. Acad. pol. Sci.* **20**: 837–843.

Dilger, W. C., 1962. The behavior of lovebirds. *Scient. Am.* **206**: 88–98.

Dilger, W. C., and J. C. Wallen, 1966. The pecking responses of peafowl chicks. *Living Bird* **5**: 115–125.

Dimond, S. J., 1968. Effects of photic stimulation before hatching on the development of fear in chicks. *J. comp. physiol. Psychol.* **65**: 320–324.

Din, N. A., and K. R. Eltringham, 1974. Breeding of the Pink-backed Pelican *Pelecanus rufescens* in Rwenzori National Park, Uganda, with notes on a colony of Marabou Storks *Leptoptilus crumeniferus*. *Ibis* **116**: 477–493.

Dittus, W. P. J., and R. E. Lemon, 1969. Effects of song tutoring and acoustic isolation on the song repertoires of Cardinals. *Anim. Behav.* **17**: 523–533.

Dixon, C. L., 1978. Breeding biology of the Savannah Sparrow on Kent Island. *Auk* **95**: 235–246.

Dobkin, D. S., 1979. Functional and evolutionary relationships of vocal copying phenomena in birds. *Z. Tierpsychol.* **50**: 348–363.

Douthwaite, R. J., 1976. Fishing techniques and foods of the Pied Kingfisher on Lake Victoria in Uganda. *Ostrich* **47**: 153–160.

Drent, R. H., 1965. Breeding biology of the Pigeon Guillemot, *Cepphus columba*. *Ardea* **53**: 99–160.

Drent, R. H., 1970. Functional aspects of incubation in the Herring Gull (*Larus argentatus* Pont.). *Bheaviour* Suppl. **17**: 1–132.

Drent, R. H., 1973. The natural history of incubation. In D. S. Farner (ed.), *Breeding Biology of Birds*, National Academy of Sciences, Washington, D.C., pp. 262–311.

Drent, R. H., and S. Daan, 1980. The prudent parent: energetic adjustments in avian breeding. *Ardea* **68**: 225–252.

Driver, P. M., 1965. 'Clicking' in the egg-young of nidifugous birds. *Nature, Lond.* **206**: 315.

Dunn, E. H., 1975a. The timing of endothermy in the development of altricial birds. *Condor* **77**: 288–293.

Dunn, E. H., 1975b. Growth, body components and energy content of nestling Double-crested Cormorants. *Condor* **77**: 431–438.

Dunn, E. H., 1975c. Caloric intake of nestling Double-crested Cormorants. *Auk* **92**: 553–565.

Dunn, E. H., 1976. The relationship between brood size and age of effective homeothermy in nestling House Wrens. *Wilson Bull.* **88**: 478–482.

Dunn, E. H., 1979. Time-energy use and life history strategies of northern seabirds. In J. C. Bartonek and D. N. Nettleship (eds), *Conservation of Marine Birds of Northern North America*, Wildlife Research Report no. 11, Fish and Wildlife Service, Washington, D.C., pp. 141–166.

Dunn, E. H., 1980. On the variability of energy allocation of nestling birds. *Auk* **97**: 19–27.

Dunn, E. H., and I. L. Brisbin, 1981. Age-specific changes in the major body components and caloric values of Herring Gull chicks. *Condor* **82**: 398–401.

Dunn, E. K., 1972. Effect of age on the fishing ability of Sandwich Terns *Sterna sandvicensis*. *Ibis* **114**: 360–366.

Dunn, E. K., 1975. The role of environmental factors in the growth of tern chicks. *J. Anim. Ecol.* **44**: 743–754.

Dunn, E. K., 1977. Predation of weasels (*Mustela nivalis*) on brooding tits (*Parus* spp.) in relation to the density of tits and rodents. *J. Anim. Ecol.* **46**: 634–652.

Dyer, M. J., J. Pinowski, and B. Pinowska, 1977. Population dynamics. In J. Pinowski and S. C. Kendeigh (eds), *Granivorous Birds in Ecosystems*, Cambridge University Press, Cambridge, pp. 53–105.

East, M., 1981. Alarm calling and parental investment in the Robin *Erithacus rubecula*. *Ibis* **123**: 223–230.

Eastman, R., 1969. *The Kingfisher*, Collins, London.

Eltringham, S. K., 1974. The survival of broods of the Egyptian Goose in Uganda. *Wildfowl* **25**: 41–48.

Emlen, S. T., 1967. Migratory orientation in the Indigo Bunting, *Passerina cyanea*. Part II: Mechanism of celestial orientation. *Auk* **84**: 463–489.

Emlen, S. T., 1969. Bird migration: influence of physiological state upon celestial orientation. *Science, N.Y.* **165**: 716–718.

Emlen, S. T., 1972. The ontogenetic development of orientation capabilities. In S. R. Galler, K. Schmidt-Koenig, G. J. Jacobs, and R. E. Belleville (eds), *Animal Orientation and Navigation*, NASA Sp-262, National Aeronautics and Space Administration, Washington, D.C., pp. 191–222.

Evans, M. E., 1975. Breeding behaviour of captive Bewick's Swans. *Wildfowl* **26**: 117–130.

Evans, R. M., 1970a. Imprinting and mobility in young Ring-billed Gulls, *Larus delawarensis*. *Anim. Behav. Monogr.* **3**: 195–248.

Evans, R. M., 1970b. Parental recognition and the 'mew call' in Blackbilled Gulls (*Larus bulleri*). *Auk* **87**: 503–513.

Evans, R. M., and M. E. Mattson, 1972. Development of selective responses to individual maternal vocalizations in young *Gallus gallus*. *Can. J. Zool.* **50**: 777–780.

Ficken, M. S., 1962. Maintenance activities of the American Redstart. *Wilson Bull.* **74**: 153–165.

Ficken, M. S., 1977. Avian play. *Auk* **94**: 573–582.

Finch, D. M., 1981. Nest predation of Albert's Towhees by coachwhips and Roadrunners. *Condor* **83**: 389.

Finlay, J. C., 1976. Some effects of weather on Purple Martin activity. *Auk* **93**: 231–244.

Finney, G., and F. Cooke, 1978. Reproductive habits in the Snow Goose: the influence of female age. *Condor* **80**: 147–158.

Fisher, H. J., 1962. The hatching muscle in Franklin's Gull. *Wilson Bull.* **74**: 166–172.

Fjeldstra, J., 1977. *Guide to the Young of European Precocial Birds*, Skarv Nature Publications, Tisvildeleje.

Fogden, M. P. L., 1972. The seasonality and population dynamics of equatorial forest birds in Sarawak. *Ibis* **114**: 307–343.

Foster, M. S., 1974a. A model to explain molt–breeding overlap and clutch size in some tropical birds. *Evolution* **28**: 182–190.

Foster, M. S., 1974b. Rain, feeding, behavior, and clutch size in tropical birds. *Auk* **91**: 722–726.

Franks, E. C., 1967. The responses of incubating Ringed Turtle Doves (*Streptopelia risoria*) to manipulated egg temperatures. *Condor* **69**: 268–279.

Fredrickson, L. H., and M. W. Weller, 1972. Responses of Adelie Penguins to colored eggs. *Wilson Bull.* **84**: 309–314.

Freeman, B. M., 1965. The relationship between oxygen consumption, body temperature and surface area in the hatching and young chicks. *Br. Poult. Sci.* **6**: 67–72.

Freeman, B. M., 1971. Body temperature and thermoregulation. In D. J. Bell and B. M. Freeman (eds), *Physiology and Biochemistry of the Domestic Fowl*, Academic Press, London, pp. 1115–1141.

Freeman, B. M., and M. A. Vince, 1974. *Development of the Avian Embryo*, Chapman and Hall, London.

Fretwell, S. D., D. E. Bowen, and H. Hespenheide, 1974. Growth rates of young passerines and the flexibility of clutch size. *Ecology* **55**: 907–909.

Frith, H. J., 1959. Breeding of the Mallee Fowl, *Leipoa ocellata* Gould (Megapodiidae). *C.S.I.R.O. Wildl. Res.* **4**: 31–60.

Gallagher, J., 1976. Sexual imprinting: effects of various regimens of social experience on mate preference in Japanese Quail *Coturnix coturnix japonica*. *Behaviour* **57**: 91–115.

Garnett, M. C., 1981. Body size, its heritability and influence on juvenile survival among Great Tits *Parus major*. *Ibis* **123**: 31–41.

Gavrilov, E. J., 1968. A possible regulation mechanism of the sex ratio in the *Passer hispaniolensis* Temm. *Int. Studies Sparrows* **2**: 20–24.

Gibb, J., 1955. Feeding rates of Great Tits. *Br. Birds* **48**: 49–58.

Gochfeld, M., 1975. Developmental defects in Common Terns of western Long Island, New York. *Auk* **92**: 58–65.

Gochfeld, M., 1977. Intraclutch egg variation: the uniqueness of the Common Tern's third egg. *Bird-Banding* **48**: 325–322.

Gofman, D. N., 1977. [Spring temperatures inside the trunks of trees in connection with breeding time of some hole-nesting birds.] *Ornitologiya* **13**: 62–66 (in Russian).

Goodwin, D., 1965. A comparative study of captive Blue Waxbills (Estrildidae). *Ibis* **107**: 285–315.

Gorman, M. L., and H. Milne, 1972. Creche behaviour in the Common Eider *Somateria m. mollissima* L. *Ornis scand.* **3**: 21–26.

Gottfried, G. M., 1979. Anti-predator aggression in birds nesting in old field habitats:

an experimental analysis. *Condor* **81**: 251–257.

Gottlieb, G., 1965. Components of recognition in ducklings. *Nat. Hist.* **74**: 12–19.

Gottlieb, G., 1968. Prenatal behaviour of birds. *Quart. Rev. Biol.* **43**: 148–174.

Gottlieb, G., 1971. *Development of Species Identification in Birds,* University of Chicago Press, Chicago.

Gottlieb, G., and P. H. Klopfer, 1962. The relation of developmental age to auditory and visual imprinting. *J. comp. physiol. Psychol.* **55**: 821–826.

Graves, H. B., and P. B. Siegel, 1968. Prior experience and the approach response in domestic chicks. *Anim. Behav.* **16**: 18–23.

Greenwood, J. J. D., 1964. The fledging of the Guillemot *Uria aalge* with notes on the Razorbill *Alca torda. Ibis* **106**: 469–481.

Greenwood, P. J., P. H. Harvey, and C. M. Perrins, 1979. The role of dispersal in the great tit (*Parus major*): the causes, consequences and heritability of natal dispersal. *J. Anim. Ecol.* **48**: 123–142.

Greenwood, R. J., 1969. Mallard hatching from an egg cracked by freezing. *Auk* **86**: 752–754.

Grimes, L. G., 1976. The occurrence of co-operative breeding behaviour in African birds. *Ostrich* **47**: 1–15.

Grohmann, J., 1939. Modification oder Funktionreifung? *Z. Tierpsychol.* **2**: 132–144.

Groves, S, 1978. Age-related differences in Ruddy Turnstone foraging and aggressive behavior. *Auk* **95**: 95–103.

Haartman, L. von, 1949. Der Trauerfliegenschnapper. I. *Acta zool. fenn.* **56**: 1–104.

Haftorn, S., 1978. Weight increase and feather development in the Goldcrest *Regulus regulus. Ornis scand.* **9**: 117–123.

Hahn, D. C., 1981. Asynchronous hatching in the Laughing Gull: cutting losses and reducing rivalry. *Anim. Behav.* **29**: 421–427.

Hailman, J., 1967. The ontogeny of an instinct. *Behaviour* Suppl. **15**: 1–159.

Hailman, J. P., and . H. Klopfer, 1962. On measuring 'critical learning periods' in birds. *Anim. Behav.* **10**: 233–234.

Hails, C. J., and D. M. Bryant, 1979. Reproductive energetics of a freeliving bird. *J. Anim. Ecol.* **48**: 471–482.

Hall, M. R., 1979. The ontogeny of thermoregulation in the Herring Gull (*Larus argentatus* Pont.), PhD thesis, University of Wales.

Hamilton, W. J., 1964. The genetical evolution of social behaviour. I, II. *J. theor. Biol.* **7**: 1–52.

Harris, M. A., and R. G. Lemon, 1974. Songs of Song Sparrows: reactions of males to songs of different localities. *Condor* **76**: 33–44.

Harris, M. P., 1976. Lack of a 'desertion period' in the nestling life of the Puffin *Fratercula arctica. Ibis* **118**: 115–118.

Harris, M. P., 1978. Supplementary feeding of young Puffins, *Fratercula arctica. J. Anim. Ecol.* **47**: 15–24.

Harrison, C., 1975. *A Field Guide to the Nests, Eggs and Nestlings of European Birds,* Collins, London.

Hartshorne, J. M., 1962. Behavior of the Eastern Blackbird at the nest. *Living Bird* **1**: 131–149.

Hartwick, E. B., 1976. Foraging strategy of the Black Oystercatcher. *Can. J. Zool.* **54**: 142–155.

Harvey, J. M., B. C. Lieff, C. D. MacInnes, and J. P. Prevett, 1968. Observations on behavior of Sandhill Cranes. *Wilson Bull.* **80**: 421–425.

Haycock, K. A., and W. Threlfall, 1975. The breeding biology of the Herring Gull in Newfoundland. *Auk* **92**: 678–697.

Hays, H., 1970. Sand-kicking camouflages young Skimmers. *Wilson Bull.* **82**: 100–101.

Hedgren, S., 1979. Seasonal variation in fledging weight of Guillemots *Uria aalge. Ibis* **121**: 356–361.

Hess, E. H., 1956a. Natural preferences of chicks and ducklings for objects of different colours. *Psychol. Rep.* **2**: 477–483.

Hess, E. H., 1956b. Space perception in the chick. *Scient. Am.* **195**: 71–80.

Hess, E. H., 1959. Imprinting. *Science, N.Y.* **130**: 133–141.

Hess, E. H., 1962. Imprinting and the 'critical period' concept. In E. L. Bliss (ed.), *Roots of Behavior*, Harper and Row, New York, pp. 254–263.

Hilden, O., 1964. Ecology of duck populations in the island group of Valassaaret, Gulf of Bothnia. *Annotnes. zool. fenn.* **1**: 153–279.

Hilden, O., 1973. Breeding system of Temminck's Stint *Calidris temmincki. Ornis fenn.* **52**: 117–146.

Hochbaum, H. A., 1955. *Travels and Tradition of Waterfowl*, University of Minnesota Press, Minneapolis.

Hoffman, H. S., D. Schiff, J. Adams, and J. L. Searle, 1966. Enhanced distress vocalization through selective reinforcement. *Science, N.Y.* **151**: 352–354.

Hogan, J. A., 1965. An experimental study of conflict and fear: an analysis of the behavior of young chicks toward a mealworm. I. The behavior of chicks which do not eat the mealworm. *Behaviour* **25**: 45–97.

Högstedt, G., 1981. Effect of additional food on reproductive success in the Magpie (*Pica pica*). *J. Anim. Ecol.* **50**: 219–229.

Holcomb, L. C., 1966a. The development of grasping and balancing coordination in nestlings of seven species of altricial birds. *Wilson Bull.* **78**: 57–63.

Holcomb, L. C., 1966b. Red-winged Blackbird nestling development. *Wilson Bull.* **78**: 283–288.

Holcomb, L. C., and G. Twiest, 1968. Red-winged Blackbird nestling growth compared to adult size and differential development of structures. *Ohio J. Sci.* **68**: 277–284.

Holleback, M., 1974. Behavioral interactions and the dispersal of the family in Black-capped Chickadees. *Wilson Bull.* **86**: 466–468.

Holyoak, D., 1969. The function of pale egg colour in the Jackdaw. *Bull. Br. Orn. Club* **89**: 159.

Hoogland, J. L., and P. W. Sherman, 1976. Advantages and disadvantages of Bank Swallo *Riparia riparia* coloniality. *Ecol. Monogr.* **46**: 33–58.

Hori, J., 1964. Parental care in the Shelduck. *A. Rept. Wildfowl Trust* **15**: 100–103.

Hori, J., 1969. Social and population studies in the Shelduck. *Wildfowl* **20**: 5–22.

Horn, G., A. L. D. Horn, P. P. G. Bateson, and S. P. R. Rose, 1971. Effects of imprinting on uracil incorporation into brain RNA in the 'split-brain' chick. *Nature, Lond.* **229**: 131–132.

Horn, G., S. P. R. Rose, and P. P. G. Bateson, 1973a. Experience and plasticity in the central nervous system. *Science, N.Y.* **181**: 506–514.

Horn, G., S. P. R. Rose, and P. P. G. Bateson, 1973b. Monocular imprinting and regional incorporation of tritiated uracil into the brains of intact and 'split-brain' chicks. *Brain Res.* **56**: 227–237.

Houston, D. C., 1979. Why do Fairy Terns *Gygis alba* not build nests? *Ibis* **121**: 102–104.

Howe, H. F., 1976. Egg size, hatching synchrony, sex, and brood reduction in the Common Grackle. *Ecology* **57**: 1195–1207.

Howell, T. R., and G. A. Bartholomew, 1962. Temperature regulation in the Red-tailed Tropic Bird and the Red-footed Booby. *Condor* **62**: 6–18.

Hoyt, D. F., D. Vleck, and C. M. Vleck, 1978. Metabolism of avian embryos: ontogeny and temperature effects in the Ostrich. *Condor* **80**: 265–271.

Hunt, G. L., and S. C. McLoon, 1975. Activity patterns of gull chicks in relation to feeding by parents: their potential significance for density-dependent mortality. *Auk* **92**: 523–527.

Hunter, R. A., H. A. Ross, and A. J. S. Ball, 1976. A laboratory simulation of predator-induced incubation interruption using ring-billed gull eggs. *Can. J. Zool.* **54**: 628–633.

Immelmann, K., 1970. The influence of early experience upon the development of social behaviour in estrildine finches. *Proc. 15th int. Orn. Congr.*: 316–338.

Immelmann, K., 1975a. Ecological significance of imprinting and early learning. *A. Rev. Ecol. Syst.* **6**: 15–37.

Immelmann, K., 1975b. The evolutionary significance of early experience. In G. Baerends, C. Beer, and A. Manning (eds), *Function and Evolution in Behaviour*, Clarendon Press, Oxford, pp. 243–253.

Impekoven, M., 1976. Response of Laughing Gull chicks (*Larus atricilla*) to parental attraction- and alarm-calls, and effects of prenatal auditory experience on the responsiveness to such calls. *Behaviour* **56**: 250–278.

Ingram, C., 1959. The importance of juvenile cannabalism in the breeding biology of certain birds of prey. *Auk* **76**: 218–226.

Ingram, C., 1962. Cannabalism by nestling Short-eared Owls. *Auk* **79**: 715.

Jehl, J. R., 1973. Breeding biology and systematic relationships of the Stilt Sandpiper. *Wilson Bull.* **85**: 115–147.

Johnsgard, P. A., 1973. *Grouse and Quails of North America*, University of Nebraska, Lincoln, Neb.

Johnsgard, P. A., and J. Kear, 1968. A review of parental carrying of young by waterfowl. *Living Bird* **7**: 89–102.

Johnson, D. H., 1979. Estimating nest success: the Mayfield method and an alternative. *Auk* **96**: 651–661.

Johnson, S. R., and G. C. West, 1975. Growth and development of heat regulation in nestlings, and metabolism of adult Common and Thick-billed Murres. *Ornis scand.* **6**: 109–115.

Kahl, M. P., 1962. Bioenergetics of growth in nestling Wood Storks. *Condor* **64**: 169–183.

Karplus, M., 1952. Bird activity in the continuous daylight of Arctic summer. *Ecology* **33**: 129–134.

Kear, J., 1962. Food selection in finches with special reference to interspecific differences. *Proc. zool. Soc. Lond.* **138**: 163–204.

Kear, J., 1964. Colour preference in young Anatidae. *Ibis* **106**: 361–369.

Kear, J., 1966. The pecking response of young Coots *Fulica atra* and Moorhens *Gallinula chloropus*. *Ibis* **108**: 118–122.

Kear, J., 1966b. Experiments with young nidifugous birds on a visual cliff. *A. Rept. Wildfowl Trust* **18**: 122–124.

Kear, J., 1966c. Further congenital malformation in birds at Slimbridge. *A. Rept. Wildfowl Trust* **17**: 35.

Kear, J., 1968. The calls of very young Anatidae. *Vogelwelt* **1**: 93–113.

Kear, J., 1970a. The adaptive radiation of parental care in waterfowl. In J. H. Crook (ed.), *Social Behaviour in Birds and Mammals*, Academic Press, London, pp. 357–392.

Kear, J., 1970b. Studies on the development of young Tufted Duck. *Wildfowl* **21**: 123–132.

Kear, J., 1973. Notes on the nutrition of young waterfowl, with special reference to slipped wing. *Int. Zoo Yb.* **13**: 97–100.

Keeton, W. T., 1979. Avian orientation and navigation: a brief overview. *Br. Birds* **72**: 451–470.

Kendeigh, S. C., 1939. The relation of metabolism to the development of temperature regulation in birds. *J. exp. Zool.* **82**: 419–438.

Kendeigh, S. C., 1952. Parental care and its evolution in birds. *Illinois biol. Monogr.* **22**: 1–356.

Kendeigh, S. C., 1963. Thermodynamics of incubation in the House Wren *Troglodytes aedon*. *Proc. 13th int. Orn. Congr.*: 884–904.

Kendeigh, S. C., and S. P. Baldwin, 1928. Development of temperature control in nestling House Wrens. *Am. Nat.* **62**: 249–278.

Kendeigh, S. C., V. R. Dolnik, and V. M. Gavrilov, 1977. Avian energetics. In J. Pinowski and S. C. Kendeigh (eds), *Granivorous Birds in Ecosystems*, Cambridge University Press, Cambridge, pp. 129–204.

Khaskin, V. V., 1961. Heat exchange in birds' eggs on incubation. *Biophysics* **6**: 97–107.

Khayatin, S., and L. Dmitrieva, 1978. [Development acceleration of sensory and motor mechanisms of nestlings as factors of some bird species.] *Zh. obshch. Biol.* **34**: 289–296 (in Russian).

Kilham, L., 1977. Altruism in nesting Yellow-bellied Sapsuckers. *Auk* **94**: 613–614.

Kitchell, R. L., L. Strom, and Y. Zotterman, 1959. Electrophysiological studies of thermal and taste reception in chickens and pigeons. *Acta physiol. scand.* **46**: 133–151.

Klopfer, P., 1963. Behavioral aspects of habitat selection: the role of early experience. *Wilson Bull.* **75**: 15–22.

Klopfer, P., 1965. Behavioral aspects of habitat selection: a preliminary report on stereotypy in foliage preferences of birds. *Wilson Bull.* **77**: 376–381.

Klopfer, P. H., 1967. Stimulus preferences and imprinting. *Science, N.Y.* **156**: 1394–1396.

Klopfer, P. H., 1967b. Behavioral sterotypy in birds. *Wilson Bull.* **79**: 290–300.

Klopfer, P. H., and G. Gottlieb, 1962. Imprinting and behavioral polymorphism: auditory and visual imprinting in domestic ducks (*Anas platyrhynchos*) and the involvement of the critical period. *J. comp. physiol. Psychol.* **55**: 126–130.

Kluyver, H. N., 1933. Bijdrage tot de biologie en de ecologie van de Spreeuw (*Sturnus vulgaris vulgaris* L.) gerdurende zijn voortplantingstijd. *Publ. Plantenziektenk. Dienst, Wageningen.*

Kluyver, H. N., 1966. Regulation of a bird population. *Ostrich* Suppl. **6**: 389–396.

Kluyver, H. N., 1971. Regulation of numbers in populations of Great Tits (*Parus m. major*). *Proc. Adv. Study Inst. Dynam. Numbers Popul., Wageningen*: 507–523.

Koelink, A. F., 1972. Bioenergetics of growth in the Pigeon Guillemot, *Cepphus columba*. MS thesis, University of British Columbia, Vancouver.

Konishi, M., 1963. The role of auditory feedback in the vocal behavior of the domestic fowl. *Z. Tierpsychol.* **20**: 349–367.

Konishi, M., 1965. The role of auditory feedback in the control of vocalizations in the White-crowned Sparrow. *Z. Tierpsychol.* **22**: 770–783.

Koskimies, J., 1950. The life of the Swift, *Micropus apus* (L.), in relation to the weather. *Ann. Acad. Sci. Fenn. A, IV Biol.*: 1–151.

Koskimies, J., 1955. Juvenile mortality and population balance in the Velvet Scoter (*Melaniita fusca*) in maritime conditions. *Proc. 11th int. Orn. Congr.*: 476–479.

Koskimies, J., 1957. Vehalten und Oekologie der Jungen und jungenfuhrenden Weibchen der Samtente. *Ann. zool. Soc. Vanamo* **18**: 1–69.

Koskimies, J., and L. Lahti, 1964. Cold-hardiness of the newly hatched young in relation to ecology and distribution in ten species of European ducks. *Auk* **81**: 281–307.

Kostelcka-Myrcha, A., 1976. Variations in the red cell picture during growth of goslings and chickens. *Br. Poult. Sci.* **17**: 93–101.

Kroodsma, D. E., 1979. Vocal dueling among male Marsh Wrens: evidence for ritualized expressions of dominance/subordinance. *Auk* **96**: 605–515.

Kroodsma, D. E., and R. Pickert, 1980. Environmentally dependent sensitive periods for avian and vocal learning. *Nature, Lond.* **288**: 477–479.

Kruuk, H., 1964. Predators and anti-predator behaviour of the Blackheaded Gull (*Larus ridibundus* L.). *Behaviour Suppl.* **11**: 1–130.

Kruuk, H., 1976. The biological function of gulls' attraction towards predators. *Anim. Behav.* **24**: 146–153.

Kushlan, J. A., 1976. Feeding rhythm in nestling White Ibis. *Wilson Bull.* **88**: 656–658.

Kushlan, J. A., 1977. Differential growth of body parts in the White Ibis. *Auk* **94**: 164–167.

Lack, D., 1947. The significance of clutch size. *Ibis* **89**: *302–352.*

Lack, D., 1954. *The Natural Regulation of Animal Numbers*, Oxford University Press, Oxford.

Lack, D., 1956. *Swifts in a Tower*, Methuen, London.

Lack, D., 1958. The significance of the colour in Turdine eggs. *Ibis* **100**: 145–166.

Lack, D., 1966. *Population Studies of Birds*, Oxford University Press, Oxford.

Lack, D., 1968. *Ecological Adaptations for Breeding in Birds*, Methuen, London.

Lack, D., 1971. *Ecological Isolation in Birds*, Harvard University Press, Cambridge.

Lack, D., and E. Lack, 1951. The breeding biology of the Swift *Apus apus*. *Ibis* **93**: 501–546.

Lack, D., and E. Lack, 1958. The nesting of the Long-tailed Tit. *Bird Study* **5**: 1–19.

Lanyon, W. E., 1957. The comparative biology of the Meadowlarks (*Sturnella*) in Wisconsin. *Publ. Nuttall Orn. Club* **1**: 1–67.

Lanyon, W. E., 1960. The ontogeny of vocalizations in birds. In W. E. Lanyon and W. N. Tavolga (eds), *Animal Sounds and Communication*, American Institute of Biological Science Publications, No. 7, American Institute of Biological Science, Washington, D.C., pp. 321–347.

Lasiewski, R. C., and W. R. Dawson, 1967. A re-examination of the relation between standard metabolic rate and body weight in birds. *Condor* **69**: 13–23.

Lavery, H. J., 1970. The comparative ecology of waterfowl in north Queensland. *Wildfowl* **21**: 69–77.

Leck, C., 1972. The impact of some North American migrants at fruiting trees in Panama. *Auk* **89**: 842–850.

LeCroy, M., 1972. Young Common and Roseate Terns learning to fish. *Wilson Bull.* **84**: 201–202.

Lehrman, D. S., 1964. The reproductive behavior of Ring Doves. *Scient. Am.* **211**: 48–54.

Lemmetyinen, R., 1971. Nest defence behaviour of Common and Arctic Terns and its effects on the success achieved by predators. *Ornis fenn.* **48**: 13–24.

Lemmetyinen, R., 1972. Growth and mortality in the chicks of Arctic Terns in the Kongsfjord area, Spitzbergen, in 1970. *Ornis fenn.* **49**: 45–53.

Lemon, R. E., 1973. Nervous control of the syrinx in the White-throated Sparow (*Zonotrichia albicollis*) *J. Zool., Lond.* **171**: 131–140.

Lemon, R. E., 1975. How birds develop song-dialects. *Condor* **77**: 385–405.

Le Morvan, P., J. L. Mougin, and J. Prevost, 1967. Ecologie du skua antartique (*Stercorarius skua maccormicki*) dans l'Archipel de Ponte Geolgoie (Terre Adeliea). *L'Oiseau et R.F.O.* **37**: 193–220.

Ligon, J. D., and D. J. Martin, 1974. Piñon seed assessment by the Piñon Jay, *Gymnorhinus cyanocephalus*. *Anim. Behav.* **22**: 421–429.

Ligon, J. D., and J. L. White, 1974. Molt and its timing in the Piñon Jay *Gymmorhinus cyanocephalus*. *Condor* **76**: 274–287.

Lind, H., 1961. *Studies on the Behaviour of the Black-tailed Godwit* (Limosa limosa L), Munksgaard, Copenhagen.

Lindenmaier, P., and M. R. Kare, 1959. The taste end-organs of the chicken. *Poult. Sci.* **38**: 545–550.

Ljunggren, L., 1971. Liver diseases and pigeon pox in South Swedish Wood Pigeon *Columba p. palumbus* populations. *Ornis scand.* **2**: 5–11.

Löhrl, H., 1959. Zur Frage des Zeitpunktes euner Prägung auf die Heimatregion beim Halsbandschnäpper (*Ficedula albiocollis*). *J. Orn., Lpz.* **100**: 132–140.

Löhrl, H., 1968. Das Nesthakchen als biologisches Problem. *J. Orn., Lpz.* **109**: 383–395.

Löhrl, H., 1975. Brutverhalten und Jugendentwicklung beim Mauerläufer (*Tichodroma muraria*). *J. Orn. Lpz.*, **116**: 229–262.

Lorenz, K., 1935. Der Kumpan in der Umwelt des Vogels. *J. Orn., Lpz.* **83**: 137–213, 289–413.

Lorenz, K., 1939. Vergleichende Verhaltensforschung. *Zool. Anz.* Suppl. 12: 69–102.

Lundy, H., 1969. A review of the effects of temperature, humidity, turning and gaseous environment in the incubator on the hatchability of the hens egg. In T. C. Carter and B. M. Freeman (eds), *The Fertility and Hatchability of the Hen's Egg,* Oliver and Boyd, Edinburgh, pp. 143–176.

MacInnes, C. D., R. G. Davis, R. N. Jones, B. C. Lieff, and A. H. Pakulak, 1974. Reproductive efficiency of McConnell River small Canada Geese. *J. Wildl. Mgmt* **38**: 686–707.

McLannahan, H. M. C., 1973. Some aspects of the ontogeny of cliff nesting behaviour in the Kittiwake (*Rissa tridactyla*) and the Herring Gull (*Larus argentatus*). *Behaviour* **44**: 36–88.

MacLean, G. L., 1969. A study of Seedsnipe in southern South America. *Living Bird* **8**: 33–80.

Maclean, G. L., 1974. Egg-covering in the Charadrii. *Ostrich* **45**: 167–174.

Maclean, G. L., 1975. Belly-soaking in the Charadriiformes. *J. Bombay nat. Hist. Soc.* **72**: 74–82.

MacLean, S. F., 1974. Lemming bones as a source of calcium for arctic sandpipers (*Calidris* spp.). *Ibis* **116**: 552–557.

McNabb, F. M. A., and R. A. McNabb, 1977. Thyroid development in precocial and altricial avian embryos. *Auk* **94**: 736–742.

McNabb, R. A., R. L. Stouffer, and F. M. A. McNabb, 1972. Thermoregulatory ability and the thyroid gland: their development in embryonic Japanese Quail (*Coturnix c. japonica*). *Comp. Biochem. Physiol* **43A**: 187–193.

Mahelka, B., 1973. Zum Vergleich des postnatalen Wachstums und der Entwicklung der Wild- und Hausente. *Acta Scient., Nat. Brno* **7**: 1–50.

Mahony, S. P., and W. Threlfall, 1981. Notes on the eggs, embryos and chick growth of Common Guillemots *Uria aalge* in Newfoundland. *Ibis* **123**: 211–218.

Makatsch, W., 1950. *Der Vogel und sein Ei,* Akademische Verlagsgesellschaft, Leipzig.

Marcstrom, V., 1960. Studies on the physiological and ecological background to the reproduction of the Capercaillie (*Tetrao urogallus* Linn.). *Viltrevy* **2**: 1–69.

Marler, P., 1959. Developments in the study of animal communication. In P. R. Bell (ed.), *Darwin's Biological Work,* Cambridge University Press, Cambridge, pp. 150–206.

Marler, P., 1967. Comparative study of song development in sparrows. *Proc. 14th int. Orn. Congr.*: 231–244.

Marler, P., 1975. On strategies of behavioural development. In G. Baerends, C. Beer, and A. Manning (eds), *Function and Evolution in Behaviour,* Clarendon Press, Oxford, pp. 254–276.

Marler, P., and S. Peters, 1977. Selective vocal learning in a sparrow. *Science* **198**: 519–521.

Marler, P., and M. Tamura, 1964. Culturally transmitted patterns of vocal behavior in sparrows. *Science, N.Y.* **146**: 1483–1486.

Marler, P., M. Kreith, and M. Tamura, 1962a. Song-development in handraised Oregon Juncos. *Auk* **79**: 12–80.

Marler, P., M. Kreith, and E. Willis, 1962b. An analysis of testosterone-induced crowing in young domestic cockerels. *Anim. Behav.* **10**: 48–54.

Martin, D. D., and A. H. Meier, 1973. Temporal synergism of corticosterone and prolactin in regulating orientation in the migratory White-throated Sparrow (*Zonotrichia albicollis*). *Condor* **75**: 369–374.

Matthews, G. V. T., and M. E. Evans, 1975. On the behaviour of the Whiteheaded Duck with especial reference to breeding. *Wildfowl* **25**: 56–66.

Maunder, J. E., and W. Threlfall, 1972. The breeding biology of the Black-legged Kittiwake in Newfoundland. *Auk* **89**: 789–816.

Maxwell, G. R., and L. S. Putnam, 1963. The maintenance behavior of the Black-crowned Night Heron. *Wilson Bull.* **80**: 467–478.

Mayfield, H., 1961. Nesting success calculated from exposure. *Wilson Bull.* **73**: 255–261.

Mayfield, H., 1975. Suggestions for calculating nest success. *Wilson Bull.* **87**: 456–466.

Maynard Smith, J., and G. A. Parker, 1976. The logic of asymmetric contests. *Anim. Behav.* **24**: 159–175.

Meinertzhagen, R., 1954. The education of young Ospreys. *Ibis* **96**: 153–155.

Mendenhall, V. M., 1978. Brooding of young ducklings by female Eiders *Somateria mollissima*. *Ornis scand.* **10**: 94–99.

Mertens, J. A. L., 1969. The influence of brood size on the energy metabolism and water loss of nestling Great Tits *Parus major*. *Ibis* **111**: 11–16.

Mertens, J. A. L., 1977a. Thermal conditions for successful breeding in Great Tits (*Parus major* L.). I. Relation of growth and development of temperature regulation in nestling Great Tits. *Oecologia* **28**: 1–29.

Mertens, J. A. L., 1977b. Thermal conditions for successful breeding in Great Tit (*Parus major* L.). 2. Thermal properties of nests and nestboxes and their implicatons for the range of temperature tolerance of Great Tit broods. *Oecologia* **28**: 31–56.

Mertens, J. A. L., 1979. Eco-physiological research. *A. Rept. Inst. Ecol. Res.*, Arnhem **1978**: 321–324.

Messmer, F., and E. Messmer, 1956. Die Entwicklung der Lautäusserungen und einiger Verhaltenweisen der Amsel (*Turdus merula merula* L.) unter natürlichen Bedigungen und nach Einzelaufzucht in schalldachten Raümen. *Z. Tierpsychol.* **13**: 341–441.

Meyburg, B. U., 1974. Sibling aggression and mortality among nestling eagles. *Ibis* **116**: 224–228.

Millikan, G. C., and R. J. Bowman, 1967. Observations on Galapagos tool-using finches in captivity. *Living Bird* **6**: 23–41.

Mills, J. A., 1973. The influence of age and pair-bond on the breeding biology of the Red-billed Gull *Larus novaehollandiae scopulinus*. *J. Anim. Ecol.* **42**: 147–162.

Montevecchi, W. A., 1974. Eggshell removal and nest sanitation in Ring Doves. *Wilson Bull.* **86**: 136–143.

Montevecchi, W. A., 1976a. Field experiments on the adaptive significance of avian eggshell pigmentation. *Behaviour* **58**: 26–39.

Montevecchi, W. A., 1976b. Egg size and the predatory behaviour of Crows. *Behaviour* **58**: 307–320.

Morton, E. S., 1973. On the evolutionary advantages and disadvantages of fruit eating in tropical birds. *Am. Nat.* **107**: 8–22.

Morton, M. L., and C. Carey, 1971. Growth and development of endothermy in the Mountain White-crowned Sparrow (*Zonotrichia leucophrys oriantha*). *Physiol. Zool.* **44**: 179–189.

Moss, W. W., and J. H. Camin, 1970. Nest parasitism, productivity, and clutch size in Purple Martins. *Science, N.Y.* **168**: 1000–1002.

Moynihan, M., 1959. Notes on the behavior of some North American gulls. IV. The ontogeny of hostile behavior and display patterns. *Behaviour* **14**: 214–239.

Mueller, H. C., 1974. The development of prey recognition and predatory behaviour in the American Kestrel *Falco sparverius*. *Behaviour* **49**: 313–324.

Mulligan, J. A., 1966. Singing behavior and its development in the Song Sparrow *Melospiza melodia*. *Univ. Calif. Publ. Zool.* **81**: 1–76.

Mundinger, P. C., 1970. Vocal imitation and individual recognition of finch calls. *Science, N.Y.* **168**: 480–482.

284

Murton, R. K., and N. G. Westwood, 1977. *Avian Breeding Cycles*, Clarendon Press, Oxford.

Murton, R. K., N. G. Westwood, and A. J. Isaacson, 1974. Factors affecting egg-weight, body-weight and moult of the Wood-pigeon *Columba palumbus. Ibis* **116**: 52–73.

Myrcha, A., J. Pinowski, and T. Tonek, 1972. Energy balance of nestlings of Tree Sparrow, *Passer m. montanus* (L.), and House Sparrows, *Passer d. domesticus* (L.), In S. C. Kendeigh and J. Pinowski (eds), *Productivity, Population Dynamics and Systematics of Granivorous Birds*, PWN, Warsaw, pp. 59–82.

Nethersole-Thomson, D., 1951. *The Greenshank*, Collins, London.

Nethersole-Thomson, C., and D. Nethersole-Thomson, 1942. Egg-shell disposal by birds. *Br. Birds* **35**: 162–169, 190–200, 214–224, 241–250.

Newton, I., 1967. The feeding ecology of the Bullfinch (*Pyrrhula pyrrhula* L.) in southern England. *J. Anim. Ecol.* **36**: 721–744.

Newton, I., 1968. The temperatures, weights and body composition of molting Bullfinches. *Condor* **70**: 323–332.

Newton, I., 1979. *Population Ecology of Raptors*, Poyser, Berkhamsted.

Nice, M. M., 1937. Studies in the life history of the Song Sparrow. *Trans. Linn. Soc. N.Y.* **4**: 1–247.

Nice, M. M., 1943. Studies in the life history of the Song Sparrow. *Trans. Linn. Soc. N.Y.* **6**: 1–328.

Nice, M. M., 1962. Development of behavior in precocial birds. *Trans. Linn. Soc. N.Y.* **8**: 1–211.

Nice, M. M., and J. ter Pelkwijk, 1941. Enemy recognition by the Song Sparrow. *Auk* **58**: 195–214.

Nicolai, J., 1959. Familie tradition in de Gesangentwicklung des Gimpels (*Pyrrhula pyrrhula* L.). *J. Orn. Lpz.*, **100**: 39–64.

Nisbet, I. C. T., 1973. Courtship-feeding, egg-size and breeding success in Common Terns. *Nature, Lond.* **241**: 141–142.

Nisbet, I. C. T., 1978. Dependence of fledging success on egg-size, parental performance and egg-composition among Common and Roseate Terns, *Sterna hirundo* and *S. dougalli. Ibis* **120**: 207–215.

Nisbet, I. C. T., and M. E. Cohen, 1975. Asynchronous hatching in Common and Roseate Terns, *Sterna hirundo* and *S. dougalli. Ibis* **117**: 374–379.

Nisbet, I. C. T., and W. H. Drury, 1972. Post-fledging survival in Herring Gulls in relation to brood-size and date of hatching. *Bird-Banding* **43**: 161–240.

Nisbet, I. C. T., K. J. Wilson, and W. A. Broad, 1978. Common Terns raise young after death of their mates. *Condor* **80**: 106–109.

Norton, D. W., 1973. Ecological energetics of calidridine sandpipers breeding in northern Alaska. PhD thesis, University of Alaska.

Norton-Griffiths, M., 1969. The organization, control and development of parental feeding in the Oystercatcher (*Haematopus ostralegus*). *Behaviour* **34**: 55–114.

Nottebohm, F., 1968. Auditory experience and song development in the Chaffinch *Fringilla coelebs. Ibis* **110**: 549–568.

Nottebohm, F., 1969. The 'critical period' for song learning. *Ibis* **111**: 386–388.

Nottebohm, F., 1970. Ontogeny of bird song. *Science, N.Y.* **167**: 950–956.

Nottebohm, F., and M. E. Nottebohm, 1971. Vocalizations and breeding behaviour of surgically deafened Ring Doves (*Streptopelia risoria*). *Anim. Behav.* **19**: 322–327.

Nye, P. A., 1964. Heat loss in wet ducklings and chicks. *Ibis* **106**: 189–197.

Oatley, T. B., 1971. The functions of vocal imitation by African coccyphas. *Ostrich* Suppl. 8: 85–89.

O'Connor, R. J., 1975a. An adaptation for early growth in tits, *Parus* spp. *Ibis* **117**: 523–526.

O'Connor, R. J., 1975b. Initial size and subsequent growth in passerine nestlings. *Bird-Banding* **46**: 329–340.

O'Connor, R. J., 1975c. Growth and metabolism in nestling passerines. *Symp. zool. Soc. Lond.* **35**: 277–306.

O'Connor, R. J., 1975d. The influence of brood size upon metabolic rate and body temperature in nestling Blue Tits *Parus caeruleus* and House Sparrows *Passer domesticus. J. Zool. Lond.* **175**: 391–403.

O'Connor, R. J., 1975e. Nestling thermolysis and developmental change in body temperature. *Comp. Biochem. Physiol.* **52A**: 419–422.

O'Connor, R. J., 1976. Weight and body composition in nestling Blue Tits *Parus caeruleus. Ibis* **118**: 108–112.

O'Connor, R. J., 1977. Differential growth and body composition in altricial passerines. *Ibis* **119**: 147–166.

O'Connor, R. J., 1978a. Growth strategies in nestling passerines. *Living Bird* **16**: 209–238.

O'Connor, R. J., 1978b. Structure in avian growth patterns: a multivariate study of passerine development. *J. Zool., Lond.* **185**: 147–172.

O'Connor, R. J., 1978c. Brood reduction in birds: selection for fratricide, infanticide and suicide? *Anim. Behav.* **26**: 79–96.

O'Connor, R. J., 1978d. Nest-box insulation and the timing of laying in the Wytham Woods population of Great Tits *Parus major. Ibis* **120**: 534–537.

O'Connor, R. J., 1979. Egg weights and brood reduction in the European Swift (*Apus apus*). *Condor* **81**: 133–145.

O'Connor, R. J., 1980. Energetics of reproduction in birds. *Proc. 17th int. Orn. Congr.*: 306–311.

O'Connor, R. J., and R. A. Morgan, 1982. Some effects of weather conditions on the breeding of the Spotted Flycatcher *Muscicapa striata* in Britain. *Bird Study* **29**: 41–48.

Odum, E. P., 1942. Muscle tremors and the development of temperature regulation in birds. *Am. J. Physiol.* **136**: 618–622.

Ollason, J. C., and G. M. Dunnet, 1978. Age, experience and other factors affecting the breeding success of the Fulmar, *Fulmarus glacialis*, in Orkney. *J. Anim. Ecol.* **47**: 961–976.

Olsen, M. W., and S. L. Haynes, 1948. Effect of age and holding temperatures. *Poultry Sci.* **27**: 420–426.

Oppenheim, R. W., 1972. Prehatching and hatching behaviour in birds: a comparative study of altricial and precocial species. *Anim. Behav.* **20**: 644–655.

Oppenheim, R. W., 1973. Prehatching and hatching behavior: comparative and physiological considerations. In G. Gottlieb (ed.), *Behavioral Embryology*, Academic Press, New York, pp. 163–224.

Orians, G. H., 1973. The Red-winged Blackbird in tropical marshes. *Condor* **75**: 28–42.

Owen, R. W., and R. T. Pemberton, 1962. Helminth infection of the Starling (*Sturnus vulgaris* L.) in northern England. *Proc. zool. Soc. Lond.* **139**: 557–587.

Parkes, K. C., and G. A. Clarke, 1964. Additional records of avian egg teeth. *Wilson Bull.* **76**: 147–154.

Parmalee, D. F., 1970. Breeding behavior of the Sanderling in the Canadian High Arctic. *Living Bird* **9**: 97–146.

Parsons, J., 1970. Relationship between egg size and post-hatching chick mortality in the Herring Gull (*Larus argentatus*). *Nature, Lond.* **228**: 1221–1222.

Partridge, L., 1974. Habitat selection in titmice. *Nature, Lond.* **247**: 573–574.

Partridge, L., 1976a. Some aspects of the morphology of blue tits (*Parus caeruleus*) and coal tits (*P. ater*) in relation to their behaviour. *J. Zool., Lond.* **179**: 121–123.

Partridge, L., 1976b. Field and laboratory observations on the foraging and feeding techniques of Blue Tits (*Parus caeruleus*) and Coal Tits (*P. ater*) in relation to their habitats. *Anim. Behav.* **24**: 534–544.

Partridge, L., 1979. Differences in behaviour between Blue and Coal Tits reared under identical conditions. *Anim. Behav.* **27**: 120–127.

Peakall, D. B., 1960. Nest records of the Yellowhammer. *Bird Study* **7**: 94–102.

Pearson, F. H., 1968. The feeding biology of sea-bird species breeding on the Farne Islands, Northumberland. *J. Anim. Ecol.* **37**: 521–552.

Peek, F. W., E. Franks, and D. Case, 1972. Recognition of nest, eggs, nest site, and young in female Redwinged Blackbirds. *Wilson Bull.* **84**: 243–249.

Pefaur, J. E., 1974. Egg neglect in the Wilson's Storm Petrel. *Wilson Bull.* **86**: 16–22.

Penney, J. G., and E. D. Bailey, 1970. Comparison of the energy requirements of fledgling Black Ducks and American Coots. *J. Wildl. Mgmt* **34**: 105–114.

Penney, R. L., 1962. Voices of the Adelie. *Nat. Hist.* **71**: 16–25.

Perdeck, A. C., 1958. Two types of orientation in migrating Starlings, *Sturnus vulgaris* L., and Chaffinches, *Fringilla coelebs* L., as revealed by displacement experiments. *Ardea* **46**: 1–37.

Perdeck, A. C., 1964. An experiment on the ending of autumn migration by Starlings. *Ardea* **52**: 133–139.

Perdeck, A. C., 1967a. Orientation of Starlings after displacement to Spain. *Ardea* **55**: 194–202.

Perdeck, A. C., 1967b. The Starling as a passage migrant in Holland. *Bird Study* **14**: 129–152.

Perdeck, A. C., 1974. An experiment on the orientation of juvenile Starlings during spring migration. *Ardea* **62**: 190–195.

Perrins, C. M., 1965. Population fluctuations and clutch size in the Great Tit, *Parus major* L. *J. Anim. Ecol.* **64**: 601–647.

Perrins, C. M., 1976. Possible effects of qualitative changes in the insect diet of avian predators. *Ibis* **118**: 580–584.

Perrins, C. M., 1977. The role of predation in the evolution of clutch size. In B. Stonehouse and C. Perrins (eds), *Evolutionary Ecology*, Macmillan, London, pp. 181–192.

Perrins, C. M., 1979. *British Tits*, Collins, London.

Petersen, A. E., 1953. Orienteringsforsøg med Haettemåge (*Larus r. ridibundus* L.) og Stormmåge (*Larus c. canus* L.) i vinterkvarteret. *Dansk. orn. Foren. Tidsskr.* **47**: 153–178.

Peterson, A., and H. Young, 1950. A nesting study of the Bronzed Grackle. *Auk* **67**: 466–476.

Pettingill, O. S., 1960. Creche behavior and individual recognition in a colony of Rockhopper Penguins. *Wilson Bull.* **72**: 213–221.

Picman, J., and A. K. Picman, 1980. Destruction of nests by the Shortbilled Marsh Wren. *Condor* **82**: 176–179.

Picozzi, N., 1978. Dispersion, breeding and prey of the Hen Harrier *Circus cyaneus* in Glen Dye, Kincardineshire. *Ibis* **120**: 498–508.

Pinkowski, B. C., 1977. Breeding adaptations in the Eastern Bluebird. *Condor* **79**: 289–302.

Pinowski, J., and A. Myrcha, 1977. Biomass and production rates. In J. Pinowski and S. C. Kendeigh (eds), *Granivorous Birds in Ecosystems*, Cambridge University Press, Cambridge, pp. 107–128.

Pitelka, F. A., P. Q. Tomich, and G. W. Treichel, 1955. Ecological relations of jaegers and owls as lemming predators near Barrow, Alaska. *Ecol. Monogr.* **25**: 85–117.

Platt, S. W., 1975. The Mexican chicken bug as a source of raptor mortality *Wilson Bull.* **87**: 557.

Potts, G. R., 1969. The influence of eruptive movements, age, population size and other factors on the survival of the Shag (*Phalacrocorax aristotelis* (L.)). *J. Anim. Ecol.* **38**: 53–102.

Potts, G. R., 1971. Moult in the Shag *Phalacrocorax aristotelis*, and the ontogeny of the 'Staffelmauser'. *Ibis* **113**: 298–305.

Potts, G. R., 1973. Factors governing the chick survival rate of the Grey Partridge (*Perdix perdix*). *Int. Congr. Game Biol.* **10**: 85–96.

Poulsen, H., 1959. Song learning in the Domestic Canary. *Z. Tierpsychol.* **16**: 173–178.

Powlesland, R. G., 1978. Behaviour of the haematophagous mite *Ornithonyssus bursa* in Starling nest boxes in New Zealand. *New Zealand J. Zool.* **5**: 395–399.

Prince, P. A., 1980. The food and feeding ecology of Grey-headed Albatross *Diomedea chrysostoma* and Black-browed Albatross *D. melanophris*. *Ibis* **122**: 476–488.

Prys-Jones, R. P., 1977. Aspects of Reed Bunting ecology, with comparisons with the Yellowhammer. DPhil thesis, Oxford University.

Proctor, D. L. C., 1975. The problem of chick loss in the South Polar Skua *Catharacta maccormicki*. *Ibis* **117**: 452–459.

Rabinowitch, V. E., 1968. The role of experience in the development of food preferences in gull chicks. *Anim. Behav.* **16**: 425–428.

Rabinowitch, V. E., 1969. The role of experience in the development and retention of seed preferences in Zebra Finches. *Behaviour* **33**: 222–236.

Radesäter, T., 1974. On the ontogeny of orienting movements in the triumph ceremony in two species of geese (*Anser anser* L. and *Branta canadensis* L.). *Behaviour* **50**: 1–15.

Rahn, H., and A. Ar, 1974. The avian egg: incubation time and water loss. *Condor* **76**: 147–152.

Rahn, H., C. V. Paganelli, and A. Ar, 1975. Relation of avian egg to body weight. *Auk* **92**: 750–765.

Ralph, C. J., and L. R. Mewaldt, 1975. Timing of site fixation upon wintering grounds in sparrows. *Auk* **92**: 698–705.

Ralph, C. J., and L. R. Mewaldt, 1976. Homing success in wintering sparrows. *Auk* **93**: 1–14.

Ramsay, A. O., and E. H. Hess, 1954. A laboratory approach to the study of imprinting. *Wilson Bull.* **66**: 196–206.

Ratcliffe, D. A., 1967. Decrease in eggshell weight in certain birds of prey. *Nature, Lond.* **215**: 208–210.

Rautenfeld, D. von, 1978. Bemerkungen zur Austauschbarkeit von Küken der Silbermöwe (*Larus argentatus*) nach der ersten Lebenswoche. *Z. Tierpsychol.* **47**: 180–181.

Recher, H. F., and J. A. Recher, 1969. Comparative foraging efficiency of adult and immature Little Blue Herons (*Florida caerula*). *Anim. Behav.* **17**: 320–322.

Reinecke, R. J., 1979. Feeding ecology and development of juvenile Black Ducks in Maine. *Auk* **96**: 737–745.

Richdale, L. E., 1957. *A Population Study of Penguins*, Clarendon Press, Oxford.

Ricklefs, R. E., 1965. Brood reduction in the Curve-billed Thrasher. *Condor* **67**: 505–510.

Ricklefs, R. E., 1966. Behavior of young Cactus Wrens and Curve-billed Thrashers. *Wilson Bull.* **78**: 47–56.

Ricklefs, R. E., 1967a. A graphical method of fitting equations to growth curves. *Ecology* **48**: 978–983.

Ricklefs, R. E., 1967b. Relative growth, body constituents and energy content of nestling Brown Swallows and Red-winged Blackbirds. *Auk* **84**: 560–570.

Ricklefs, R. E., 1968a. Patterns of growth in birds. *Ibis* **110**: 419–451.

Ricklefs, R. E., 1968b. Weight recession in nestling birds. *Auk* **85**: 30–35.

Ricklefs, R. E., 1968c. On the limitation of brood size in passerine birds by the ability to nourish their young. *Proc. natn. Acad. Sci. U.S.A.* **61**: 847–851.

Ricklefs, R. E., 1969a. An analysis of nesting mortality in birds. *Smithson. Contr. Zool.* **9**: 1–48.

Ricklefs, R. E., 1969b. Preliminary models for growth rates of altricial birds. *Ecology*

288

50: 1031–1039.

Ricklefs, R. E., 1973. Patterns of growth in birds. II. Growth rate and mode of development. *Ibis* **115**: 177–201.

Ricklefs, R. E., 1974. Energetics of reproduction in birds. *Publ. Nuttall Orn. Club* **15**: 152–292.

Ricklefs, R. E., 1975. Patterns of growth in birds. III. Growth and development of the Cactus Wren. *Condor* **77**: 34–45.

Ricklefs, R. E., 1976. Growth rates of birds in the humid New World Tropics. *Ibis* **118**: 179–207.

Ricklefs, R. E., 1977a. Composition of eggs of several bird species. *Auk* **94**: 350–356.

Ricklefs, R. E., 1977b. On the evolution of reproductive strategies of birds: reproductive effort. *Am. Nat.* **111**: 453–478.

Ricklefs, R. E., 1979a. Adaptation, constraint, and compromise in avian postnatal development. *Biol. Rev.* **54**: 269–290.

Ricklefs, R. E., 1979b. Patterns of growth in birds. V. A comparative study of development in the Starling, Common Tern, and Japanese Quail. *Auk* **96**: 10–30.

Ricklefs, R. E., 1984. Prenatal development. In D. S. Farner and J. R. King (eds), *Avian Biology*, Vol. V, Academic Press, New York, in press.

Ricklefs, R. E., and F. R. Hainsworth, 1968. Temperature regulation in nestling Cactus Wrens: development of homeothermy. *Condor* **70**: 121–127.

Ricklefs, R. E., and F. R. Hainsworth, 1969. Temperature regulation in nestling Cactus Wrens: the nest environment. *Condor* **71**: 32–37.

Ricklefs, R. E., and W. A. Montevecchi, 1979. Size, organic composition and energy content of North Atlantic Gannet *Morus bassanus* eggs. *Comp. Biochem. Physiol.* **64A**: 161–165.

Ricklefs, R. E., and S. Weremiuk, 1977. Dynamics of muscle growth in the Starling and Japanese Quail: a preliminary study. *Comp. Biochem. Physiol.* **56A**: 419–423.

Ricklefs, R. E., D. C. Hahn, and W. A. Montevecchi, 1978. The relationship between egg size and chick size in the Laughing Gull and Japanese Quail. *Auk* **95**: 135–144.

Ricklefs, R. E., S. C. White, and J. Cullen, 1980a. Energetics of postnatal growth in Leach's Storm-petrel. *Auk* **97**: 566–575.

Ricklefs, R. E., S. C. White, and J. Cullen, 1980b. Postnatal development of Leach's Storm-petrel. *Auk* **97**: 768–781.

Rohwer, S., 1977. Status signalling in Harris Sparrows: some experiments in deception. *Behaviour* **61**: 107–129.

Rolnik, V. V., 1970. *Bird Embryology*, Israel Program for Scientific Translations, Jerusalem.

Romanoff, A. L., 1960. *The Avian Embryo*, Macmillan, New York.

Romanoff, A. L., 1967. *Biochemistry of the Avian Embryo*, Wiley, New York.

Romanoff, A. L., and A. J. Romanoff, 1949. *The Avian Egg*, Wiley, New York.

Romijn, C., and W. Lokhorst, 1955. Chemical heat regulation in the chick embryo. *Poult. Sci.* **34**: 649–654.

Romijn, C., and W. Lokhorst, 1960. Foetal heat production in the fowl. *J. Physiol., Lond.* **150**: 239–249.

Ross, H., 1980. Growth of nestling Ipswich Sparrows in relation to season, habitat, brood size, and parental age. *Auk* **97**: 721–732.

Rothstein, S. J., 1975. Mechanisms of avian egg-recognition: do birds know their own eggs? *Anim. Behav.* **23**: 268–278.

Royama, T., 1966a. Factors governing feeding rate, food requirement and brood size of nestling Great Tit *Parus major*. *Ibis* **108**: 313–347.

Royama, T., 1966b. A re-interpretation of courtship feeding. *Bird Study* **13**: 116–129.

Safriel, U. N., 1975. On the significance of clutch size in nidifugous birds. *Ecology* **56**: 703–708.

Salomonsen, F., 1968. The moult migration. *Wildfowl* **19**: 5–24.

Schaller, G. B., and J. T. Emlen, 1961. The development of visual discrimination patterns in the crouching reactions of nestling grackles. *Auk* **78**: 125–137.

Scharf, W. A., and E. Balfour, 1971. Growth and development of nestling Hen Harriers. *Ibis* **113**: 323–329.

Schiermann, G., 1927. Untersuchungen an Nestern des Haubentauchers *Podiceps cristatus*. *J. Orn., Lpz.* **75**: 619–638.

Schifferli, A., 1948. Über Markscheidbildung im Gehirn von Huhn und Star. *Rev. suisse Zool.* **55**: 117–122.

Schifferli, L., 1973. The effect of egg-weight on subsequent growth in nestling Great Tits. *Ibis* **115**: 549–558.

Schiller, E. L., 1954. Studies on the helminth fauna of Alaska. XVIII. Cestode parasites in young anseriformes on the Yukon Delta nesting grounds. *Trans. Am. Microscopical Soc.* **73**: 194–201.

Schleidt, W., 1961a. Über die Auslösung der Flucht vor Raubvögeln bei Truthühnern. *Naturwissenschaft* **48**: 141–142.

Schleidt, W., 1961b. Reaktionen von Truthühnern auf fliegende Raubvogel und Versuche zur Analyse ihrer AAMs. *Z. Tierpsychol.* **18**: 534–560.

Schüz, E., 1951. Überblick über die Orientierungsversuche der Vogelwarte Rossitten. *Proc. 10th int. Orn. Congr.*: 249–269.

Schüz, E., 1963. On the northwestern migration divide of the White Stork. *Proc. 13th int. Orn. Congr.*: 475–480.

Schwartz, P., 1963. Orientation experiments with Northern Waterthrushes wintering in Venezuela. *Proc. 13th int. Orn. Congr.*: 481–484.

Scott, M. L., 1973. Nutrition in reproduction – direct effects and predictive functions. In D. S. Farner (ed.), *Breeding Biology of Birds*, National Academy of Sciences, Washington, D.C., pp. 46–68.

Sealy, S. G., 1973. Adaptive significance of post-hatching development patterns in the Alcidae. *Ornis scand.* **4**: 113–121.

Sealy, S. G., 1976. Biology of nesting Ancient Murrelets. *Condor* **78**: 294–306.

Sealy, S. G., 1980. Reproductive responses of Northern Orioles to a changing food supply. *Can. J. Zool.* **58**: 221–227.

Seastedt, T. R., and S. F. MacLean, Jr, 1977. Calcium supplements in the diet of nestling Lapland Longspurs *Calcarius lapponicus* near Barrow, Alaska. *Ibis* **119**: 531–533.

Seel, D. C., 1960. The behaviour of a pair of House Sparrow while rearing young. *Br. Birds* **53**: 303–310.

Seel, D., 1965. The breeding of the House Sparrow (*Passer domesticus*). DPhil thesis, Oxford University.

Selander, R. K., and D. R. Giller, 1960. First-year plumages of the Brown-headed Cowbird and Redwinged Blackbird. *Condor* **62**: 202–214.

Shettleworth, S. J., 1972. The role of novelty in learned avoidance of unpalatable 'prey' by domestic chicks (*Gallus gallus*). *Anim. Behav.* **20**: 29–35.

Shilov, J. A., 1973. *Heat Regulation in Birds*, Amerind Publishing Co., New Delhi.

Siegel-Causey, D., 1980. Progenicide in Double-crested Cormorants. *Condor* **82**: 101.

Siegfried, W. R., 1972. Aspects of the feeding ecology of Cattle Egrets (*Ardeola ibis*) in South Africa. *J. Anim. Ecol.* **41**: 71–78.

Siegfried, W. R., 1974. Climbing ability of some cavity-nesting waterfowl. *Wildfowl* **25**: 74–80.

Siegfried, W. R., A. E. Burger, and P. G. H. Frost, 1976. Energy requirements for breeding in the Maccoa Duck. *Ardea* **64**: 171–191.

Simkiss, K., 1974. The air space of the avian egg: an embryonic "cold nose". *J. Zool., Lond.* **173**: 225–232.

Skar, H. J., A. Hagen, and E. Ostbye, 1972. Caloric values, weight, ash and H_2O content of the body of the Meadow Pipit. *Norw. J. Zool.* **20**: 51–59.

290

Skead, D. M., 1975. Ecological studies of four Estrildines in the Central Transvaal. *Ostrich* Suppl. 11: 1–55.

Skowron, C., and M. Kern, 1980. The insulation in nests of selected North American songbirds. *Auk* 97: 816–824.

Skutch, A. F., 1949. Do tropical birds rear as many young as they can nourish? *Ibis* 91: 430–455.

Skutch, A. F., 1961. Helpers among birds. *Condor* 63: 198–226.

Skutch, A. F., 1976. *Parent Birds and their Young*, University of Texas Press, Austin, Tx.

Sladen, W. J. L., 1955. Some aspects of the behaviour of Adelie and Chinstrap Penguins. *Proc. 11th int. Orn. Congr.*: 241–246.

Slonaker, J. R., 1921. The development of the eye and its accessory parts in the English Sparrow. *J. Morph.* 35: 263–357.

Smith, H. M., 1943. Size of breeding populations in relation to egglaying and reproductive success in the Eastern Red-wing (*Agelaius p. phoeniceus*). *Ecology* 24: 183–207.

Smith, S. M., 1972. The ontogeny of impaling behaviour in the Loggerhead shrike, *Lanius ludovicianus* L. *Behaviour* 44: 113–141.

Smith, S. M., 1973a. A study of prey attack behaviour in young Loggerhead Shrikes, *Lanius ludovicianus*. *Behaviour* 44: 113–141.

Smith, S. M., 1973b. Factors directing prey attack by the young of three passerine species. *Living Bird* 12: 55–67.

Smith, S. M., 1975. Innate recognition of coral snake pattern by a possible avian predator. *Science, N.Y.* 187: 759–760.

Smith, S. M., 1976. Predatory behaviour of young Turquoise-browed Motmots *Eumomota superciliosa*. *Behaviour* 56: 309–320.

Smith, J. M. N., 1978. Division of labour by Song Sparrow feeding fledged young. *Can. J. Zool.* 56: 187–191.

Snow, B. K., 1970. A field study of the Bearded Bell Bird in Trinidad. *Ibis* 112: 299–329.

Snow, D. W., 1958. *A Study of Blackbirds*, Methuen, London.

Snow, D. W., 1962. The natural history of the Oilbird, *Steatornis caripensis*, in Trinidad, W.I. Part 2: Population, breeding ecology and food. *Zoologica, N.Y.* 47: 199–221.

Sossinka, R., E. Pröve, and H. H. Kalberlah, 1975. Der Einfluss von Testosteron auf den Gesangsbeginn beim Zebrafinken (*Taeniopygia guttata castanotis*). *Z. Tierpsychol.* 39: 259–264.

Southern, W. E., 1969. Orientation behavior of Ring-billed Gull chicks and fledglings. *Condor* 71: 418–425.

Southern, W. E., 1971. Gull orientation by magnetic cues: a hypothesis revisited. *Ann. N.Y. Acad. Sci.* 188: 295–311.

Southern, W. E., 1972. Influence of disturbances in the earth's magnetic field on Ring-billed Gull orientation. *Condor* 74: 102–105.

Spaans, A. L., 1977. Are Starlings faithful to their individual winter quarters? *Ardea* 65: 83–87.

Spidso, T. K., 1980. Food selection by Willow Grouse *Lagopus lagopus* chicks in Northern Norway. *Ornis scand.* 11: 99–105.

Stallcup, J. A., and G. E. Woolfenden, 1978. Family status and contributions to breeding by Florida Scrub Jays. *Anim. Behav.* 26: 1144–1156.

Stokes, A. W., 1950. Breeding behavior of the Goldfinch. *Wilson Bull.* 62: 107–127.

Stokkan, K. A., and J. B. Steen, 1980. Age determined feeding behaviour in Willow Ptarmigan chicks *Lagopus lagopus lagopus*. *Ornis scand.* 11: 75–76.

Storey, A. E., and L. J. Shapiro, 1979. Development of preferences in white Peking ducklings for stimuli in the natural post-hatch environment. *Anim. Behav.* 27: 411–416.

Street, M., and T. MacDonald, 1977. Amey Roadstone Wildfowl project: 1976. *A. Rev. Game Conservancy* **1976**: 35–43.

Sturkie, P. D., 1946. The production of twins in *Gallus domesticus*. *J. exp. Zool.* **101**: 51–63.

Sugden, L. S., and L. E. Harris, 1972. Energy requirements and growth of captive Lesser Scaup. *Poult. Sci.* **51**: 625–633.

Swingland, I. R., 1977. The social and spatial organization of winter communal roosting in Rooks (*Corvus frugilegus*). *J. Zool., Lond.* **182**: 509–528.

Thielcke, G., 1973. On the origin of divergence of learned signals (songs) in isolated populations. *Ibis* **115**: 511–516.

Thielcke, G., 1974. Stabilität erlernter Singvogel-Gesänge trotz vollständiger geographischer Isolation. *Die Vogelwarte* **27**: 209–215.

Thielcke-Poltz, H., and G. Thielcke, 1960. Akustisches Lerner verschieden alter schallisolierter Amseln (*Turdus merula* L.) und die Entwicklung erlernter Motive ohne und mit künstlichem Einfluss von Testosteron. *Z. Tierpsychol.* **17**: 211–244.

Thorpe, W. H., 1958. The learning of song patterns by birds, with especial reference to the song of the Chaffinch *Fringilla coelebs*. *Ibis* **100**: 535–570.

Thomas, R. H., 1946. A study of Eastern Bluebirds in Arkansas. *Wilson Bull.* **58**: 143–183.

Tinbergen, J., 1981. Foraging decisions in Starlings (*Sturnus vulgaris* L.). *Ardea* **69**: 1–67.

Tinbergen, N., 1951. *The Study of Instinct*, Oxford University Press, London.

Tinbergen, N., 1963. The shell menace. *Nat. Hist.* **72**: 28–35.

Tinbergen, N., 1972. *The Animal in its World*, Vol. I, *Field Studies*, George Allen and Unwin, London.

Tinbergen, N., 193. *The Animal in its World*, Vol. II, *Laboratory Experiments and General Papers*, George Allen and Unwin, London.

Tinbergen, N., and D. J. Kuenen, 1939. [The releasing and directing stimulus situations of the gaping response in young Blackbirds and Thrushes (*Turdus m. merula* L. and *T. e. ericetorum* Turton).] *Z. Tierpsychol.* **3**: 37–60 (in German; English translation in Tinbergen (1973)).

Tinbergen, N., and A. C. Perdeck, 1951. On the stimulus situation releasing the begging response in the newly hatched Herring Gull chick (*Larus argentatus* Pont.), *Behaviour* **3**: 37–60.

Tinbergen, N., G. H. Broekhuysen, F. Feekes, J. C. W. Houghton, H. Kruuk, and E. Szulc, 1962. Egg shell removal by the Black-headed Gull, *Larus ridibundus*; a behaviour component of camouflage. *Behaviour* **19**: 74–117.

Tretzel, E., 1965. Über das Spotten der Singvögel, insbesondere ihre Fähigkeit zu spontaner Nachahmung. *Verh. dt. Zool. Ges., Kiel* **1964**: 556–565.

Trivers, R. L., 1974. Parent–offspring conflict. *Am. Zool.* **14**: 249–269.

Tschanz, B., 1968. Trottellummen. *Z. Tierpsychol.* Suppl. **4**: 1–103.

Tschanz, B., and M. Hirsbrunner-Scharf, 1975. Adaptations to colony life on cliff ledges: a comparative study of Guillemot and Razorbill chicks. In G. Baerends, C. Beer, and A. Manning (eds), *Function and Evolution in Behaviour*, Clarendon Press, Oxford, pp. 358–380.

Tuck, L. M., and H. J. Squires, 1955. Food and feeding habits of Brunnich's Murre (*Uria lomvia lomvia*) on Atpatok Island. *J. Fish Res. Bd. Can.* **12**: 781–792.

Turner, R. A., 1964. Social feeding in birds. *Behaviour* **24**: 1–46.

Tyrvainen, H., 1969. The breeding biology of the Redwing (*Turdus iliacus* L.). *Annotnes. zool. fenn.* **6**: 1–46.

Utter, J. M., and E. A. Le Febvre, 1973. Daily energy expenditure of Purple Martin (*Progne subis*) during the breeding season: estimates using D_2O^{18} and time budget methods. *Ecology* **54**: 597–605.

Veselovsky, Z., 1973. The breeding biology of Cape Barren geese. *Int. Zoo Yb.* **13**: 48–55.

Vestjens, W. J. M., 1975. Breeding behaviour of the Darter at Lake Cowal, NSW. *Emu* **75**: 121–131.

Victoria, J. K., 1972. Clutch characteristics and egg discriminative ability of the African Village Weaverbird *Ploceus cucullatus*. *Ibis* **114**: 367–376.

Vince, M. A., 1960. Developmental changes in responsiveness in the Great Tit (*Parus major*). *Behaviour* **15**: 219–243.

Vince, M. A., 1966. Potential stimulation produced by avian embryos. *Anim. Behav.* **14**: 34–40.

Vince, M. A., 1968. Retardation as a factor in the synchronization of hatching. *Anim. Behav.* **16**: 332–335.

Vince, M. A., 1969. Embryonic communication, respiration and the synchronization of hatching. In R. A. Hinde (ed.), *Bird Vocalizations: Their Relations to Current Problems in Biology and Psychology*, Cambridge University Press, Cambridge, pp. 233–260.

Vince, M. A., 1972. Communication between quail embryos and the synchronization of hatching. *Proc. 15th int. Orn. Congr.*: 357–362.

Vince, M. A., and R. Cheng, 1970. The retardation of hatching in Japanese Quail. *Anim. Behav.* **18**: 210–214.

Vince, M. A., and S. Chinn, 1972. Effects of external stimulation on the domestic chick's capacity to stand and walk. *Br. J. Psychol.* **63**: 89–99.

Vleck, C. M., and G. J. Kenagy, 1980. Embryonic metabolism of the Forktailed Storm Petrel: physiological patterns during prolonged and interrupted incubation. *Physiol. Zool.* **53**: 32–42.

Vleck, C. M., D. F. Hoyt, and D. Vleck, 1979. Metabolism of avian embryos: patterns in altricial and precocial birds. *Physiol. Zool.* **52**: 363–377.

Voronov, N., 1974. [Rate of development of the digestive system in the Jackdaw *Corvus monedula* and Rook *Corvus frugilegus* in the postnatal period.] *Z. obsch. Biol.* **35**: 934–943 (in Russian; English summary).

Wada, M., J. Goto, H. Nishinyama, and K. Nobukuni, 1979. Colour exposure of incubating eggs and colour preference of chicks. *Anim. Behav.* **27**: 359–364.

Walker, J. E. S., 1972. Attempts at fledging of a runt Great Tit. *Bird Study* **19**: 250–251.

Walkinshaw, L. H., 1953. Life-history of the Prothonotary Warbler. *Wilson Bull.* **65**: 152–168.

Walkinshaw, L. H., 1963. Some life history studies of the Stanley Crane. *Proc. 13th int. Orn. Congr.*: 344–353.

Ward, L. D., and J. Burger, 1980. Survival of Herring Gull and Domestic Chicken embryos after simulated flooding. *Condor* **82**: 142–148.

Warham, J., 1975. The Crested Penguins. In B. Stonehouse (ed.), *The Biology of Penguins*, MacMillan Press, London, pp. 189–269.

Weeks, H. P., 1978. Clutch size variation in the Eastern Phoebe in Southern Indiana. *Auk* **95**: 656–666.

Wehr, E. E., and C. M. Herman, 1954. Age as a factor in acquisition of parasites by Canada Geese. *J. Wildl. Mgmt* **18**: 239–247.

Wehrlin, J., 1977. Verhaltenpassungen junger Trottellummen ans Felskippen- und Koloniebruten. *Z. Tierpsychol.* **44**: 45–79.

Weller, M. W., 1957. Growth, weights, and plumages of the Redhead *Aythya americana*. *Wilson Bull.* **69**: 5–38.

Weller, M. W., 1961. Breeding biology of the Least Bittern. *Wilson Bull.* **73**: 11–35.

Welty, J. C., 1975. *The Life of Birds*, 2nd edn, Saunders, Philadelphia.

Werschkul, D. F., 1979. Nestling mortality and the adaptive significance of early locomotion in the Little Blue Heron. *Auk* **96**: 116–130.

West, G. C., 1965. Shivering and heat production in wild birds. *Physiol. Zool.* **38**: 111–120.

Westerterp, K., 1973. The energy budget of the nestling Starling *Sturnus vulgaris*: a field study. *Ardea* **61**: 137–158.

Wetherbee, D. K., 1957. Natal plumages and downy pteryloses of passerine birds of North America. *Bull. Am. Mus. nat. Hist.* **113**: 345–436.

Wetherbee, D. K., and L. M. Bartlett, 1962. Egg teeth and shell rupture of the American Woodcock. *Auk* **79**: 117.

White, F. N., and J. L. Kinney, 1974. Avian incubation. *Science, N.Y.* **186**: 107–115.

Whittow, G. C., and A. J. Berger, 1977. Heat loss from the nest of the Hawaiian Honeycreeper, 'Amahiki'. *Wilson Bull.* **89**: 480–483.

Williams, M., 1974. Creching behaviour of the Shelduck *Tadorna tadorna* L. *Ornis scand.* **5**: 131–143.

Wiltschko, W., R. Wiltschko, and W. T. Keeton, 1976. Effects of a 'permanent' clock-shift on the orientation of young homing pigeons. *Behavl. Ecol. Sociobiol.* **1**: 229–243.

Winkel, W., and R. Berndt, 1972. Beobachtungen und Experimente zur Dauer der Huderperiode bein Trauerschnapper (*Ficedula hypoleuca*). *J. Orn. Lpz.* **113**: 9–20.

Witschi, E., 1956. *Development of Vertebrates*, Saunders, Philadelphia.

Wood, R. C., 1971. Population dynamics of breeding South Polar Skuas. *Auk* **88**: 805–814.

Wooler, R. D., and J. C. Coulson, 1977. Factors affecting the age of first breeding of the Kittiwake *Rissa tridactyla*. *Ibis* **119**: 339–349.

Woolfenden, G. E., 1975. Florida Scrub Jay helpers at the nest. *Auk* **92**: 1–15.

Woolfenden, G. E., and J. W. Fitzpatrick, 1978. The inheritance of territory in group-breeding birds. *Bioscience* **28**: 104–108.

Wurdinger, J., 1974. Die Entwicklung der Objektfixierung einiger Verhaltensweisen während der ersten vier Lebenstage bei der Streifengans *Answer indicus* (*Eulabeia indica* Latham). *Z. Tierpsychol.* **35**: 209–218.

Wynne-Edwards, V. C., 1962. *Animal Dispersion in Relation to Social Behaviour*, Oliver and Boyd, Edinburgh.

Yarbrough, C. G., 1970. The development of endothermy in nestling Graycrowned Rosy Finches *Leucosticte tephrocotis griseonucha*. *Comp. Biochem. Physiol.* **34**: 917–925.

Yom-Tov, Y., 1974. The effect of food and predation on breeding density and success, clutch size and laying date of the Crow (*Corvus corvone* L.). *J. Anim. Ecol.* **43**: 479–498.

Yom-Tov, Y., 1975. Food of nestling Crows in Northeast Scotland. *Bird Study* **22**: 47–51.

Young, E. C., 1963. The breeding behaviour of the South Polar Skua. *Ibis* **105**: 203–233.

Young, H., 1963. Age specific mortality in the eggs and nestlings of Blackbirds. *Auk* **80**: 145–155.

Zwickel, F. C., 1967. Some observations of weather and brood behaviour in blue grouse. *J. Wildl. Mgmt* **31**: 563–568.

Appendix

Scientific names of species mentioned in the text

STRUTHIONIDAE	*Struthio camelus*	Ostrich
RHEIDAE	*Rhea americana*	Common Rhea
CASUARIIDAE	*Casuarius casuarius*	Common Cassowary
SPHENISCIDAE	*Aptenodytes forsteri*	Emperor Penguin
	Aptenodytes patogonica	King Penguin
	Pygoscelis papua	Gentoo Penguin
	Pygoscelis adeliae	Adelie Penguin
	Eudyptes crestatus	Rockhopper Penguin
	Megadyptes antipodes	Yellow-eyed Penguin
PODICIPEDIDAE	*Podiceps cristatus*	Great Crested Grebe
	Podiceps nigricollis	Black-necked or Eared Grebe
DIOMEDEIDAE	*Diomeda nigripes*	Black-footed Albatross
	Diomede immutabilis	Laysan Albatross
PROCELLARIIDAE	*Fulmarus glacialis*	Northern Fulmar
	Puffinus puffinus	Common or Manx Shearwater
HYDROBATIDAE	*Oceanodroma furcata*	Fork-tailed Storm Petrel
	Oceanodroma leucorhoa	Leach's Storm Petrel
PHAETHONTIDAE	*Phäethon rubricauda*	Red-tailed Tropicbird
SULIDAE	*Sula sula*	Red-footed Booby
	Sula dactylatra	Masked or Blue-faced Booby
	Sula variegata	Peruvian Booby
	Sula bassana	(Northern or Atlantic) Gannet
	Sula capensis	Cape Gannet
PHALACROCOR-ACIDAE	*Phalacrocorax auritus*	Double-crested Cormorant
	Phalacrocorax aristotelis	Shag
ANHINGIDAE	*Anhinga melanogaster*	Darter

PELECANIDAE	*Pelecanus onocrotalus*	Great White or Rosy Pelican
	Pelecanus rufescens	Pink-backed Pelican
	Pelecanus occidentalis	Brown Pelican
ARDEIDAE	*Ixobrychus exilis*	Least Bittern
	Bubulcus ibis	Cattle Egret or Buff-backed Heron
	Hydranassa caerulea	Little Blue Heron
	Ardea herodias	Great Blue Heron
CICONIIDAE	*Ciconia ciconia*	White Stork
THRESKIOR-NITHIDAE	*Eudocimus albus*	White Ibis
ANATIDAE	*Anseranas semipalmata*	Magpie Goose
	Dendrocygna arcuata	Wandering Whistling Duck
	Dendrocygna autumnalis	Black-bellied Tree Duck
	Cygnus olor	Mute Swan
	Cygnus columbianus	Bewick's Swan
	Anser indicus	Bar-headed Goose
	Anser caerulescens	Snow Goose and Blue Goose
	Anser canagicus	Emperor Goose
	Branta canadensis	Canada Goose
	Cereopsis novaehollandiae	Cape Barren Goose
	Alocopochen aegyptiacus	Egyptian Goose
	Tadorna tadorna	Shelduck
	Aix sponsa	Wood Duck
	Anas americana	Baldplate or American Wigeon
	Anas strepera	Gadwall
	Anas crecca	Teal
	Anas platyrhynchos	Mallard
	Anas rubripes	Black Duck
	Anas acuta	Pintail
	Anas discors	Blue-winged Teal
	Aythya valisineria	Canvasback
	Aythya ferina	Pochard
	Aythya americana	Redhead
	Aythya australis	Australasian White-eye
	Aythya baeri	Baer's Pochard
	Aythya nyroca	Ferruginous Duck or White-eyed Pochard
	Aythya fuligula	Tufted Ducks
	Aythya marila	(Greater) Scaup
	Aythya affinis	Lesser Scaup
	Somateria mollissima	Eider
	Clangula hyemalis	Long-tailed Duck or Oldsquaw
	Melanitta nigra	Common or Black Scoter
	Melanitta fusca	Velvet or White-winged Scoter
	Bucephala islandica	Barrow's Goldeneye
	Bucephala clangula	(Common) Goldeneye
	Mergus serrator	Red-breasted Merganser
	Oxyura jamaicensis	Ruddy Duck
	Oxyura leucocephala	White-headed Duck

ACCIPITRIDAE	*Milvus migrans*	Black Kite
	Circus cyaneus	Hen Harrier or Marsh Hawk
	Accipiter nisus	Sparrowhawk
	Aquila pomarina	Lesser Spotted Eagle
	Aquila heliaca	Imperial Eagle
PANDIONIDAE	*Pandion haliaetus*	Osprey
FALCONIDAE	*Falco sparverius*	American Kestrel or American Sparrowhawk
	Falco mexicanus	Prairie Falcon
MEGAPODIIDAE	*Leipoa ocellata*	Mallee Fowl
	Alectura lathami	Brush Turkey
TETRAONIDAE	*Dendrogapus obscurus*	Blue Grouse
	Bonasa umbellus	Ruffed Grouse
	Lagopus lagopus	Willow Grouse (Willow Ptarmigan)
	Lagopus mutus	Ptarmigan or Rock Ptarmigan Black Grouse (Blackcock)
	Tetrao tetrix	Capercaillie
	Tetrao urogallus	
PHASIANIDAE	*Oreortyx picta*	Mountain Quail
	Colinus virginianus	Bobwhite
	Perdix perdix	Partridge
	Coturnix japonica	Japanese Quail
	Phasianus colchicus	Common Pheasant
RALLIDAE	*Crex crex*	Corncrake
	Gallinula chloropus	Moorhen or Common Gallinule
	Porphyrio porphyrio	Purple Gallinule (Pukeko)
	Fulica atra	European Coot
	Fulica cornuta	Horned Coot
MELEAGRIDIDAE	*Meleagris gallopavo*	Turkey
GRUIDAE	*Grus canadensis*	Sandhill Crane
	Anthropoides paradisea	Stanley Crane
HAEMATOPODIDAE	*Haematopus ostralegus*	Oystercatcher
	Haematopus bachmani	American Black Oystercatcher
BURHINIDAE	*Burhinus oedicnemus*	Stone Curlew
GLAREOLIDAE	*Pluvianus aegyptius*	Egyptian Plover
CHARADRIIDAE	*Charadrius hiaticula*	Ringed Plover
	Vanellus vanellus	Lapwing

SCOLOPACIDAE	*Calidris alba*	Sanderling
	Calidris pusilla	Semipalmated Sandpiper
	Micropalama himantopus	Stilt Sandpiper
	Philomachus pugnax	Ruff
	Scolopax rusticola	Eurasian Woodcock
	Scolopax minor	American Woodcock
	Limosa limosa	Black-tailed Godwit
	Numenius arquata	Eurasian Curlew
	Actitis macularia	Spotted Sandpiper
	Catoptrophorus semipalmatus	Willet
	Arenaria interpres	Turnstone or Ruddy Turnstone
THINOCORIDAE	*Thinocorus rumicivorous*	Least or Patagonian Seedsnipe
STERCORARIIDAE	*Stercorarius pomarinus*	Pomarine Skua or Jaeger
	Stercorarius parasiticus	Arctic Skua or Parasitic Jaeger
	Stercorarius longicaudus	Long-tailed Skua or Jaeger
	Stercorarius skua	Great Skua (Brown Skua)
	Catharacta antarctica	Antarctic Skua
	Stercorarius maccormicki	Maccormick's or South Polar Skua
LARIDAE	*Larus bulleri*	Black-billed Gull
	Larus novaehollandiae	Red-billed Gull
	Larus atricilla	Laughing Gull
	Larus pipixcan	Franklin's Gull
	Larus sabini	Sabine's Gull
	Larus ridibundus	Black-headed Gull
	Larus delawarensis	Ring-billed Gull
	Larus canus	Common or New Gull
	Larus fuscus	Lesser Black-backed Gull
	Larus argentatus	Herring Gull
	Larus marinus	Great Black-backed Gull
	Rissa tridactyla	(Black-legged) Kittiwake
STERNIDAE	*Sterna caspia*	Caspian Tern
	Sterna maxima	Royal Tern
	Sterna sandvicensis	Sandwich Tern
	Sterna dougallii	Roseate Tern
	Sterna hirundo	Common Tern
	Sterna paradisaea	Arctic Tern
	Sterna fuscata	Sooty Tern (Wideawake Tern)
	Chlidonias nigra	Black Tern
	Anous tenuirostris	Lesser or Black Noddy
	Gygis alba	White or Fairy Tern
RYNCHOPIDAE	*Rynchops nigra*	Black Skimmer
ALCIDAE	*Uria aalge*	Guillemot or Common Murre
	Uria lomvia	Brunnich's Guillemot or Thick-billed Murre

	Alca torda	Razorbill
	Cepphus grylle	Black Guillemot
	Cepphus columba	Pigeon Guillemot
	Synthliboramphus antiquus	Ancient Murrelet
	Fratercula arctica	Common Puffin
PETEROCLIDIDAE	*Pterocles namaqua*	Namaqua Sandgrouse
	Pterocles orientalis	Black-bellied or Imperial Sandgrouse
COLUMBIDAE	*Columba livia*	Rock Dove or Domestic Pigeon
	Columba palumbus	Woodpigeon
	Streptopelia risoria	Ringed Turtle Dove
	Zenaida macroura	Mourning Dove
	Scardafella inca	Inca Dove
CUCULIDAE	*Cuculus canorus*	Cuckoo
TYTONIDAE	*Tyto alba*	Barn Owl
	Nyctea scandiaca	Snowy Owl
	Strix aluco	Tawny Owl
	Asia flammeus	Short-eared Owl
STEATORNITHIDAE	*Steatornis caripensis*	Oilbird
CAPRIMULGIDAE	*Caprimulgus europaeus*	Nightjar
POPIDAE	*Chaetura pelagica*	Chimney Swift
	Apus apus	Swift
TROCHILIDAE	*Selasphorus platycerus*	Broad-tailed Hummingbird
ALCEDINIDAE	*Alcedo atthis*	Kingfisher
	Ceryle rudis	Pied Kingfisher
MOTMOTIDAE	*Eumomota superciliosa*	Turquoise-browed Motmot
BUCEROTIDAE	*Tockus erythrorhynchus*	Red-billed Hornbill
PICIDAE	*Asyndesmus lewis*	Lewis' Woodpecker
	Sphyrapicus varius	Yellow-bellied Sapsucker
PIPRIDAE	*Manacus manacus*	White-bearded Manakin
TYRRANNIDAE	*Sayornis phoebe*	Eastern Phoebe
	Onychorhynchus mexicanus	Northern Royal Flycatcher
	Tolmomyias sulphurescens	Sulphury Flatbill
ALAUDIDAE	*Galerida cristata*	Crested Lark
	Alauda arvensis	Skylark
	Eremophila alpestris	Shore Lark or Horned Lark

HIRUNDINIDAE	*Riparia riparia*	Sand Martin or Bank Swallow
	Stelgidopteryx ruficollis	American Rough-winged Swallow
	Tachycineta bicolor	Tree Swallow
	Progne subis	Purple Martin
	Hirundo rustica	Swallow or Barn Swallow
	Delichon urbica	House Martin
MOTACILLIDAE	*Anthus campestris*	Tawny Pipit
	Anthus trivialis	Tree Pipit
	Anthus pratensis	Meadow Pipit
PYCNONOTIDAE	*Pycnonotus eotilus*	Crested Brown Bulbul
TROGLODYTIDAE	*Campylorhynchus brunneica-pillus*	Cactus Wren
	Campylorhynchus zonatus	Banded-backed Wren
	Cistothorus platensis	Short-billed Marsh Wren
	Cistothorus palustris	Marsh Wren or Long-billed Marsh Wren
	Troglodytes aëdon	House Wren
MIMIDAE	*Mimus polyglottos*	Northern Mockingbird
	Toxostoma curvirostre	Curve-billed Thrasher
	Dumetella carolinensis	Northern or Gray Catbird
PRUNELLIDAE	*Prunella collaris*	Alpine Accentor
TURDIDAE	*Erithacus rubecula*	Robin or European Robin
	Hylocichla mustelina	Wood Thrush
	Turdus merula	Blackbird
	Turdus philomelos	Song Thrush
	Turdus migratorius	American Robin
SYLVIIDAE	*Acrocephalus scirpaceus*	Reed Warbler
	Sylvia borin	Garden Warbler
MUSCICAPIDAE	*Muscicapa striata*	Spotted Flycatcher
	Ficedula albiocollis	Collared Flycatcher
	Ficedula hypoleuca	Pied Flycatcher
TIMALIIDAE	*Turdoides striatus*	Jungle Babbler
AEGITHALIDAE	*Aegithalos caudatus*	Long-tailed Tit
PARIDAE	*Parus palustris*	Marsh Tit
	Parus atricapillus	Black-capped Chickadee
	Parus ater	Coal Tit
	Parus caeruleus	Blue Tit
	Parus major	Great Tit

SITTIDAE	*Sitta europaea*	Common or Eurasian Nuthatch
TICHODROMA-DIDAE	*Tichodroma muraria*	Wall Creeper
CERTHIIDAE	*Certhia brachydactyla*	Short-toed Tree Creeper
REMIZIDAE	*Remiz pendulinus*	Penduline Tit
LANIDAE	*Lanius collurio*	Red-backed Shrike
	Lanius ludovicianus	Loggerhead Shrike
CORVIDAE	*Cyanocorax cristata*	Blue Jay
	Aphelocoma ultramarina	Mexican or Gray-breasted Jay
	Aphelocoma caerulescens	Scrub Jay
	Gymnorhinus cyanocephalus	Piñon Jay
	Gymnorhinus dorsalis	Australian Magpie
	Pica pica	Magpie or Black-billed Magpie
	Nucifraga caryocatactes	Nutcracker
	Corvus mondedula	Jackdaw
	Corvus frugilegus	Rook
	Corvus brachyrynchos	Crow (Carrion Crow and Hooded Crow)
	Corvus corax	Ragen or Northern Raven
STURNIDAE	*Onychognatus morio*	Red-winged Starling
	Sturnus vulgaris	Starling
PASSERIDAE	*Passer domesticus*	House Sparrow
	Passer hispaniolensis	Spanish or Willow Sparrow
	Passer montanus	(Old World) Tree Sparrow
PLOECIDAE	*Philetairus socius*	Sociable Weaver
	Ploceus alienus	Strange Weaver
	Ploceus cucullatus	Village Weaver
	Malimbus rubriceps	Red-headed Weaver
FRINGILLIDAE	*Fringilla coelebs*	Chaffinch
	Fringilla montifringilla	Brambling
	Serinus canaria	Canary
	Carduelis spinus	Siskin
	Carduelis tristis	American Goldfinch
	Carduelis flammea	Common Redpoll
	Loxia tephrocotis	American or Gray-crowned Rosy Finch
	Pyrrhula pyrrhula	Bullfinch or Eurasian Bullfinch
ESTRILDIDAE	*Lagonosticta senegala*	Red-billed or Senegal Firefinch

	Lonchura striata	Bengalese Finch
	Poephila guttata	Zebra Finch
	Uraeginthus angolensis	Blue Waxbill
	Estrilda erythronotus	Black-headed Waxbill
PARULIDAE	*Setophaga ruticilla*	American Redstart
	Seiurus aurocapillus	Ovenbird
	Seiurus noveboraeensis	Northern Waterthrush
	Geothlypsis trichas	Yellowthroat
COEREBIDAE	*Coerba flaveola*	Bananaquit
EMBERIZIDAE	*Aimophila carpalis*	Rufous-winged Sparrow
	Spizella passerina	Chipping Sparrow
	Spizella pusilla	Field Sparrow
	Pooecetes gramineus	Vesper Sparrow
	Ammodramus sandwichensis	Savannah Sparrow
	Zonotrichia melodia	Song Sparrow
	Zonotrichia georgiana	Swamp Sparrow
	Zonotrichia querula	Harris's Sparrow
	Zonotrichia leuchophrys	White-crowned Sparrow
	Zonotrichia albicollis	White-throated Sparrow
	Junco hyemalis	Dark-eyed or Northern Junco (Oregon Junco)
	Junco phaenonotus	Yellow-eyed or Mexican Junco (Arizona Junco)
	Calcarius lapponicus	Lapland Bunting or Lapland Longspur
	Emberiza citrinella	Yellowhammer
	*Emberiza pusilla| *	Little Bunting
	Emberiza schoeniclus	Reed Bunting
	Cardinalis cardinalis	Cardinal or Red or Northern Cardinal
	Passerina cyanea	Indigo Bunting
ICTERIDAE	*Molothrus ater*	Brown-headed Cowbird
	Euphagus cyanocephalus	Brewer's Blackbird
	Quiscalus quiscula	Common Grackle
	Sturnella magna	Eastern Meadowlark
	Agelaius phoeniceus	Red-winged Blackbird
	Icterus galbula	Northern Oriole (Baltimore Oriole)

Index

304

312